Tributes
Volume 34

Models: Concepts, Theory, Logic, Reasoning, and Semantics
Essays Dedicated to Klaus-Dieter Schewe on the Occasion of his 60th Birthday

Volume 25
The Facts Matter. Essays on Logic and Cognition in Honour of Rineke Verbrugge
Sujata Ghosh and Jakub Szymanik, eds.

Volume 26
Learning and Inferring. Festschrift for Alejandro C. Frery on the Occasion of his 55th Birrthday
Bruno Lopes and Talita Perciano, eds.

Volume 27
Why is this a Proof? Festschrift for Luiz Carlos Pereira
Edward Hermann Haeusler, Wagner de Campos Sanz and Bruno Lopes, eds.

Volume 28
Conceptual Clarifications. Tributes to Patrick Suppes (1922-2014)
Jean-Yves Béziau, Décio Krause and Jonas R. Becker Arenhart, eds.

Volume 29
Computational Models of Rationality. Essays Dedicated to Gabriele Kern-Isberner on the Occasion of her 60th Birthday
Christoph Beierle, Gerhard Brewka and Matthias Thimm, eds.

Volume 30
Liber Amicorum Alberti. A Tribute to Albert Visser
Jan van Eijck, Rosalie Iemhoff and Joost J. Joosten, eds.

Volume 31
"Shut up," he explained. Essays in Honour of Peter K. Schotch
Gillman Payette, ed.

Volume 32
From Semantics to Dialectometry. Festschrift in Honour of John Nerbonne.
Martijn Wieling, Martin Kroon, Gertjan van Noord, and Gosse Bouma eds.

Volume 33
Logic and Computation. Essays in Honour of Amílcar Sernadas
Carlos Caleiro, Fransciso Dionísio, Paula Gouveia, Paulo Mateus and João Rasga, eds.

Volume 34
Models: Concepts, Theory, Logic, Reasoning, and Semantics. Essays Dedicated to Klaus-Dieter Schewe on the Occasion of his 60th Birthday
Atif Mashkoor, Qing Wang and Bernhrd Thalheim, eds.

Tributes Series Editor
Dov Gabbay   dov.gabbay@kcl.ac.uk

Models: Concepts, Theory, Logic, Reasoning, and Semantics

Essays Dedicated to Klaus-Dieter Schewe on the Occasion of his 60th Birthday

edited by

Atif Mashkoor

Qing Wang

and

Bernhard Thalheim

© Individual authors and College Publications 2018. All rights reserved.

ISBN 978-1-84890-276-3

College Publications
Scientific Director: Dov Gabbay
Managing Director: Jane Spurr

http://www.collegepublications.co.uk

Cover design by Laraine Welch

Printed by Lightning Source, Milton Keynes, UK

All rights reserved. No part of this publication may be reproduced, stored in a retrieval system or transmitted in any form, or by any means, electronic, mechanical, photocopying, recording or otherwise without prior permission, in writing, from the publisher.

Preface

We are pleased to dedicate this Festschrift to Prof. Klaus-Dieter Schewe on his 60th Birthday. This Festschrift contains a collection of papers contributed by his former students, collaborators and colleagues – most are researchers whose academic careers have been strongly influenced by him. We offer this Festschrift as a token of acknowledgment of his scholarly achievements, services and contributions to teaching.

Prof. Klaus-Dieter Schewe received a Ph.D. in Mathematics from the University of Bonn in 1985. His Ph.D. thesis was in the area of group representation theory. After that, he worked in various ICT establishments in different roles such as IT consult, industrial developer, and industrial researcher. In 1990, he decided to return to an academic career, and subsequently obtained a Habilitation degree (Dr.rer.nat.habil.) in Computer Science, from the Brandenburg University of Technology in 1995. His habilitation thesis was in the area of formal semantics, in particular topos theory and intuitionistic logic. Since then, Prof. Klaus-Dieter Schewe has worked in a number of universities and scientific institutions over the world, including: Technical University of Clausthal in Germany, Massey University in New Zealand, and Software Competence Center Hagenberg GmbH in Austria. This helped him build up a wide research spectrum covering on one side theoretical foundations and on the other side their practical applications.

This Festschrift contains three themes: formal methods, databases and conceptual modeling, because Prof. Klaus-Dieter Schewe has made important contributions in all these research areas.

- In the area of formal methods, the early work of Prof. Klaus-Dieter Schewe was on the integration of algebraic and model-based specifications and types expressing semantics via topoi, which was the major theme in his habilitation thesis. While his research was originally inspired by guarded commands, predicate transformers and thus the B language, Prof. Klaus-Dieter Schewe later became more interested in Abstract State Machines (ASMs). This change was motivated by the theoretical foundations within ASMs as reported by Gurevich's seminal work on the sequential ASM thesis, and the potential of ASMs for supporting real-world applications as exemplified in the area of software engineering. Together with his collaborators, Prof. Klaus-Dieter Schewe has studied behavioral theory and logic of different classes of algorithms such as (synchronous) parallel algorithms, concurrent algorithms, and reflective sequential algorithms. One of his articles on "Concurrent Abstract State Machines" was listed in the 21st Annual Best of Computing in 2016 under the category "F. Theory of Computation" by ACM.

- In the area of databases, Prof. Klaus-Dieter Schewe concentrated on object-oriented databases or more abstractly speaking databases that use rational trees as building blocks. One of his major interests concerns integrity constraints and in particular data dependencies, and the largely neglected area of updates. With his students and collaborators, axiomatizations for several classes of dependencies in presence of various constructors have been achieved. In addition to these, he also stimulated research on a physical architecture for object-oriented databases emphasizing distribution design, linguistic reflection, a stack-based approach to the integration of query and programming languages, persistent storage, and multi-level transactions.

- In the area of conceptual modeling, Prof. Klaus-Dieter Schewe created the co-design approach to Web Information Systems development, which combines semi-formal usage-oriented modeling with formal reasoning techniques derived from dynamic and deontic logics, and algebraic term rewriting.

Although Prof. Klaus-Dieter Schewe has conducted his research on diversified topics, almost all of his scientific contributions are directly or indirectly linked to formal, mathematical foundations.

In addition to his many contributions to the aforementioned research areas, Prof. Klaus-Dieter Schewe has been an awesome colleague and mentor, reading and commenting on the work of us and giving encouragement to young students. He has always been the key driver for inspiring his colleagues and students to pick up problems and to collaborate on them. Most importantly, he has empowered his colleagues and students to grow and realize their potential.

March 2018
Hagenberg, Canberra and Kiel

Atif Mashkoor
Qing Wang
Bernhard Thalheim

Table of Contents

I Conceptual Modeling

Development of Conceptual Models and the Knowledge Background
Provided by the Rigor Cycle in Design Science 3
Ajantha Dahanayake, Bernhard Thalheim

Integrating Social Media Information into the Digital Forensic
Investigation Process 29
Antje Raab-Düsterhöft

Normal Models and Their Modelling Matrix 44
Bernhard Thalheim

II Databases

Extremal Combinatorics of SQL Keys 75
Sven Hartmann, Markus Kirchberg, Henning Koehler, Uwe Leck, Sebastian Link

Polynomially Bounded Valuations in Higher-Order Logics over
Relational Databases 92
Flavio Ferrarotti, Loredana Tec, José María Turull-Torres

Tolerant Constraint-Preserving Snapshot Isolation: Extended
Concurrency for Interactive Transactions 122
Stephen J. Hegner

Viewpoint-Oriented Data Management in Collaborative Research
Projects 146
Yannic Ole Kropp, Bernhard Thalheim

III Rigorous Methods

Cyberphysical Systems: A Behind-the-Scenes Foundational View 177
Richard Banach, Wen Su

The Role of Validation in Refinement-Based Formal Software
Development 202
Jean-Pierre Jacquot, Atif Mashkoor

Hagenberg Business Process Modelling Method – Towards a Homogeneous
Framework for Integrating Process, Actor, Dialogue, and Data Models ... 220
*Verena Geist, Felix Kossak, Christine Natschläger, Christa Illibauer,
Thomas Ziebermayr, Atif Mashkoor*

Closing the gap between the specification and the implementation:
the ASMETA way 242
Paolo Arcaini, Angelo Gargantini, Elvinia Riccobene

Addressing Client Needs for Cloud Computing using Formal Foundations 264
Andreea Buga, Sorana Tania Nemeş, Atif Mashkoor

From Concepts in Non-Monotonic Reasoning to High-Level Implementations
Using Abstract State Machines and Functional Programming 286
Christoph Beierle, Steven Kutsch, Gabriele Kern-Isberner

IV Miscellaneous

Recent Developments in Armstrong Codes 311
Attila Sali

Misunderstandings and Their Overwhelming Success 330
Alexander Bienemann

QoS-Aware Web Service Composition Using Graph Databases 336
Hui Ma, Zhaojiang Chang, Alexandre Sawczuk da Silva, Sven Hartmann

Author Index 352

Part I

Conceptual Modeling

Development of Conceptual Models and the Knowledge Background Provided by the Rigor Cycle in Design Science

Ajantha Dahanayake[1] and Bernhard Thalheim[2]

[1] School of Business and Management, Lappeenranta University of Technology, Finland
`Ajantha.Dahanayake@lut.fi`
[2] Department of Computer Science, Christian Albrechts University Kiel, D-24098 Kiel, Germany
`thalheim@is.informatik.uni-kiel.de`

Abstract. Models, modeling languages, modeling frameworks and their background have dominated conceptual modeling research and information systems engineering for four decades. Conceptual models are mainly used as mediators between the application world and the implementation or system world. Modelling is still conducted as the work of an artisan, involving skilled workmanship. While a general notion of the model and of the conceptual model already has been developed, the modelling process is not investigated so well. It is currently based on modelling methodologies. Each modeling step considers specific work products, orients towards specific aspects of the system or application, involves different partners, and uses a variety of resources. The reasoning process in a modeling step involves a different style of reasoning.

Modelling has to be based on principles and a general theory of modelling activities. The modelling activities need however a conceptualization. Design science distinguishes the relevance cycle as the iterative process that re-inspects the application and the model, the design cycle as the iterative model development process, and the rigor cycle that aims in grounding and adding concepts developed to the knowledge base. This separation of concerns into requirements engineering, model development and conceptualization is the starting point for this paper. We combine approaches developed in design science, ontology engineering, decision processes, and conceptual modelling for the development of general stages, phases and steps of modelling. The main elements of our approach discussed in this paper are the way in which a modelling decision is made and which phases and steps are commonly observed during modelling.

Keywords: Conceptual modelling, modelling actions, modelling decisions, steps and phases of modelling, design science, rigor cycle, support for modelling decisions, modelling as an apprenticeship and art

1 Introduction

Conceptual modelling is one of the central activities in Computer Science. Conceptual models are mainly used as intermediate artefact or mediator for system construction, i.e. describing an understanding of an application domain and prescribing the system that is going to be developed, to be modernised, to be re-engineered, or to be integrated. Other functions of conceptual models are documentation of an existing information system, communication and negotiation among stakeholders, control of parts of a system, simulation of system behaviour visualisation of structure or behaviour of information systems, or explanation and reasoning on a system.

The notion of the conceptual model is not yet commonly accepted within the database community. There are several competing notions. The *conceptual model of an information system* consists of a conceptual schema and of a collection of conceptual views that are associated (in most cases tightly by a mapping facility) to the conceptual schema [67]. In a nutshell, a *conceptual model* is an enhancement of a model by concepts from a concept(ion) space.

1.1 Perspectives on Conceptual Modelling

Based on the notions in the Encyclopedia Britannica [47], we distinguish between the conception of a *model*, the conception of a *model activity*, and the conception of systematic, reflected and well-organised *modelling* (models-model-activities-modelling (briefly MMM[1]) [66]). These three 'M's' are elaborated as follows::

The model as an artefact: A model is a well-formed, adequate, and dependable instrument that represents origins [4, 11, 63, 64]. The model is something set or held for guidance or imitation of an origin and is a product at the same time. Models are enduring, justified and adequate artefacts from one side. From the other side, models represent the state of comprehension or knowledge of a user.

To model as an activity: 'To model' is a scientific or engineering activity beside theoretical or experimental investigation. The activity is an additive process. Corrections are possible during this activity. Modelled work may be used for construction of systems, for exploration of a system, for definition and negotiation, for communication, for understanding and for problem solving.

Modelling as a systematically performed, reflected, technological process: Modelling is a technique for systematically using knowledge from computer science and engineering to introduce technological innovations into the planning and development stages of a system. At each stage the modeller is

---

[1] In [66], the MMM approach to modelling is forwarded. The name MMM comes from the German Modelle (the plural noun models), modellieren (the verb model), and Modellierung (the noun modelling).

likely to ask both why and how, rather than merely how. Modelling is thus based on paradigms and principles.

This MMM approach to modelling has been investigated for models in agriculture, archaeology, arts, biology, chemistry, computer science, economics, electrotechnics, environmental sciences, farming, geosciences, historical sciences, languages, mathematics, medicine, ocean sciences, pedagogical science, philosophy, physics, political sciences, sociology, and sports (see [66]). The MMM approach provides a foundation for the *modelling method*.

While the notion of the model has already attracted a lot of research, activities of modelling, modelling methodologies, and systematic modelling have not attracted the same research effort. So, we face the need to systematically develop an approach to modelling activities. Modelling is typically performed with background knowledge and experience. Any modelling step is a result of decisions made by the modeller in accordance to the requirements to the forthcoming system or to the modernisation of a system. Most textbooks and monographs on conceptual modelling introduce a modelling language and use this language for modelling in a 'cookbook' manner. Modelling techniques and methods are less systematically introduced. Experienced modellers use their previous models, learned in the process of development of these models, abstracted from those models, developed some kind of reference models that might be of use for later development, and developed their approach to development of conceptual models. So, we might ask whether there is *an explicit and sophisticated engineering of models* based on the knowledge somebody or a community has already gained. If engineering approaches exist then we might generalise these approaches to *systematic guidance for performing modelling*.

Comparing this separation with systematic engineering [24], we reconsider these three perspectives in the layers behind the W*H specification pattern [62, 65]. which includes:

- a plan or purpose dimension (concept: wherefore, why, to what place or end[2], for when, for which reason),
- an artefact dimension (content: what, with or by means of with[3]),
- an annotation dimension (topic: with or by which[4]),
- a development and deployment dimension (how, whence, what in, what out),
- a language dimension (representation)
- a context dimension (at or towards which[5], where about, to what place or situation[6], when)
- a user or stakeholder dimension (who, by whom, to whom, whichever)

---
[2] In the sense of the more archaic: whereto.
[3] In the sense of the more archaic: wherewith.
[4] In the sense of the more archaic: whereof.
[5] In the sense of the more archaic: whereat.
[6] In the sense of the more archaic: whither.

- an application dimension (in what particular or respect [7], from which[8], for what, where, whence), and
- additionally, the added value dimension (evaluation).

The rhetoric framework [10] developed by Hermagoras of Temnos[9] distinguishes five modelling perspectives:

Purpose, function, goal perspective: Models and model development serve a certain purpose in some utilisation scenarios. The model has to function in these scenarios and should thus be of certain quality. At the same time it is embedded into the context and is acceptable by a community of practice with its rules and understandings.

Product perspective: Models are products that are requested, have been developed, are delivered according to the first perspective, are potentially applicable within the scenarios, and have their merits and problems.

Engineering perspective: Models are mastered within an engineering process. Modelling is a systematically performed process that uses methods, techniques, preparations, and experience already gained in former modelling processes.

Science perspective: Model development and utilisation is a systematic, well-founded process that allows one to reason on the capacity and potential of the model, to handle adequacy and dependability of models in a proper way, and the reason on the model and its origins that it represents. A science also answers the by-what-means question beside providing the background.

Culture and society perspective: Modelling is a well-accepted, well-structured, and well-founded process that is commonly applied and accepted in a community of practice as a culture of models, of modelling, and common understanding. The modelling culture [10] [21] is the basis of the MMM method and thus a well-accepted, learned, socially practised system of shared values, approaches, techniques, and attitudes with its rules of the 'game' within a social community.

These perspectives have a common underpinning that could be considered as the 'ground-zero' perspective:

---

[7] In the sense of the more archaic: wherein.

[8] In the sense of the more archaic: wherefrom.

[9] The work of Hermagoras of Temnos is almost lost. He have had a great influence on orality due to his proposals. For instance, Cicero has intensively discussed his proposals and made them thus available. It consists of seven questions: Quis, quid, quando, ubi, cur, quem ad modum, quibus adminiculis (W7: Who, what, when, where, why, in what way, by what means). Later, E. Koziol [32] and may be independently J. A. Zachman [75] rephrased it.

[10] According to Hofstede [20] et al., culture can be defined as "a system of shared values, which distinguishes members of one group or category of people from those of another group; culture is therefore intrinsic in the mind of individuals and it can be measured".

Collaboration and modelling background: The social community of the MMM method [66] has intentionally set-up its understanding of collaboration. The community communicates, knows languages, explains, recognizes, accept the grounding behind the models, has been introduced to the basis and is common with it.

The background is typically considered to be given and not explicitly explained. It consists of an undisputable grounding from one side (paradigms, postulates, restrictions, theories, culture, foundations, conventions, authorities) and of a disputable and adjustable basis from other side (assumptions, concepts, practices, language as carrier, thought community and thought style, methodology, pattern, routines, commonsense).

1.2 The Design Science Contribution to Modelling

Design science, according to [28], "is the scientific study and creation of artefacts as they are developed and used by people with the goal for solving practical problems of general interest." Conceptual modeling is about describing (syntax,) semantics (, and pragmatics) of software applications at a high level of abstraction[11]. Specifically, conceptual modelers (1) describe models that represent structures in terms of entities, relationships, and constraints; (2) describe behavior or functional models in terms of states, transitions among states, and actions performed in states and transitions; and (3) describe interactions and user interfaces in terms of messages sent and received, information exchanged, and look-and-feel navigation and appearance [12].

Comparing these general statements, we observe a good overlap whenever information systems are the target of development. The design of an IT artefact includes explication of the problem, definition of requirements, development of the artefact, demonstration of the artefact, and evaluation of the artefact. A similar flow of activities can be distinguished for modelling. Both approaches to development of artificial artefacts [52] may be based on methodological frameworks, on general paradigms and principles, and on general ethical, economical, ecological etc. principles.

Therefore, it is beneficial to integrate the two approaches. In general, design and modelling are partially different and - at the same time - largely similar activities[12]. It seems that the two development approaches share many issues and can benefit from each other.

The controversy [2, 29, 34, 41] discuss the differences between design science and information systems research. It seems that the two research directions are

---

[11] The original citation has been concentrating on semantics. We added the two other semiotics dimensions.

[12] *To design* means, for instance, (1) to create, fashion, execute, or construct according to a plan, (2) to conceive and plan out in the mind driven by a purpose and devised for a specific function or end, and (3) to develop an artefact. *To model* means, for instance, (1) to plan or form after a pattern, (2) to shape or fashion, and (3) to construct a model guided by an origin. [47].

completely different and do not have too much in common. Design science has its background in industrial and interior design and in psychology. Conceptual modelling started with database modelling and is more directly influenced by computer engineering.

Conceptual modelling is a specific form of modelling. Models become conceptualised due to incorporation of concepts - or more generally, conceptions - into the model. These concepts are commonly shared within a community of practice that is involved in the modelling process. Models are a universal vehicle (or better instrument) in almost all sciences and engineering. They can be understood as the 'third' dimension of science [7, 66].

Therefore, we can compare design science research for information systems development with conceptual modelling of information systems. Design science distinguishes the relevance cycle, the modelling cycle, and finally the rigor cycle in Figure 1.

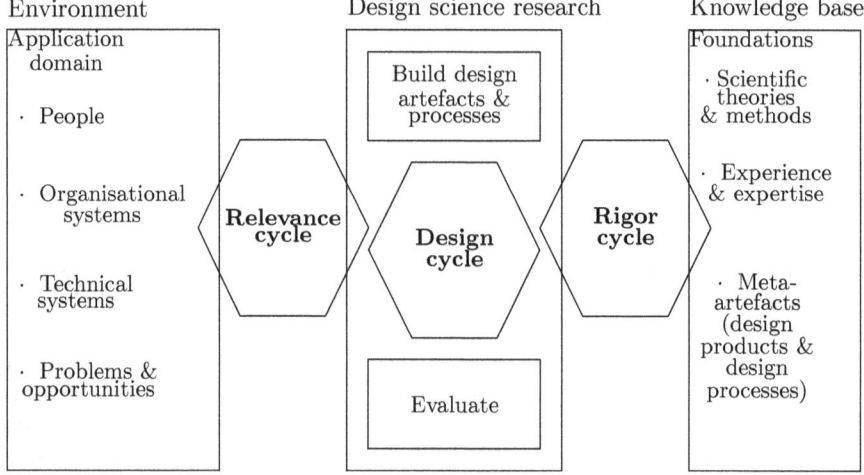

Fig. 1. Design science research cycles (as introduced by A.R. Hevner [19])

This approach can be considered as a specific development process that seeks to enhance the model development process by experience gained so far. The design and development process may by guided by a separation of activities into requirements and objectivity settlement, model development activities (called design cycle), and a reuse of existing development experience from some kind of knowledge base.

In this paper we look more specifically into the rigor cycle and use it for development principles that can be incorporated into for conceptual modelling. Since neither the rigor cycle nor the principles of conceptual modelling have led to an accepted theory, we start our research with one specific aspect of systematic modelling: support for design decisions.

1.3 Objectives and the Storyline of the Paper

In this paper, we discuss modelling foundational principles and theoretical underpinnings for purpose-oriented models and modelling. We continue the research in [60]: "A general theory of modelling also considers modelling *as an apprenticeship* and *as a technology*." The art of modelling should consider all relevant aspects of an application. A typical example is the co-design methodology for design and development of structuring, functionality, and supporting systems such as view and interaction support [57]. The community of practice iteratively obtains a deeper insight and understanding about the necessities and conditions of the problem and the strengths, weaknesses, opportunities and threats of the solution depending on the purpose of the modelling within a modelling process. The central research problems we investigate here are: What are the ingredients of a technology of modelling? Can we approach the modelling process in a systematic way? Can be the modelling process be structured? What is the basis of decisions made in the modelling process?

Our approach extends the MMM approach to modelling activities based on the three cycles of design science research activities of research artefact creation. We thus combine conceptual modelling, design science approaches, decision processes, ontology engineering, and the theory of information system models. We do not intend to review all relevant literature in the rich body of knowledge developed in design science research or conceptual modelling. There are conference series such as DESRIST, ER, and Models etc. and journals such as DKE, EJIS, and MISQ etc. Instead, we follow the approach presented in [73] and use design science research for conceptual modelling of information systems.

Section 2 provides an account of design science, its position on modelling, and the stages of design. Section 3 describes the modelling decisions and its parallels to systematic decision support and the modelling act leading to models and solution imperfection. Section 4 gives an account of systematic conceptual modelling, exploration, and model amalgamation leading to formal model foundation. Section 5 summarises the conclusions of this research.

2 Design Science and Modelling

Design science originated in the area of IT development. It aims at creation on novel and innovative artefacts in the form of models, methods, and systems that support people while developing, using, maintaining, reconsidering, and migrating IT solutions. It considers four perspectives [28]: (1) people, practices and problems; (2) artefacts as solutions to problems in IT practices; (3) the context and anatomy of artefacts; and (4) the study of artefacts. It contributes to improvement by providing new solutions for known problems, to preadaptation of known solutions extended to new problems, to invention of new solutions to new problems, and to routine design of known solutions for known problems.

2.1 Separation of Concern in Design Science through Main Cycles

The Relevance Cycle. Design science research requires the creation of an innovative, purposeful artefact for a special problem domain. The artefact must be evaluated in order to ensure its utility for the specified problem. According to [19], the *relevance cycle* is to initiate design science research within the context of an application to provide the requirements for the research as inputs as well as to define the acceptance criteria for the eventual evaluation of the research results.

The Modelling or Design Cycle. Modelling is a crucial activity in the creation of the design, the artefact. The models and modelling itself implies an ethical change from describing and explaining of the existing world to shaping it. One can question the values of this type of models and modelling oriented design research, i.e. whose values and what values dominate it, emphasizing that research may openly or latently serve the interests of particular dominant groups. The interests served may be those of the host organization as perceived by its top management, those of IS users, those of IS professionals or potentially those of other stakeholder groups in society. Therefore, in order to define the acceptance criteria for ultimate evaluation of the research, modelling and models need to be mapped to a theoretical foundation.

The Rigor Cycle. The *rigor cycle* is considered as the conceptualisation and generalisation or knowledge development cycle [70]. The rigor cycle also aims at the development of knowledge about the application domain and the model. This part of the rigor cycle is conceptualisation. The second target of the rigor cycle is the derivation of abstract knowledge and experience, of scientic theories that can be applied in similar application cases, of (pragmatical) experience for modelling, and of meta-artefact or reference models based on model-driven development (MDD) approaches. Design science aims at another kind of model refinement by adding more rigor after evaluation of a model. This refinement is essentially model evolution and model evaluation. Another refinement is the enhancement of models by concepts. This refinement is essentially a 'semantification' or conceptualization of the model.

We observe that the rigor cycle is orthogonal to the modelling and relevance cycles. The modelling cycle may be broken into a description stage that relates the application domain to the model and a prescription stage that uses the model for system construction. The rigor cycle which is somehow orthogonal has at least two facets: one facet that is important for the model and one facet that is important for generalisation of the model, e.g., for derivation of patterns or reference models and for extraction of model and modelling knowledge beyond the actual modelling activity. In this paper, we concentrate on the rigor cycle of conceptualization and the knowledge development for modelling foundational principles and theoretical underpinning to validate the purposeful values of models and modelling within the design science research activities.

2.2 Revisiting the Three Cycles for Design Science Research

The three-cycle view of design science suggests that relevance is attained through identification of requirements (or business needs) and field testing of an artefact within an environment, while rigor is achieved by appropriately grounding the research in existing foundations, methodologies and design theories and subsequently making contributions that add to the existing knowledge base. The design aspect is achieved through a Design Cycle in which the artefact must be built and evaluated thoroughly before "releasing it" to the Relevance Cycle and before the knowledge contribution is output into the Rigor Cycle [17, 19].

Design science combines the social view and the technical view of information systems. Information systems are purposefully designed and developed. The combined view considers technologies, technical solutions, structuring and behaviour or information systems, and people involved. The model is an artefact with its constructs, variety of representations, methods, and instantiations. Model development is a process that also contains building the model and evaluating it. It becomes thus a complex collaboration effort of a community of practice. It can be guided by principles such as: design/model as an artefact, problem relevance, evaluation, extracting the research contribution, rigor, considering design as a manifold of activities such as search, and scholarship of obtained results.

Design science is thus a contribution to *apprenticeship for modelling activities* within a system construction scenario. This apprenticeship also orients on development of some kind of knowledge as shown in Figure 1. This modelling insight may be applied in another model development activity. Design science research provides a guideline for knowledge transfer and has thus an agenda for development and delivery.

2.3 Models in Design Science

The knowledge transfer in design science is based on systematic exploration of the perception and domain-situation models. Perception and domain-situation models are specific mental models either of one member or of the community of practice within one application area. It is not the real world or the reality that is represented. It is the common consensus, world view and perception which is represented. Domain-situation models represent a settled perception within a context, especially an application. Perception models might differ from the domain-situation model. They are personal perceptions and judgements of a member of the community of practice. Both model kinds represent observations and phenomena for the community of practice. Typical elements are classifications, categorisations, ontologies and catalogs, background especially the grounding, practices and principles, pattern and solutions, and a commonly accepted basis from the modelling background.

These models also reuse a commonly accepted basis from the modelling background such as potentially available constructions or conceptions as definitional knowledge, signs from a language (symbols, indexes and icons), language-based semiotics and semiology, commonly accepted methods and techniques, guidelines and development approaches, approaches to realisation of models.

Design theory is a theory in another sense than typically considered [4]. Essentially, it is an offer, i.e. a scientific, an explicit and systematic discussion of foundations and methods, with critical reflection, and as a system of assured conceptions providing a holistic understanding. Many scientific and engineering disciplines use this constructive understanding of the notion of theory. At the same time, it is constructive theory with a collection of settled instruction conceptions (e.g. concepts, rules, laws, conditions) for (system) development within practical (technical) and quality (esthetic) norms, according to the goals of construction, and guided by some background. Following this understanding, a theory is understood as the underpinning of engineering principles, similar to theory as understood in the architecture discipline [49] and approaches by Vitruvius [71] and L.B. Alberti [1].

Models, modelling languages, modelling frameworks and their background have dominated conceptual modelling research and information systems engineering for the last four decades. Design science research considers artefacts. It is understood as an object or thing made by humans with the intention that it will be used to address a practical problem. Artefacts are, for instance, physical objects, drawings or blueprints. Models are also artefacts whenever they are not virtual. Artefacts are used in development scenarios. Their functions are what they can do for members in their community of practice, what role they can play for them, and how they can support them in their activities.

Conceptual models are mediators between the application world and the implementation or system world [8]. Design science distinguishes the relevance cycle as the iterative process that re-inspects the application and the model, the design cycle as the iterative model development process, and the rigor cycle that aims in grounding and adding concepts developed to the knowledge base [19]. Research in design science and on conceptual modelling has resulted in a large body of knowledge, practices, and techniques. Modelling is based on modelling activities. Each modelling step considers specific work products, orients towards specific aspects of the system or application, involves different partners, and uses a variety of resources [61] used for system development in computer engineering. The separation into the application world, the modelling and model world and the knowledge or design science world [18, 59] supporting an assessment of the results of modelling and an evaluation of the results of research on modelling.

Conceptual modelling has been oriented in the past mainly to clarification on languages, on methods for deployment of such languages, on (mathematical) theories as foundations of syntactic, semantics and pragmatics of model, and on evaluation and quality guaranteeing methods [24, 40, 43, 59]. The application world is used as a starting point for the development of systems that solve some problems of the application domain under consideration. By analyzing these two directions we come to a conclusion similar to [76]. In reality design science research and research on conceptual modelling are two research issues that may benefit from each other. The two communities are already engaged in a discussion of the added value of each side [5, 6, 33, 35, 42, 44, 45, 53, 73, 72, 74].

We may now compare design science approaches and information systems approaches. Figure 2 is a revision of a comparison given in [9]. It correlates design science notions with approaches to professional system design used in information systems research. The different ways of acting nicely correspond to the three cycles in design science.

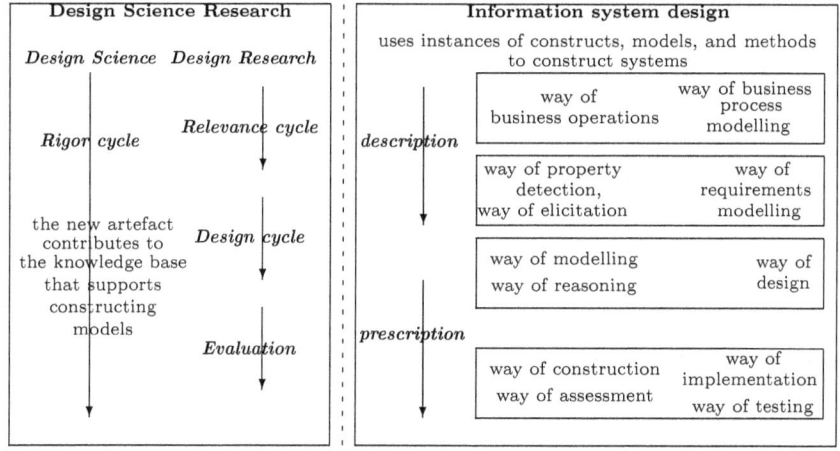

Fig. 2. Framework for understanding Design Science Research and Information Systems Design & Construction

2.4 Modelling as an Activity

Modelling includes two different kinds of activities:

Model deployment is based on activities such as
- adaption, concept enrichment, optimisation, specialisation, instantiation, refinement, grinding,
- applicability studies (evaluation, assurance, composition for application),

- integration, selection, renovation, modernisation, evolution and revolution, migration,
- problem solution, classification, practice, understanding, theory or paradigm evolution (or revolution), and
- explanation.

Model development is typically based on another set of activities such as
- abstraction of origin, scoping, validation, verification, testing, optimisation,
- construction, composition, definition, integration, classification, invention,
- enrichment, adaption, mutation, recombination, refinement, reuse, preparation for deployment, and
- understanding, theory or paradigm injection.

Based on foundations of conceptual modelling [57], ontology engineering [56], and design science for information systems development (e.g. [38]) and summarising, we distinguish three stages of modelling activities:

Stage I: *Model development* is based on four phases: description, formulation, ramification, and validation. In the description phase, individual perception and situation models involved into the modelling situation, are isolated and the corresponding primary properties are identified and represented. We realise in the next sections that this phase includes exploration and model amalgamation. In the formulation phase, properties are interrelated, integrated and combined into a preliminary, initial model. This model is analysed in a ramification phase in order to check whether the model is a proper solution and to interpret and to consider its implications. Finally, the model and its capability and capacity are assessed in a validation phase.

Stage II: *Model deployment* considers the developed model within the given application situation, assesses this model in other application contexts in order to evaluate its stability and plasticity, and derives its added value.

Stage III: The rigor cycle also investigates the experience we have gained during developing the given model. Conceptual modelling uses this experience as a hidden intuitive basis for further development. We may however use this experience within a *paradigmatic synthesis* for recapitulation and consolidation of conceptualisation concept gathering, ontologisation, grounding and tagging, i.e. for knowledge acquisition.

Design science has been oriented on the second kind of activities and on stage I. We follow this restriction in this paper. Stage II and stage III are however equally important for a sophisticated modelling approach. Maturity of models can be guided by maturity approaches depending on whether the perception and domain-situation models are mature or not and whether the model engineering became mature or not, e.g. CMM and SPICE [23, 58]. We may distinguish development maturity at the specification, at the control or technical, at the application or technology, at the establishment or organisational, at the value or prediction, or at the optimisation levels. These maturity levels use experiences, skills, knowledge etc. gained for the innovation and adaptation of other processes and products that have not yet reached this maturity.

3 Modelling Decisions

The main question is now *how, when, why, on what, in which way* and *why* design decisions are made beside the organisation of the design process itself, its flow of activities, and the involvement of actors into the design process. Model development is performed in teams. They make their decisions based on some common understanding, based on the perception and domain-specific models, being guided by the principles of modelling agreed in advance, and based on some background for models. These decisions must be properly supported. We thus briefly investigate which kind of support is necessary and how this support can be provided.

3.1 Systematic Decision Support

According to [30], modelling and modelling decisions enhancement (DE) activities are encouraged within the studio concept. A DE studio has five main components:
  - *Studio style*: Learning, Enquiry and Participative.
  - *Decision process coordinators*: these include facilitators, domain experts, and suite.
  - *Scripting*: the balance between improvisation and formalized methods.
  - *Suites and development and support expertise.*
  - *Location and rooms*: the options here range from fixed point to distributed Web conferencing and from simple technology infrastructure to multimedia heaven.

[30] consider the mix of skills required to fulfill the demands of a studio for modelling decisions listed as landscaping, facilitation, recipes, suites and process as a means of a complete package of developing an architecture - a solution. Let us combine this approach with the technology proposal for change management in [27].

- *Landscaping* is the domain of expertise of the business strategist and domain expert. In terms of both understanding the decision issues and decision-makers, information resources, processes and the basics of what to model, why and how. In addition, the landscaper has to have some credibility, whether as an insider or outside adviser, with senior managers and stakeholders. Otherwise, the studio is just an exercise or a "pilot", "prototype" or "lab", all of which are euphemisms for "do not take this too seriously."
- *Facilitation*: Behavioral knowledge and process skills are a key for the process of arriving at a solution.

- *Recipes* apply wherever possible proven recipes that include effective scripts. Recipes are proven, repeatable and transferable, specify ingredients and sequencing, permit variations and innovations, and result in something people eat and are likely to come back for another meal. Building recipes requires research and writing and the willingness to place "secrets" and "methodology" in the public domain. It demands teaching as well: developing a body of knowledge and building a critical mass of skilled practitioners. Since technology moves so fast, each new generation of software draws on a new generation of developers and there is little passing on of experience and knowledge.
- *Suites* ensure that tools are designed and implemented within an overall distributed architecture. The goal of suite development is to make the "system" as transparent, easy to access, reliable as the electrical system, where any breakdown is a news item and crisis.
- *Processes* make commitment to a decision of the explicit target and agenda.

The obstruction here is organizational culture, management style, stakeholder relationships and legacy of existing decision processes.

3.2 The Modelling Action and Design Decisions

The modelling action is similar to the speech act [3, 48] and consists of

1. a selection and construction of an appropriate model depending on the task and purpose and depending on the properties we are targeting and the context of the intended system and thus of the language appropriate for the system,
2. a workmanship on the model for detection of additional information about the original and of improved model,
3. an analogy conclusion or other derivations on the model and its relationship to the real world, and
4. a preparation of the model for its use in systems, to future evolution and to change.

Therefore, the DE studio approach provides specific tactics to modelling.

3.3 Modelling Knowledge and Decisions Imperfection

In the case of conceptual modelling, the rigor cycle can be based on knowledge obtained within the five consecutive phases [14]:

1. *exploration*,
2. *model amalgamation and adduction*,
3. *model formulation*,
4. *model deployment*, and
5. *paradigmatic synthesis*.

Modelling decisions have to be based on transparent and realistic objectives [14, 66, 68, 69]. The correspondences of elements of the conceptual model to particular pattern in the real world or the perception models must be based on conformity criteria. Modelling can be considered as progressive cognition within the context, for the purpose of development and within the concept space. Models cannot be developed in its full scientific rigor and are thus objects of evolution. Modelling actions balance between exploratory decisions (description, explanation, prediction) and inventive aspects (reification, refinement). Modelling kits are supporting the quality of modelling decisions. Modelling actions suffer from the breadth-depth paradox. We want to have as much detail as necessary and want to be as broad as sufficient. Modelling decisions must be continuously evaluated within a modelling process by either mode or all three modes of assessment (coherence, correspondence, commensurability). They are conditionally anchored to the experience, knowledge and intuition gained so far. Modelling also includes negotiation within the community of practice and with the stakeholders of the information system.

Adequacy of models is based on analogy, focusing, and purposefulness of the model. Focusing provides a means for explicit modelling of the divergence from the real world with incompleteness, open issues and potential errors [22]. Therefore, a model is imperfect [25] due to exceptional states that are not considered, incompleteness due to limitations of the modelling language and the scope of modelling, and due to errors, which are either based on real errors or exceptional states or on biases by the community of practice.

4 Systematic Development of Conceptual Models

Let us now consider the modelling phases and steps and highlight the decisions that must be made during modelling. We concentrate in this paper on the first two model development phases: description and formulation. Ramification and validation extend the approach in [69]. The two next stages (model deployment and paradigmatic synthesis) are deferred to a forthcoming paper due to space limitations. The section is based on our entire experience on conceptual modelling and on the experiences of several decades of database realisation[13]. The body of knowledge developed so far and used in real practice is very large. It

---

[13] Due to involvement of the second author into the development and the service for the CASE workbenchs $(DB)^2$ and ID^2 we have collected a large number of real life applications. Some of them have been really large or very large, i.e., consisting of more than 1.000 attribute, entity and relationship types. The largest schema in our database schema library contains of more than 19.000 entity and relationship types and more than 60.000 attribute types that need to be considered as different. Another large database schema is the SAP R/3 schema. It has been analysed in 1999 by a SAP group headed by the second author during his sabbatical at SAP. At that time, the R/3 database used more than 16.500 relation types, more than 35.000 views and more than 150.000 functions. The number of attributes has been estimated by 40.000. Meanwhile, more than 21.000 relation types are

needs however a systematisation, categorisation and generalisation. There are very few publications (e.g. [13, 15, 39, 51, 50, 54]) that provide such systematisation of the experience gained so far. The generalisation and the categorisation is however an open research field so far.

Modelling of structures is a systematically performed technological process. It is a technique for applying knowledge from other branches of engineering and disciplines of science in effective combination to solve a multi-faceted engineering problem. In addition to structure development, it is important to define databases systems themselves. The systems are first of all man-made. migration-resistant. Modelling and especially information system modelling[14,15] is a creation and production process, an explanation and exploration process, an optimisation and variation process, and a verification process. This distinction allows to relate the specific purpose with macro-steps of modelling and with criteria for approval or refusal of modelling results. Modelling is thus at the same time problem solving and engineering.

---

used. The schema has a large number of redundant types which redundancy is only partially maintained. The SAP R/3 is a very typical example of a partially documented system. Many of the design decisions are now forgotten. The high type redundancy is mainly caused by the incomplete knowledge on the schema that has been developed in different departments of SAP over several decades.

[14] We develop our approach here on the approach that is established and widely practised and taught in almost all textbooks. We leave out the more sophisticated approach in [67]. At the business user level, user viewpoints can be represented by user viewpoint schemata or more generally views. At the conceptual level, these viewpoints are going to be harmonised and mapped to a conceptual schema. It is assumed that the user viewpoints are then sub-schemata of the conceptual schema. This conceptual schema is mapped to a logical and later to a physical schema. The viewpoints are cut down to logical views which typically consist of single-table definitions on the basis of a query to the logical schema. A user viewpoint is then called external view. The query might be more complex and thus not be based on a sub-schema of the conceptual schema. The database structure architecture consists of the logical schema, external views defined on top of the logical schema and an implementation or physical schema. With the introduction of the conceptual model, the architecture description has been changed by considering the logical and the physical as an implementation schema and using the conceptual schema as the mediator between views and the implementation schemata. It creates a mismatch since the views are defined on top of the implementation schema. [26] breaks with this three-layer architecture by proposing the conceptual view tower mechanism where business user viewpoints are represented by conceptual views. [67] rounds off this approach by considering the conceptual model to consist of a conceptual schema and a collection of conceptual views.

[15] We concentrate the investigation of the modelling process to 'greenfield' modelling called modelling from scratch and to model gardening called evolutionary modelling. We do not investigate 'brownfield' modelling and modernisation for already operating database systems based on modelling and redevelopment for legacy systems based on macro-modelling methods and especially migration strategies [31]

4.1 The Model Description Phase

In this paper we concentrate our investigation of the model description phase on two (macro-)steps: model exploration and model amalgamation.

Main Phases of Model Description for Starting from Scratch. The exploration step is based on state-of-affairs and the functions a model should play during information system development. The state-of-affairs is typically represented by a reality model that already abstracts from the state-of-affairs and perception models that are used in the community of modellers and business users. We may distinguish in this step a number of activities: the situation and the perception models are disassembled. Later, manifestation may be applied to some situation model. This situation model is negotiated within the community of practice and users and represented by a nominal model.
The next step is model amalgamation. The result is a real model. Amalgamation is oriented on the justification of the model and on the quality criteria for the model. It integrates also criteria for well-formedness of models.

Modelling typically also results in modelling experience that can be elicited during or after model development. This modelling experience elicitation and acquisition is part of paradigmatic synthesis and therefore of the rigor cycle.

Evolutionary Model-Based and Background-Aware Modelling. Modelling is often not performed from scratch. Rather we start with an explication of experience that uses stereotypical or generic models. We may also start with existing models for an already existing information system. We thus elaborate artefacts of interest, e.g. reference or existing models. We explicitly extract the background of these models. Next we explicate their purpose, their background, their context and compare the result with the reality models and the objectives of development. The result is again a situation model.
This situation model is now assessed and evaluated. It is typically reformulated by specialisation and refinement. We thus explicitly describe why the model is adequate and dependable. This model can also be enhanced by formal methods.

In a similar form, the experience gained is incorporated into the body of modelling knowledge.

4.2 The Exploration (Macro-)Step

Exploration starts with a well-defined modelling task, a well-defined scope, and consider choices. It is often assumed to be based on deductive approaches. It seems better however to consider inductive and abductive approaches first.

It is based on the following three steps:

Disassemble: The perception and situation models are converted into constituent parts in dependence on the specific assumptions, specific reality and state-of-the-art properties, and specific foci and scopes. Methods are dissolution, segmentation, analysis of coherent units, refinement and categorisation, and examination.

Manifestation is the act of demonstrating, exhibiting, and demonstration. It often considers familiar situations and examples. We consider typical application situations, typical phenomena, typical system states, concepts and conceptions. During manifestation we are more interested in specific kinds of models, Galilean models oriented on improvement of the state-of-the-art. Manifestation may follow the W*H specification framework [10]. Typical questions answered are: What is the demonstrated situation about? What systems and phenomena are involved into the situation? What is the state of every system? What concepts are necessary to describe and/or explain? What is the reference system? How can these concepts be represented?
Manifestation and reflection consider model properties, model variants and model capabilities. Based on these considerations we may derive obligations for model revision.

Nominal models in the exploration step use parameters and variables that can be instantiated during further model development and progressive model refinement. The glossary, namespace, agreements on conceptions, assumptions on the model background, decisions on model structures, and composition pattern are often imported from the application domain.

The result of this step is now a nominal (or perception) model that is a generalisation of a subsidiary model. This model explicitly represents the background via its grounding (paradigms, culture, background, foundations, theories, postulates, restrictions, authorities, conventions, commonsense) and its basis (concepts, language, routine, training, infrastructure, assumptions, though community, thought style, pattern, methodology, guidelines, practices).

4.3 The Model Amalgamation (Macro-)Step

Amalgamation aims at combination or unification of model elements into one form [58]. It includes merger, consolidation, and mixing or blending of different elements. We typically focus on one real model during conceptual data modelling of information systems. We might also use several models but leave it out of scope within this paper.

Amalgamation is mainly based on inductive reasoning. It might incorporate abductive and adduction reasoning based on the association to the situation and perception models. It can be enhanced by methods of plausible reasoning. Classical development strategies (top-down, bottom-up, modular, inside-out, mixed) provide means for combination and unification of elements. In this case we can use local top-down and bottom-up development operations for ER models [57].

Model composition can follow a number of strategies and tactics, especially for unification. Since we are interested in adequacy and dependability of models we explicitly propagate these properties during amalgamation. Typical ER-based composition principles are: global-as-design, unification of viewpoints, explicit consideration of realisability, empiric evaluation by sample data, homomorphic mappings from the situation and perception models, and consideration of specific elements of the ER modelling language. This step is often governed by practical guidelines and rational constraints (general guiding principles, acceptable tolerance (approximation limits, precision intervals, data preciseness), convenient modes for logging and handling of data, appropriate mathematical or formal representations and operations, norms and corroboration for the real model and criteria for evolution (refinement, modernisation, modification, replacement); level of detail, type system and mapping style to type system, handling of exceptions and deviations (NULL, default), treatment of hierarchies, controlled redundancy, ground type system, quantity matrix (Mengengerüst), constraint enforcement, treatment of cardinality, inherent constraints, naming conventions, abbreviation rules, kind of semantics (set or pointer), weak types, translation and tolerance for complex attributes, handling of identification). We may also require that structures used correspond to natural situations (good design is functional, useful, aesthetic, innovative, good business, honest, long-lasting, minimalistic, understandable, user-oriented, unobtrusive, simple as possible, thorough down to the last detail, and focused).

Finally, maturity of models based on SPICE or CMM has to be provided [69]. The real model should be fully defined, must be well-understood, provide all semantics in an explicit form, and use explicit concepts. We might use different definitional frames but in a coherent form. General modelling principles are modularisation, abstraction, and explicit coupling. It is a good approach to use best practices within the modelling framework. They should allow to preserve also design principles that are given for the realisation environment.

4.4 The Model Formulation Phase

The model formulation phase aims at formulation of the well-formed, adequate and dependable model. This model may use several representation forms, e.g. conceptual data models combine diagrammatic and formal representations. It may also contain several sub-models that represent viewpoints of business users [67]. Finally, the assessment of the model is explicitly given.

The formulation is based on decisions such as depiction of the elements of previous models with an explicit consideration of the model function and purpose. We develop criteria for adequacy and dependability of the model and start to explicitly represent the model grounding and background. Since the model should also represent various viewpoints of business users we have to enhance it by explicit view schemata and aggregations as well as abstractions. Reasoning on justification might be based on an argument calculus [66] or argument logics [36].

The model must be completed. Typical drivers for completion are application domain requirements, the background behind the situation and perception models, the specifics of the modelling language, and the generic model behind the model. We need to adjust the scope of modelling elements. The goal is also to develop a well-formed model that fits well to the situation and perception models.

The model is also assessed by an elementary deployment and tested against the real world. This assessment is often backed by some test data based on an experimentation strategy. It might also be tested by elementary utilisation. The first main result of assessment is a justification of the model by an explanatory statement, by confirmation of rational coherence, by a validation of the model against the state-of-affairs, and by explicit consideration of stability of the model against non-essential deviations of the state-of-affairs. The second main result of assessment is an explicit statement on model quality based on quality characteristics for quality in use, external quality and internal quality. The assessment allows to reason on the model capacity and potential.

5 Combining Conceptual Modelling and Design Science

Design science and information system design and development can be seen as two different approaches to modelling. Our investigation leads however to the conclusion that these two approaches can neatly enhance each other. The process of model development starts with a model representing the state-of-affairs. The final model should represent the extended, idealised and distorted reality. Perception and domain-specific models are the basis for the first model. They are also used for evaluation of the final model. Figure 3 combines the approaches of design science (research) [18, 38] with those based on main duties for system development [16] and those typically used for conceptual modelling [57].

The rigor cycle can be understood as a knowledge and experience gathering process. The relevance cycle is based on observation of the state of affairs, scoping of the demands for system development, and describing a view of the application domain. These cycles form the y-dimension. We also use the x-dimension for explicit display of the changes imposed to the reality. Typically, information systems augment the reality.

The design or modelling cycle uses the scoped application domain description for the development of a model. The rigor cycle adds semantics, meaning and context to the model. The description of the scoped application domain may directly be used for system development. For instance, agile development is typically following this direct approach. The model may also be directly used for system development. The advantage of such approach is that all relevant elements are supported by a model and that the model may be used for understanding the system. The system is therefore defined. We may also use the model for development of a behaviour description, guidelines (e.g., for system deployment) and documentation. In this case modelling is established.

Furthermore, we might background the model by concepts. In this case, users of the model may perform system construction with a sense of groundwork behind the model and the description of the application domain. Models may

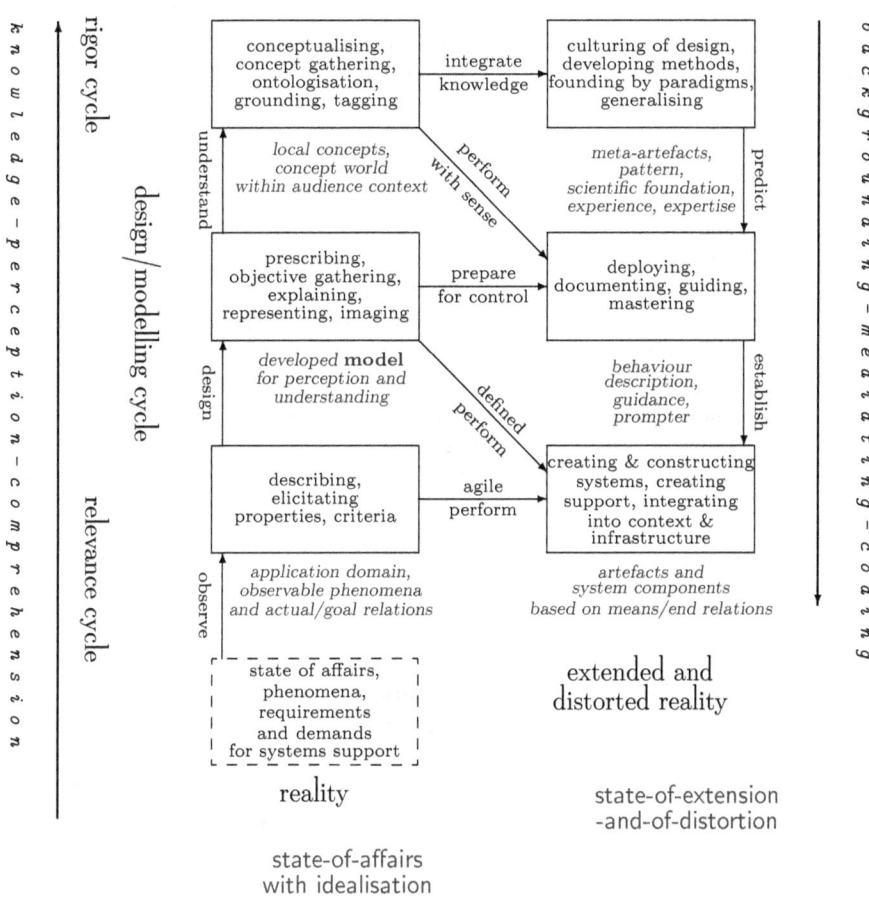

Fig. 3. The relevance/design/rigor and the state-of-affairs/augmentation dimensions [61]

also be a part of a knowledge base. In this case we integrate, generalise and found the model through concepts in the knowledge base.

The relevance, design and rigor cycles are based on comprehension of the application domain, perception of the relevant elements and knowledge or understanding development for those elements. During system development models are used as a mediating artefact. They describe and image the problems, phenomena and demand form one side and serve as a prescription for systems development from the other side. Models may also serve as a background and foundation of the system if they are integrated with concepts.

6 Conclusion

This paper shows how design science and conceptual modelling can benefit from each other. More specifically we discuss how conceptual modelling can benefit from design science research. Modelling of IT artefacts typically starts with an understanding of the state of affairs, with objectives and consideration of requirements. This perception may be described and directly used for the development of new IT artefacts without any models, such as when using the agile approaches. The modelling cycle results however in a model that can be used for IT development either directly or in a more reflected way. The rigor cycle in design science and modelling as a systematically performed, reflected and well organised process is based on an understanding of all actions undertaken. This process can also be used for development of new knowledge and its integration into the existing body of knowledge.

Therefore, we observe that the rigor cycle and systematics of modelling may enhance and complement each other.

Since design science research and conceptual modelling are tackling the same problem - proper development of (information) systems - we discussed in this paper how design science research can be used for underpinning of modelling activities. The decision steps we presented are the basis for a general stepwise procedure of systematic design.

The formalisation of this approach is delayed to a forthcoming paper. Formalisation also includes approaches to a general theory of modelling such as [14, 37, 46, 55]. The main issue so far is the development of the combined approach.

Acknowledgement. We acknowledge the constructive and fruitful proposals and suggestions, the comments, and the critics of our reviewers. All this led to a substantial revision and improvement of the paper.

References

1. L.B. Alberti. *On the art of building in ten books.* MIT Press, Cambridge. Promulgated in 1475, published in 1485, 1988.

2. R. Baskerville, K. Lyytinen, V. Sambamurthy, and D. Straub. A response to the design-oriented information systems research memorandum. *European Journal of Information Systems*, pages 1–5, 2010.
3. M. Berg, A. Düsterhöft, and B. Thalheim. Query and answer forms for sophisticated database interfaces. In *Information Modelling and Knowledge Bases*, volume XXIV, pages 161–180. IOS Press, 2013.
4. M. Bichler, U. Frank, D. Avison, J. Malaurent, P. Fettke, D. Hovorka, J. Krämer, D. Schnurr, B. Müller, L. Suhl, and B Thalheim. Theories in business and information systems engineering. *Business & Information Systems Engineering*, pages 1–29, 2016.
5. D. Bjørner. *Domain engineering*, volume 4 of *COE Research Monographs*. Japan Advanced Institute of Science and Technolgy Press, Ishikawa, 2009.
6. J. Vom Brocke and C. Buddendick. Reusable conceptual models - requirements based on the design science research paradigm. In *DESRIST*, pages 576–604, 2006.
7. S. Chadarevian and N. Hopwood, editors. *Models - The third dimension of science*. Stanford University Press, Stanford, California, 2004.
8. A. Dahanayake and B. Thalheim. Co-evolution of (information) system models. In *EMMSAD 2010*, volume 50 of *LNBIP*, pages 314–326. Springer, 2010.
9. A. Dahanayake and B. Thalheim. Enriching conceptual modelling practices through design science. In *BMMDS/EMMSAD*, volume 81 of *Lecture Notes in Business Information Processing*, pages 497–510. Springer, 2011.
10. A. Dahanayake and B. Thalheim. W*h: The conceptual model for services. In *Correct Software in Web Applications and Web Services*, Texts & Monographs in Symbolic Computation, pages 145–176, Wien, 2015. Springer.
11. D. Embley and B. Thalheim, editors. *The Handbook of Conceptual Modeling: Its Usage and Its Challenges*. Springer, 2011.
12. D. Embley and B. Thalheim. Preface. In *The Handbook of Conceptual Modeling: Its Usage and Its Challenges*, pages v–ix. Springer, Berlin, 2011.
13. M. Fowler. *Refactoring*. Addison-Wesley, Boston, 2005.
14. I.A. Halloun. *Modeling Theory in Science Education*. Springer, Berlin, 2006.
15. D. C. Hay. *Data model pattern: Conventions of thought*. Dorset House, New York, 1995.
16. L.J. Heinrich, A. Heinzl, and R. Riedl. *Wirtschaftsinformatik: Einführung und Grundlegung*. Springer, Berlin, 4 edition, 2011.
17. A. Hevner and S. Chatterjee. *Design research in formation systems*. Springer, Berlin, 2010.
18. A. Hevner, S. March, J. Park, and S. Ram. Design science in information systems research. *MIS Quaterly*, 28(1):75–105, 2004.
19. A. R. Hevner. The three cycle view of design science research. *Scandinavian J. Inf. Systems*, 19(2), 2007.
20. G. Hofstede, G.J. Hofstede, and M. Minkow. *Cultures and Organizations: Software of the Mind: Intercultural Cooperation and Its Importance for Survival*. McGraw-Hill, New York, 2010.
21. H. Jaakkola, J. Henno, B. Thalheim, and J. Mäkelä. Collaboration, distribution and culture - challenges for communication. In *MiPRO*, pages 657–664. IEEE, 2015.
22. H. Jaakkola, J. Henno, T. Welzer-Družovec, J. Mäkelä, and B. Thalheim. Why information systems modelling is so difficult. In *SQAMIA'2016*, pages 29–39, Budapest, 2016. CEUR Workshop Proceedings.

23. H. Jaakkola, T. Mäkinen, B. Thalheim, and T. Varkoi. Evolving the database co-design framework by SPICE. In *Informaton Modelling and Knowledge Bases Vol. XVII, Series Frontiers in Arificial Intelligence, volume 136*, pages 268–279. IOS Press, May 2006.
24. H. Jaakkola and B. Thalheim. A framework for high quality software design and development: A systematic approach. *IET Software*, pages 105–118, April 2010.
25. H. Jaakkola and B. Thalheim. Exception-aware (information) systems. In *Information Modelling and Knowledge Bases*, volume XXIV, pages 300–313. IOS Press, 2013.
26. H. Jaakkola and B. Thalheim. Multicultural adaptive systems. In *Information Modelling and Knowledge Bases*, volume XXVI of *Frontiers in Artificial Intelligence and Applications, 272*, pages 172–191. IOS Press, 2014.
27. K. Jannaschk, H. Jaakkola, and B. Thalheim. Technologies for database change management. In *ADBIS Vol. II*, volume 312 of *Advances in Intelligent Systems and Computing*, pages 271–286. Springer, 2014.
28. P. Johannesson and E. Perjons. *An introduction to design science*. Springer, Cham, 2014.
29. I. Junglas, B. Niehaves, S. Spiekermann, B.C. Stahl, T. Weitzel, R. Winter, and R. Baskerville. The inflation of academic intellectual capital: the case for design science research in europe. *European Journal of Information Systems*, pages 1–6, 2010.
30. P.G.W. Keen and H.G. Sol. *Decision enhancement services : rehearsing the future for decisions that matter*. IOS Press, 2008.
31. M. Klettke and B. Thalheim. Evolution and migration of information systems. In *The Handbook of Conceptual Modeling: Its Usage and Its Challenges*, chapter 12, pages 381–420. Springer, Berlin, 2011.
32. E. Koziol. *Organisation der Unternehmung*. Gabler, 1962.
33. B. Kuechler and V. Vaishnavi. On theory development in design science research: Anatomy of a research project. *European Journal of Information Systems*, 17(5):489–504, 2008.
34. K. Lyytinen, R. Baskerville, J. Iivari, and D. Téeni. Why the old world cannot publish? overcoming challenges in publishing high-impact is research. *EJIS*, 16(4):317–326, 2007.
35. L. Maciaszek. *Requirements analysis and design*. Addison-Wesley, Harlow, Essex, 2001.
36. L. Magnani, W. Carnielli, and C. Pizzi, editors. *Model-Based Reasoning in Science and Technology: Abduction, Logic, and Computational Discovery*. Springer, 2010.
37. B. Mahr. On judgements and propositions. *ECEASST*, 26, 2010.
38. S.T. March and V.C. Storey. Design science in the information systems discipline: An introduction to the special issue on design science research. *MIS Quarterly*, 4:725–730, 2008.
39. D. Marco and M. Jennings. *Universal meta data models*. Wiley Publ. Inc., 2004.
40. A. Olivé. *Conceptual modeling of information systems*. Springer, Berlin, 2007.
41. H. Österle, J. Becker, U. Frank, T. Hess, D. Karagiannis, H. Krcmar, P. Loos, P. Mertens, A. Oberweis, and E.J. Sinz. Memorandum on design-oriented information systems research. *European Journal of Information Systems*, pages 1–4, 2010.
42. K. Pohl. *Process centred requirements engineering*. J. Wiley and Sons Ltd., 1996.

43. N. Prat, I. Comyn-Wattiau, and J. Akoka. Artifact evaluation in information systems design-science research - a holistic view. In *18th Pacific Asia Conf. on Inf. Syst., PACIS 2014, Chengdu, China, June 24-28, 2014*, page 23, 2014.
44. S. Robertson and J. Robertson. *Requirements-led project management*. Pearson, Boston, 2005.
45. C. Rolland. From conceptual modeling to requirements engineering. In *Proc. ER'06*, LNCS 4215, pages 5–11, Berlin, 2006. Springer.
46. A. Rutherford. *Mathematical Modelling Techniques*. Dover publications, 1995.
47. J.E. Safra and j. Aquilar-Cauz et. al, editors. *Encyclopædia Britannica Ultimate Reference Suite*. Encyclopaedia Brittanica Inc., Chicago, fifteenth edition, Ultimate Reference Suite, Version 2015, CD-ROM / DVD, Deluxe Edition (Software Development, Magic Software, India).
48. J. Searle. *Speech Acts: An essay in the philosophy of language*. Cambridge University Press, Cambridge, England, 1969.
49. G. Semper. *Die vier Elemente der Baukunst*. Braunschweig, 1851.
50. L. Silverston. *The data model resource book. Revised edition*, volume 2. Wiley, 2001.
51. L. Silverston, W. H. Inmon, and K. Graziano. *The data model resource book*. John Wiley & Sons, New York, 1997.
52. H.A. Simon. *The science of the artificial*. MIT Press, Cambridge, 1996.
53. G. Simsion. *Data modeling essentials - Analysis, design and innovation*. Van Nonstrand Reinhold, New York, 1994.
54. G. Simsion and G.C. Witt. *Data modeling essentials*. Morgan Kaufmann, San Francisco, 2005.
55. B.Ja. Sovetov and S.A. Jakovlev. *Systems Modelling*. Vysschaja Schkola, 2005. In Russian.
56. V. Sugamaran and V. Storey. The role of domain ontologies in database design: An ontology management and conceptual modeling environment. *ACM TODS*, 31(3):1064–1094, 2005.
57. B. Thalheim. *Entity-relationship modeling – Foundations of database technology*. Springer, Berlin, 2000.
58. B. Thalheim. Towards a theory of conceptual modelling. *Journal of Universal Computer Science*, 16(20):3102–3137, 2010. http://www.jucs.org/jucs_16_20/towards_a_theory_of.
59. B. Thalheim. The theory of conceptual models, the theory of conceptual modelling and foundations of conceptual modelling. In *The Handbook of Conceptual Modeling: Its Usage and Its Challenges*, chapter 17, pages 547–580. Springer, Berlin, 2011.
60. B. Thalheim. The art of conceptual modelling. In *Information Modelling and Knowledge Bases XXII*, volume 237 of *Frontiers in Artificial Intelligence and Applications*, pages 149–168. IOS Press, 2012.
61. B. Thalheim. The science and art of conceptual modelling. In A. Hameurlain et al., editor, *TLDKS VI*, LNCS 7600, pages 76–105. Springer, Heidelberg, 2012.
62. B. Thalheim. The conception of the model. In *BIS*, volume 157 of *Lecture Notes in Business Information Processing*, pages 113–124. Springer, 2013.
63. B. Thalheim. The conceptual model ≡ an adequate and dependable artifact enhanced by concepts. In *Information Modelling and Knowledge Bases*, volume XXV of *Frontiers in Artificial Intelligence and Applications, 260*, pages 241–254. IOS Press, 2014.

64. B. Thalheim. Conceptual modeling foundations: The notion of a model in conceptual modeling. In *Encyclopedia of Database Systems*. Springer US, 2017.
65. B. Thalheim and A. Dahanayake. Comprehending a service by informative models. In *Conceptual Modeling of Services*, LNCS 10130, pages 87–108, Berlin, 2016. Springer.
66. B. Thalheim and I. Nissen, editors. *Wissenschaft und Kunst der Modellierung: Modelle, Modellieren, Modellierung*. De Gruyter, Boston, 2015.
67. B. Thalheim and M. Tropmann-Frick. The conception of the conceptual database model. In *ER 2015*, LNCS 9381, pages 603–611, Berlin, 2015. Springer.
68. B. Thalheim and M. Tropmann-Frick. Models and their capability. In C. Beierle, G. Brewka, and M. Thimm, editors, *Computational Models of Rationality*, volume 29 of *College Publications Series*, pages 34–56. College Publications, 2016.
69. B. Thalheim and M. Tropmann-Frick. Wherefore models are used and accepted? The model functions as a quality instrument in utilisation scenarios. In I. Comyn-Wattiau, C. du Mouza, and N. Prat, editors, *Ingénierie Management des Systèmes d'Information*, pages 131–143. Cépaduès, 2016.
70. J. R. Venable. Design science research post Hevner et al.: Criteria, standards, guidelines, and expectations. In *DESRIST*, volume 6105 of *Lecture Notes in Computer Science*, pages 109–123. Springer, 2010.
71. Vitruvius. *The ten books on architecture (De re aedificatoria)*. Oxford University Press, London, 1914.
72. Y. Wand, D. E. Monarchi, J. Parsons, and C. C. Woo. Theoretical foundations for conceptual modeling in information systems development. *Decision Support Systems*, 15:285 – 304, 1995.
73. R. Wieringa. *Design science methodology for information systems and software engineering*. Springer, Heidelberg, 2014.
74. R. Winter. Design science research in Europe. *European Journal of Information Systems*, 17(5):470–475, 2008.
75. J. A. Zachman. A framework for information systems architecture. *IBM Systems Journal*, 38(2/3):454–470, 1999.
76. S. Zelewski. Kann Wissenschaftstheorie behilflich für die Publikationspraxis sein? In F. Lehner and S. Zelewski, editors, *Wissenschaftstheoretische Fundierung und wissenschaftliche Orientierung der Wirtschaftsinformatik*, pages 71–120. GTO, 2007.

Integrating Social Media Information into the Digital Forensic Investigation Process

Antje Raab-Düsterhöft

Hochschule Wismar
University of Applied Sciences, Technology, Business and Design
Philipp-Müller-Strae 14
23966 Wismar
Germany
Email: antje.duesterhoeft@hs-wismar.de

Abstract

In this paper, we will illustrate how data from social media sources can be integrated into an environment for a digital forensic investigation process and the underlying model. Social media information will be combined with retrieved information from different forensic tools, e.g. X-Ways, Autopsy or database forensic investigation information. The retrieved forensic information as well as the investigation steps will be time-based stored into a database, so that the chain of custody is supported and time-based questions of an investigation can be used for an investigation documentation.

1 Introduction

Digital forensics is becoming increasingly important with the escalation of cybercrime and other network-related serious crimes. Unfortunately, there are times when an incident occurs and organizations have to be able to support the digital investigation process with the electronic data needed to conduct analysis and arrive at credible and factual conclusions.

An investigation is a systematic examination, typically with the purpose of identifying and verifying facts. The 5WH formula sets the following objectives: Who (persons involved), Where (the location of crime), What (the description of the facts), When (the time of crime), Why (the motivation of crime) and How (the crime was committed). Digital investigation and digital forensics refer to forensic science applied to digital information. Digital crimes and incidents consist of a digital event or a sequence of events.

Two most important fundamental principles of a digital investigation are digital evidence integrity and chain of custody. Digital evidence integrity refers to the preservation of evidence in its original form. Chain of custody refers to the documentation of acquisition, control, analysis, and disposition of physical and electronic evidence.

The research project described below is supported by WINGS[1] and related to the correspondence degree course of study digital forensics.

2 State of the art

Survey of digital investigation and forensics. In [25], [26] a guide to digital computer forensic is given. In the paper [18] a historical view at forensic science and topics of digital forensics are discussed. The paper [24] gives an overview of the current state in digital forensic research and addresses all aspects of the investigations process.

Modelling the forensic process. The paper [7] presents an overview of the application of design science research to the tactical management of forensic evidence processing and the problems faced by those dealing with evidence. A conceptual meta-model for a unified approach to forensic evidence is developed. Any practical application of the suggested model would be predominantly law enforcement driven; law enforcement participants in several international jurisdictions have carried out evaluation of sections of the model.
The paper [17] presents a systematic discussion of the forensic analysis protocol that allows us to identify many types of corruption events and to model them. In [23] a discussion of different investigation process models is given.

Investigation Process Environments. The paper [39] proposes a new model for digital forensic tools utilization via integrating open-source single tools into a platform and setting up into Live DVD/USB. The platform, an Integrated Open Forensic Environment (named IOFE), takes full advantage of these tools and, at the same time, elevates its power and interoperability via standardized input/output data.

Professional forensic tools. There are several professional tools based on, e.g. X-Ways [38], Encase [8] or FTK [12]. These tools have the same basic forensic functions; e.g. can clone and image disc as well as access to logical memory of running processes, have various data recovery techniques, have a powerful file carving, have a recursive view of all existing and deleted files in all subdirectories, can analyse browser artefacts, etc. and have a complete case management involving a structured documentation.

Traditional open source forensic tools. Besides that, open source tools are available e.g. The Sleuth Kit and Autopsy [32]. Autopsy is a GUI-based program that allows you analyse hard drives and smart phones efficiently. It

---

[1] WINGS - Wismar International Graduation Services GmbH

has a plug-in architecture that allows you to find add-on modules or develop custom modules in Java or Python. The Sleuth Kit is a collection of command line tools and a C library that allows you to analyse disk images and recover files from them. It is used behind the scenes in Autopsy and many other open source and commercial forensics tools.

Database forensics. The basics of database forensics are given in [36], [11] using the Oracle and the MSSQL Server database systems. Databases provide a means for storing large quantities of data, which can be of interest in forensic investigations [21].

The research in [23] considers several techniques for hiding data in a database. In particular, it demonstrates how sensitive data can be hidden within an object-relational database (ORD) using PostgreSQL queries.

The paper [5] describes a database forensic method that transforms a DBMS into the required state for a database forensic investigation. The method segments a DBMS into four abstract layers that separate the various levels of DBMS metadata and data. A forensic investigator can then analyse each layer for evidence of malicious activity. Tests performed on a compromised PostgreSQL DBMS demonstrate that the segmentation method provides a means for extracting the compromised DBMS components.

An algorithm for reconstructing a database for forensic purposes is presented in [10]. Given the current instance of a database and the log of modifying queries executed on the database over time, the database reconstruction algorithm determines the data that was present in the database at an earlier time. The algorithm employs inverse relational algebra operators along with a relational algebra log and value blocks of relations to perform database reconstruction.

In paper [36] the architecture of Oracle 10g DBMS is analysed and different queries that can be fired to retrieve the facts and details from the system log files and redo log files from the oracle database are discussed. A proposed design of a forensic tool that will detect the tamper on the content of the database and analyse when, where and who did the tamper on the database is stated.

In paper [1], a framework is projected for analysing and reconstructing the activity of any suspicious behaviour at intervals information. The aim is to spot, analyse, validate, interpret, generate rhetorical report and preserve the proof for digital investigations. To prove within the idea, the information of the MySQL information schema is studied and analysed for this projected framework.

There is an imperative need in the field of database forensics to make several redundant copies of sensitive data found in database server artefacts, audit logs, cache, table storage etc. for analysis purposes. Large volume of metadata is available in database infrastructure for investigation purposes but most of the effort lies in the retrieval and analysis of that information from computing systems. Thus, in the paper [32] primarily relevance of metadata in design of a generalized database forensics tool independent of DBMS used is focused. The various tools of database forensics along with the challenges faced are also discussed.

The paper [22] initially discusses categories of possibilities that exist to (surreptitiously) change the application schema; practical examples are used to illustrate these possibilities. The paper is based on the premise that a specific combination of DBMS layers of metadata and data should be assembled to test specific hypotheses. A process is proposed on how forensic evidence should be extracted from the application schema layer of a DBMS.

Information retrieval for text and images. Digital image forensics is a brand new research field, which aims at validating the authenticity of images by recovering information about their history. In paper [27] two main problems are addressed: the identification of the imaging device that captured the image, and the detection of traces of forgeries.

Current digital forensic text string search tools fail to group and/or order search hits in a manner that appreciably improves the investigators ability to get to the relevant hits first (or at least more quickly). The paper [26] proposes and empirically tests the feasibility and utility of post-retrieval clustering of digital forensic text string search results specifically by using Kohonen Self-Organizing Maps, a self-organizing neural network approach.

Web-Browser and Social Networks (SNSs) Forensics. The research in paper [20] is to discover residual artefacts from private and portable web browsing sessions. In addition, the artefacts must contain more than just file fragments and enough to establish an affirmative link between user and session. Certain aspects of this topic have triggered many questions, but there have never been enough authoritative answers to follow. As a result, we propose a new methodology for analysing private and portable web browsing artefacts. Our research will serve to be a significant resource for law enforcement, computer forensic investigators, and the digital forensics research community.

The paper [15] suggests a digital forensic process for digital devices using Social Networks. To analyse digital evidence about SNSs, this proposed method is composed of effective processes, classifying digital devices, collecting digital evidence, and analysis.

In papers [37], [14], a strategy for detecting trends and opinions in geo-tagged social text streams (Twitter) is explained.

Cloud data. One of the current most important data sources are clouds. Forensic questions using cloud data are discussed in [3], [16], and [9].

3 The digital forensic investigation process

The digital forensics investigation process will ensure that an investigation is forensically sound and is based on common law enforcement and the principles of digital forensics. Forensically sound means

- an exact copy of digital data is created,
- the authenticity of digital is preserved,
- the chain of custody is established, and
- actions taken by people through the different investigation phases are recorded.

A digital forensics process model is shown in fig. 1, cf. [28]. There are seven common phases of the digital investigation workflow: preparation, identification, collection, preservation, examination, analysis, and presentation.

- Preparation includes activities to ensure equipment and personnel are prepared.
- Identification involves detection of an incident or event.
- Collection comprises the collection of relevant data using approved techniques.
- Preservation establishes proper evidence gathering and the chain of custody.
- Examination evaluates digital evidence to reveal data and reduce volumes.
- Analysis examines the context and content of digital evidence to determine relevancy.
- Presentation includes preparing reporting documentation.

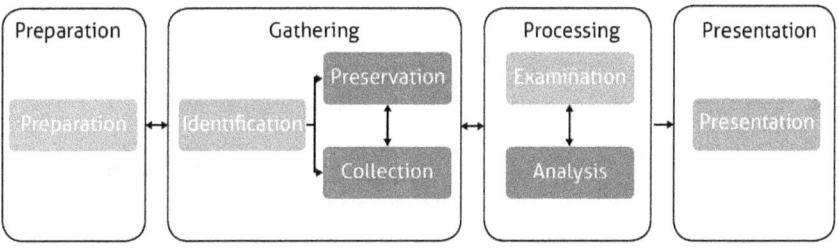

Fig. 1. High-level digital forensic process model [28]

In the collection phase, data from different sources will be collected. Volatile data (network connections, running processes, open files system dates and time, login sessions, hibernation files, cache files) and non-volatile data (account information, log files, configurations files, data files, paging or swap file, registry entries, slack space, dump files, database files) have to be recorded and analysed.

Additionally information from social networks or messengers services like Facebook, Twitter, Telegram, etc. has to be used and therefore integrated into the investigation process.
Data sources exist beyond traditional computer systems. So, data will be found

e.g. at cloud systems as well as Internet service providers. Of course, unidentified data sources exist too.

Different technical environments based on a forensic process model exist [28].

The lack of these technical environments is that only common digital data sources can be gathered and analysed. Whereas mostly information from e.g. WhatsApp can be retrieved from the mobile device, information from social media sources, especially from social networks (e.g. Facebook) or messenger tools (e.g. Twitter, Telegram), cannot be gathered and processed using the environments.

4 Extending the forensic process using social media data

In this section, we will explain an extension of the digital forensic process and the according process model for handling social media data (cf. fig. 2).

For illustrating the extension, we use an example incident. For that, we analysed the typical forensic data discussed above with the forensic tools X-Ways, The Sleuth Kit respectively Autopsy as well as forensic data from a PostgreSQL database [40], [37], and [17]. In addition we had special media data for that incident in Facebook [35],[31], Telegram [30] and Twitter [14].

The aim of the example incident was to combine and store the information retrieved in parallel from these different sources in a time-based way in a database.

Each component of the forensic process model is extended with special functions additionally to the common used forensic functions of the forensic tools.

The preparation component is extended to have special software for analysing social media. We implemented a crawler for social networks (in our case Facebook), a BOT for the messenger services Telegram and a keyword retrieval software for Twitter tweets.

In the second step, the gathering component, the specific software is used in context. The specific questions and tasks of an investigation are addressed. The software has to be configured and started to collect the data from the social media.

In the third step, the processing component is extended. The retrieved data from the social media are reduced to the relevant data. The relevant data has

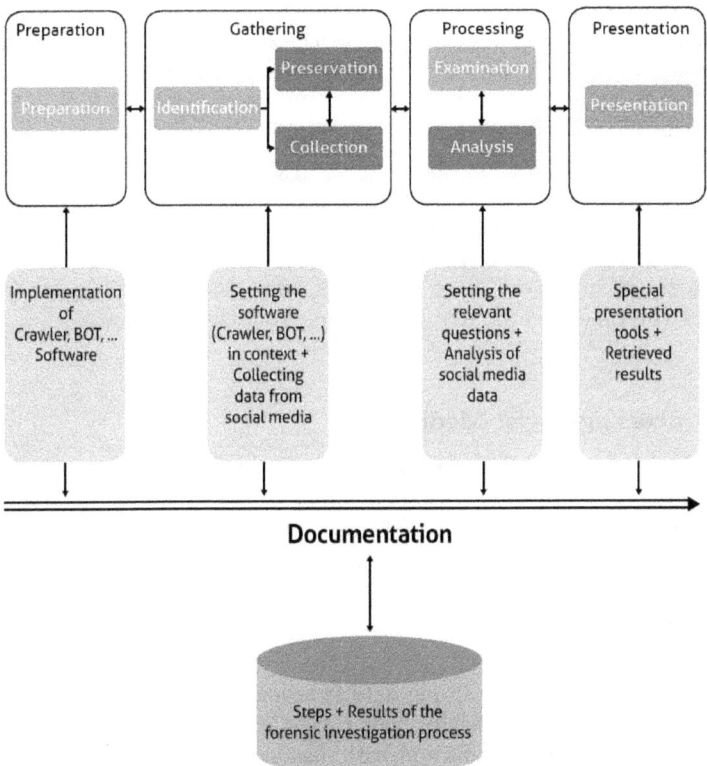

Fig. 2. Social media extension of the high-level digital forensic process model

to be analysed and the results are stored in a specific format (e.g. CSV) or in a database. Special analysing software is used to process the large resulting data collections.

The last component is the presentation component. It is important to find the right presentation for the retrieved social media data. For instance, from a list of Facebook-friends in a CSV-format the information of friends of a special person is difficult to recognize. That is why specific software (e.g. Gephi) is used for the visual presentation of large social media data (e.g. a network of Facebook friends). To document chain of custody, the resulting graphical representations are stored into the documentation database too.

During the digital forensic process all steps, software and configurations as well as the according results are documented in a special database, the documentation database. Transformation software was implemented for storing the time-based analysis results of X-Ways, The Sleuth Kit and Autopsy in that documentation database. In parallel, the analysis results from the social media data are stored in the database too. Thus, time-based questions can be taken to gather the timeline events of an incident from the database.

4.1 Processing Social Media Data

In this section, we show how media data from the messenger service Telegram and from the social network Facebook will be processed and integrated into the forensic process. The use of geo-tagged social media information from Twitter is shown in [14].

Telegram Fraud detection in big data is becoming more and more important nowadays. Several Mobile devices per person exist, everybody is chatting everywhere via email, chatrooms and with messaging services. This is generating many data over time.
What does it mean in a fraud detection context? During an investigation, big data must be analysed in the fastest and easiest way possible. For our project, the messenger service Telegram is chosen as an example for any messaging service [2].

The focus is on one hand the data extraction out of the Telegram messenger service. Which kind of data can be extracted and which information is possible?

---

[2] We assume that we have not encrypted information. Information in Telegram as well as in other messengers can be encrypted but there are different methods for the decryption, e.g. [29].

For that, the interesting topics of the investigation, the relevant Telegram threads and the context of questions has to be found. That could be such questions like:

- number of members,
- topic of the thread,
- timeliness of the postings, and
- relevance according to the forensic question.

A PHP software and a BOT is created using the BOT API of Telegram [34]. Communication- and Metadata are extracted out of the messenger service Telegram and stored in a MySQL database. The extracted data structure is shown in fig. 3.

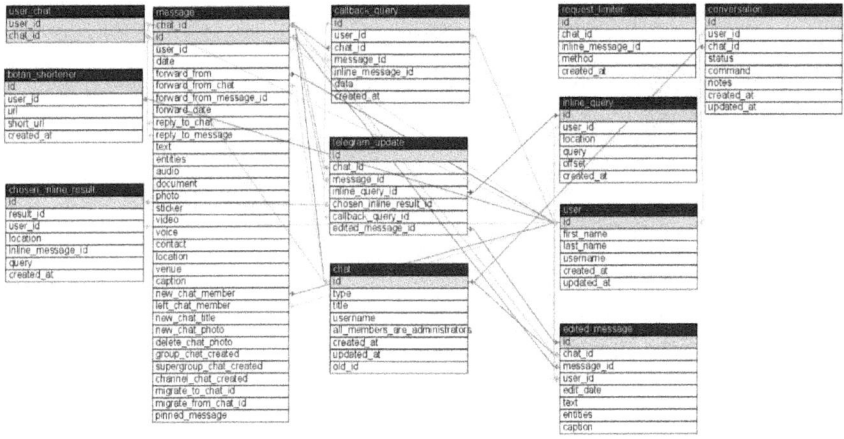

Fig. 3. Structure of the Telegram data stored in a MySQL database

The data stored in the MySQL Database can be reduced and setting up to specific questions; the focus of the examination phase in the investigation process model.

Afterwards data visualization is another focus especially for the presentation component in the investigation model. Extracted data are organized and visualized in a graph database Neo4j [19]. With Neo4j, data can be visualised easily and effective to get an overview over complex structures between communication partners. Entities and context between communication partners will be shown.
An example of Cypher question for a Neo4J database shown in fig. 4 focuses

on specific words used in cybercrime context e.g. drugs (Cypher is the query language of Neo4J.). The results include conversations and account information of suspicious persons who used that words. The result visualization is shown in fig. 5.

The Cypher questions, the results and the visualization are stored into the documentation database.

```
1 MATCH (n:Person {sex:"female"})-[:HAD]->(convo:Conversation)<-[:HAD]-(m:Person {sex:"male"}),(mWords:MaliciousWords)
2 WHERE length(FILTER(w in mWords.words where convo.text contains w )) > 0
3 AND length(FILTER(w in mWords.words where convo.text contains w )) > 0
4 RETURN m
```

Fig. 4. Cypher-Query for the use of specific word in cybercrime context

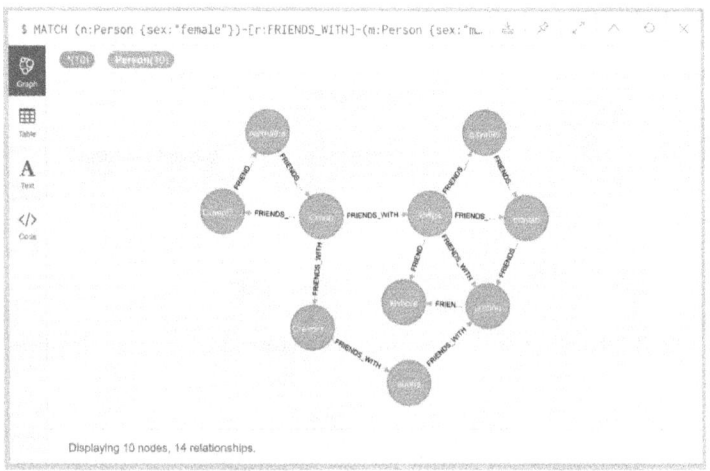

Fig. 5. Visualization of the results of the Cypher question for suspicious persons

Facebook Facebook is chosen for our project to illustrate how social networks can be analysed for digital forensic investigations. Starting from a given Facebook account the structure of friends can be extracted. The Java crawler software was developed (preparation phase) and has to be set up to the investigation context. In our example, we will use a list of Facebook-friends as interesting forensic data (identification phase).

After starting the crawler software, a list of friends is extracted automatically (the collection phase). The resulting list of friends is stored as a CSV file and in the documentation database.

In the presentation phase, we use the software Gephi [13] for visualization of the friends network. Fig. 6 shows this visualization. The different visualization parameters let us have different views at the data; groups of friends can be identified and central persons with a large number of friends too. Thus, we can have a better overview of the data and can extract logical structure that can be used to recognise steps as well as persons or objects of the incident.

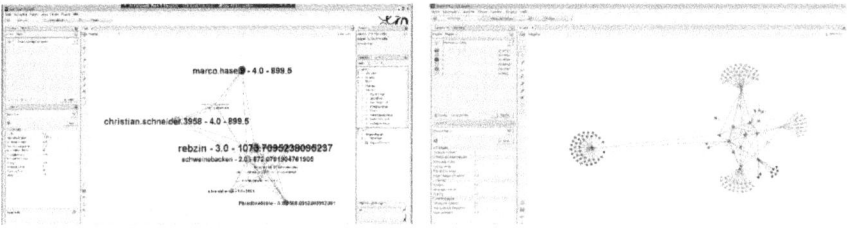

Fig. 6. Gephi visualization of the Facebook friends list

For that reason, we used a bidirectional working flow between the gathering and the processing component, especially the analysis phase. We consider that data from social network has to be often visualised in order to find helpful facts for the enlightenment of a crime. In a second step, the social media data has to be gathered again starting with the new facts from the analysis phase.

Combining analysed data from different sources In general, we realize a cycle between the gathering and the processing phases in the digital forensic process model. The results of the data analysis can be thus again input data for the gathering phases. Currently, user input is necessary to classify relevant results as new input data for the gathering phase. At present, different search templates are used for connecting the gathering and processing phases.

This general approach makes it possible to combine data from different media. For example, analysis results from a Twitter search can be applied on the data from an image file using e.g. X-Ways. The basis for this cross-media search is the storage of all data and results in a central database.

4.2 Storing social media data and analysis

The common law enforcement and the principles of digital forensics require the documentation of all steps of a forensic investigation. Thus, a documentation of the whole process of the forensic investigation is established so that each investigation step will be logged. A database to store this information and a framework for the forensic investigation process was developed.

The database structure includes tables for every investigation case and each step of a case. The logbooks of the common forensic software (e.g. X-Ways, En-Case, and FTK) are transformed onto the database structure and related to a specific case. The gathered data and the resulting investigation data from the social media are related to a specific case too. A specific new object relational data type called forensic social media was developed and integrated into the database. The new data type was used for storing e.g. the search question, the results and the pictures of the visualizations as well as the derived logical facts of the investigation.

The presentation component of the forensic investigation process model was extended for using the database information. A timeline-based documentation can be generated using SQL queries to the database.

5 Conclusion

In this paper, we illustrate an extension of a forensic investigation process for handling social media information. We have shown how forensic data from common forensic software can be combined with forensic data from social media using examples from Telegram and Facebook. Therefore, information from the social networks has to be gathered using the crawler and BOT technique and afterwards the data has to be analysed. Visualization software was used for getting an overview about the resulting data and detecting logical relations. A central database was used during the whole investigation process for storing all steps and data. The timeline-based documentation of the digital forensic investigation can be generated from the database entries.

Future work will consider the automation of the forensic process and the integration of knowledge basis as well as dictionaries. In particular, we work on integrating the analysis of biometric data.

References

1. Karen B. Alexander: Database Forensic Analysis. In: International Journal of Advance Research in Computer Science and Management Studies. Volume 2, Issue 3, March 2014.
2. Andre Arnes (Eds.): Digital forensics. John Wiley & Sons, July 2017.
3. Saad Alqahtany, Nathan Clarke, Steven Furnell and Christoph Reich: A forensic acquisition and analysis system for IaaS. In: Cluster Comput (2016) 19:439453, Springer Science+Business Media New York, 2015.
4. Nicole Lang Beebe and Jan Guynes Clark: Digital forensic text string searching: Improving information retrieval effectiveness by thematically clustering search results. In: Proc. Of The Digital Forensic Research Conference DFRWS,2007, USA Pittsburgh.
5. Hector Beyers, Martin Olivier and Gerhard Hancke: Assembling Metadata for Database Forensics. In: G. Peterson and S. Shenoi (Eds.): Advances in Digital Forensics VII, IFIP AICT 361, pp. 8999, 2011. IFIP International Federation for Information Processing, 2011.
6. Hector Beyers, Martin Oliviery and Gerhard Hancke: Database Application Schema Forensics. In: SACJ No. 55, December 2014.
7. Colin Armstrong and Helen Armstrong: Modeling Forensic Evidence Systems Using Design Science. In: J. Pries-Heje et al. (Eds.): IS Design Science Research, IFIP AICT 318, pp. 282300, 2010. IFIP International Federation for Information Processing, 2010.
8. EnCase. https://www.guidancesoftware.com/encase-forensic
9. Darren R. Hayes: Practical Guide to Computer Forensics Investigations. Pearson Prentice Hall Computing, September 2014.
10. Oluwasola Mary Fasan and Martin Olivier: Reconstruction in Database Forensics. In: G. Peterson and S. Shenoi (Eds.): Advances in Digital Forensics VIII, IFIP AICT 383, pp. 273287, 2012.
11. Kevvie Fowler: SQL Server Forensics. Pearson Education 2009.
12. FTK. http://accessdata.com/products-services/forensic-toolkit-ftk
13. Gephi. https://gephi.org/
14. Jevgenij Jakunschin, Antje Raab-Düsterhöft, Andreas Heuer: Detection of Trends and Opinions in Geo-Tagged Social Text Streams. In: Proc. of the Adbis Conference 2015, Workshop New Trends in Databases and Information Systems, pp.259-267, September 2015.
15. Yu-Jong Jang and Jin Kwak: Digital forensics investigation methodology applicable for social network services. In: Multimed. Tools Appl. (2015) 74:50295040, Springer Science+Business Media New York, 2014.
16. Da-Yu Kao: Cybercrime investigation countermeasure using created-accessed-modified model in cloud computing environments. In: J. Supercomput. (2016) 72:141160, Springer Science+Business Media New York, 2016.
17. Korinna Kruse and Antje Raab-Düsterhöft: Analyse von Email-Anhängen. Project report IT-Forensik-Projekt II, University of Applied Science Wismar, July 2017.
18. Dirk Labudde, Frank Czerner und Michael Spranger: Einführung. D. Labudde, M. Spranger (Hrsg.), Forensik in der digitalen Welt, Springer-Verlag GmbH Deutschland, 2017.
19. Neo4J. https://neo4j.com/

20. Donny J. Ohana and Narasimha Shashidhar: Do private and portable web browsers leave incriminating evidence?: a forensic analysis of residual artifacts from private and portable web browsing sessions. In: Ohana and Shashidhar EURASIP Journal on Information Security 2013, 2013:6, http://jis.eurasipjournals.com/content/2013/1/6 .
21. Martin Olivier: On metadata context in database forensics, Digital Investigation, vol. 5(3-4), pp. 115123, 2009.
22. Kyriacos Pavlou and Richard T. Snodgress: Generalizing Database Forensics, In: ACM Transactions on Database Systems, Vol. 38, No. 2, Article 12, 2013.
23. Heloise Pieterse and Martin Olivier: Data Hiding Techniques for Database Environments. G. Peterson and S. Shenoi (Eds.): Advances in Digital Forensics VIII, IFIP AICT 383, pp. 289301, 2012. IFIP International Federation for Information Processing, 2012.
24. Sriram Raghavan: Digital forensic research: current state of the art. In: CSIT (March 2013) 1(1):91114, CSI Publications 2012, Springer Science+Business Media New York, 2013.
25. Shagufta Rajguru and Deepak Sharma: Database Tamper Detection and Analysis. In: International Journal of Computer Applications (0975 8887) Volume 105 No. 15, November 2014.
26. B. K. S. P. Kumar Raju and G. Geethakumari: Event correlation in cloud: a forensic perspective. In: Computing (2016) 98:12031224, Springer-Verlag Wien, 2016.
27. Judith A. Redi, Wiem Taktak and Jean-Luc Dugelay: Digital image forensics: a booklet for beginners. In: Multimed. Tools Appl. (2011) 51:133162, Springerlink.com .
28. Jason Sachowski: Implementing Digital Forensic Readiness: From Reactive to Proactive Process, Syngress Februar 2016.
29. Hayk Saribekyan, Akaki Margvelashvili: Security Analysis of Telegram. 6.857 Final Project, May 2017, https://courses.csail.mit.edu/6.857/2017/project/19.pdf, December 2017
30. Thomas Schmalz, Rebecca Zinke, Antje Raab-Düsterhöft: Betrugserkennung in grossen Datenmengen mit dem Messengerdienst Telegram durch Einsatz einer Graph-Datenbank. Project report IT-Forensik-Projekt II, University of Applied Science Wismar, July 2017.
31. Matthias Schtz, Markus Keller, Rebecca Zinke and Antje Raab-Düsterhöft: Auslesen und Visualisieren von Freundes-Netzwerke in Facebook. Project report, University of Applied Science Wismar, January 2016.
32. Shraddha Suratkar and Harmeet Khanuja: On The Role of Log Based Metadata in Forensic Analysis of Database Attacks. International Journal of Engineering Research and Applications (IJERA), pp.25-41, International Conference on Industrial Automation and Computing, April 2014.
33. Sleuth Kit. https://www.sleuthkit.org
34. Telegram BOT API. https://core.telegram.org/api
35. Alexander Thoms and Antje Raab-Dsterhft: Automatische Auswertung von Facebook-Informationen. Project report, University of Applied Science Wismar, July 2015
36. Paul M. Wright: Oracle Forensics: Oracle Security Best practice. Rampant Tech Press, North Carolina, USA, 2007.

37. Petra Wurzler, Nastasja Napierski, Stephan Brendel and Antje Raab-Dsterhft: Erstellung eines Szenarios fr X-Ways. Project report IT-Forensik-Projekt II, University of Applied Science Wismar, July 2017.
38. X-Ways. http://www.x-ways.com/
39. Zhang Jun, Wang Lina: An Integrated Open Forensic Environment for Digital Evidence Investigation. In: Wuhan University (China), Journal of Natural Science, Vol.17 No.6, pp. 511-515, 2012.
40. Dirk Zimmermann, Matthias Schütz, Markus Keller and Antje Raab-Düsterhöft: IT-Forensische Analyse eines mit Malware befallenen Webservers - Mithilfe der IT-Forensik Software X-Ways und Autopsy. Project report IT-Forensik-Projekt II, University of Applied Science Wismar, July 2017.

Normal Models and Their Modelling Matrix

Bernhard Thalheim

Department of Computer Science, Christian Albrechts University Kiel, D-24098 Kiel
`thalheim@is.informatik.uni-kiel.de`

Abstract. Models are one of the central instruments of modern Computer Science and Computer Engineering. The notion of model is however not commonly agreed. There are - from one side - very general and universal notions and - from the other side - rather specific ones which are easy to use within some focus and scope and fail to be applicable in other sub-disciplines. Model development is typically based on an explicit and rather quick description of the 'surface' or normal model and on the mostly unconditional acceptance of a deep model. We discover that model development is based on stereotypes. The basis of a stereotype is the deep model which is tacit and latent knowledge in normal models. The deep model is the 'logos' of a normal model. The scenarios of model deployment and the functions the model plays in these scenarios are tacit and latent engineering in normal models.

Keywords: model, model notion, conceptual modelling, (modelling) matrix, deep model, normal model

1 Models, Models, Models: Everywhere but Different

1.1 101 Notions of Model Concept

Computer science and computer engineering expressively use the conception of model for daily work. Modelling is one of their four central paradigms beside structures (in the small and large), evolution or transformation (in the small and large), and collaboration (based on communication, cooperation, and coordination). E.g. [81] selected 35 of notions which are commonly used in business informatics. As a very short list we may consider the following statements:

[3]: A model is a mathematical description of a business problem.
[4]: A model is the result of a construction process for which the selected part of the origin is satisfying the purpose.
[8]: A model is the representation of an object system for the purpose of some subject. It is the result of a construction process by the modeller who addresses a representation of these objects for model user at a certain time and based on some language. A model consists of this construction, the origin, the time and a language.

[28]: A model can be simply considered to be a material or virtual artifact which is called model within a community of practice based on a judgement of appropriateness for representation of other artifacts (things in reality, systems, ...) and serving a purpose within this community.

[34]: A model is an abstraction externalised in a professional language. A model is assumed to be simpler than, resemble, and have the same structure and way of functioning as the phenomena it represents.

[42]: The model prescribes concepts as a particular kind of relation relating a subject and an entity.

[59]: A model is an object that has been developed and is used for solution of tasks which cannot be directly solved for the origin by a subject, because of its structural and behavioural analogy to an origin.

[65]: Models are governed by the purpose, are mappings of an origin and reflect some of the properties observed or envisioned for the origin. They use languages as carrier.

[86]: A model is a simplified reproduction of a planned or real existing system with its processes on the basis of a notational and concrete concept space. According to the represented purpose-governed relevant properties, it deviates from its origin only due to the tolerance frame for the purpose.

The following general notion in [78] has been combined and generalised the understanding of the concept of a model in Archeology, Arts, Biology, Chemistry, Computer Science, Economics, Electrotechnics, Environmental Sciences, Farming and Agriculture, Geosciences, Historical Sciences, Humanities, Languages and Semiotics, Mathematics, Medicine, Ocean Sciences, Pedagogical Science, Philosophy, Physics, Political Sciences, Sociology, and Sport Science.

Definition 1 *[18, 58, 74, 76, 79] A* model *is an instrument that is adequate and dependable. It has a profile (goal or purpose or function), represents artifacts and is used for some deployment scenario. As an instrument, a model has its own background (e.g. foundation (paradigms, postulates, theories, disciplinary culture, etc.) and basis (concepts, language, assumptions, practice, etc.)). It should be well-defined or well-formed.*

Adequacy is based on satisfaction of the purpose, analogy to the artifacts it represents and the focus under which the model is used. Dependability is based on a justification for its usage as a model and on a quality certificate. Models can be evaluated by one of the evaluation frameworks. A model is functional if methods for its development and for its deployment are given. A model is effective if it can be deployed according to its portfolio, i.e. according to the tasks assigned to the model. Deployment often uses some deployment model, e.g. for explanation, exploration, construction, description and prescription.

1.2 Models as the Third Dimension of Science

Models have been considered to be somewhere in the middle between state of affairs (world, situations, data etc.) and theories (concepts and conceptions, statements, beliefs, etc.) since they may describe certain aspects of a situation and may represent parts of a theory. Figure 1 displays this understanding.

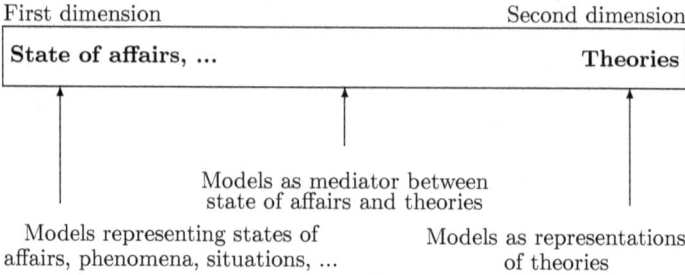

Fig. 1. Models between state of affairs and theories

"Models are partially independent of both theories and worlds." [48] The understanding of a model to be a mediator between a world and a theory is however far too restricted.

Models should be considered to be the third dimension of science [10, 78, 80][1] as depicted in Figure 2. Disciplines have developed a different understanding of the notion of model, of the function of models in scientific research and of the purpose of the model. Models are often considered to be artifacts. Models might also be mental models and thought concepts. Models are used in *utilisation scenarios* such as construction of systems, verification, optimization, explanation, and documentation. *In these scenarios* they *function* as *instruments*[2].

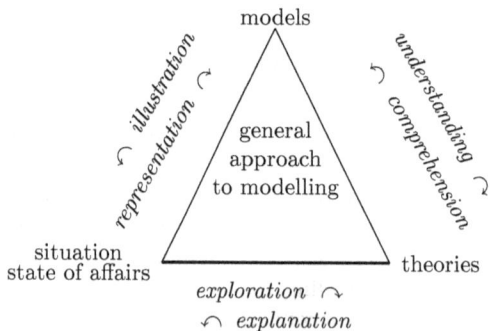

Fig. 2. Models are independent and are the third dimension of science

---

[1] The title of the book [14] has inspired this observation.
[2] An instrument is among others (1) a means whereby something is achieved, performed, or furthered; (2) one used by another as a means or aid or tool [56].

Given the utilisation scenarios, we may use models as perception models, mental models, situation models, experimentation models, formal model, mathematical models, conceptual models, computational models, inspiration models, physical models, visualisation models, representation models, diagrammatic models, exploration models, heuristic models, informative models, instructive models, etc. They are a means for some purpose (or better: function within a certain utilisation scenario), are often volatile after having been used, are useful inside and often useless outside the utilisation scenario. Models are different [10] in the four generations of science (empirical science, theory-oriented science, computational science, data science) [22].

1.3 Kuhn's Conception of Normal Science

T. Kuhn realised that the history of science consists of normal science punctuated by periods of revolution. He combined the underpinning of normal science by a notion of *paradigms*. His notion of paradigms substantially and circularly changes within his work. He did not get a satisfying definition for it[3]. He integrated normative and empirical disciplines and got a general picture on how science works. Until his work, philosophers of science made some assumptions about how science works, because they were confident in the methods of science and its success. However failures and irrationalities were not explainable. Kuhn suggested that science can only be understood "warts and all". The main conception in his book [35] is the distinction between *normal science* and (revolutionary) evolving science. Normal science is strict and governed by what he called paradigm as an object of consensus within a community and context. Education in such sciences is governed by success stories, e.g. examples, and is not governed by rules or methods. So, it is far more dogmatic [11, 21, 35]. Dogmatism, brainwashing, and indoctrination have the advantage of simplicity, fruitfulness, parsimony, and understandability within certain community and are thus enablers of success. Normal sciences maintain confidence within this community. They condition the members of a scientific community.

The investigations Kuhn has made can also be observed for modern natural sciences. Modern physics is partially normal science. It uses *standard models*. For instance, astrophysics [26] is based on the Lyndon-Bell hypothesis that assumes the existence of a black hole in galaxies with accretion layers around the black hole with broad and narrow line regions it as the central energy source. The "inner" or "deep" model behind defines the space of possible models. The modelling approach develops the final model based on a composition approach starting with parametric model fragments. Each fragment is conditioned on a background, i.e. a grounding and a basis, for instance, for approximations,

---

[3] Later [37], he revises this notion to a '*disciplinary matrix*'. His understanding of the matrix combines a wider notion of commonsense within a community of practice for relatively unproblematic disciplinary communication and for relative common agreement. We shall turn to this notion and clarify what it means in this paper. Our clarification follows [38].

perspective, granularity, and operating assumptions underlying the fragment. Mutually contradictory backgrounds are organised into background groups, and a set of coherence constraints is used to govern the use of the backgrounds [49]. Techniques such as constraint satisfaction, (causal, fitting, ...) approximations, conflict resolution graphs, and inverse modelling are then used for construction of an adequate and dependable model. Typical other deep models are the standard big bang model and the standard model of particle physics.

1.4 The Thought Style by L. Fleck: Before and Beyond T. Kuhn

L. Fleck's[4] thought style [19] is related to Kuhn's paradigms. It is however more oriented on the sociology of science. A *thought style* or *hermeneutics* is based on some kind of biased reception and perception that allows drawing conclusions from observations. The style uses forms and categories or more generally some a priori synthetic knowledge. The modelling thought style can be based on some 'logos' ($\lambda \acute{o} \gamma o \varsigma$) [13, 39] that is the rationale or first principle behind modelling.

A *thought community*[5] is a carrier and collecting tank of some sub-discipline. It arises for communication or more general for collaboration and has its peculiar thought styles that are bounded by a specific community mood. It has its opinion leaders (called 'esoteric' by Fleck) with specific leader communication such as journals, conferences, or other opinion pools. Leader influence the acceptance of thought styles and the ignorance of other thought styles. They are a source for a 'thought school'. The community also integrates followers and layman (called 'exoteric' by Fleck). Members are used to conceive and to think in a similar way. This community mutually exchanges ideas, thoughts, insights, approaches, techniques, methods, etc. within some collaboration. It maintains interpersonal interaction. The thought styles form an interpersonal space of such communities. The principles of tenacity and proliferation guide the behaviour of a thought community.

In the case of modelling communities, we shall see in the sequel that the basis for a thought community is a commonly accepted collection of deep models or, more general, of modelling matrices. A thought community also shares a common ignorance of other deep models. The accepted and practised modelling method is not neutral. It is based on some utilisation scenarios. A thought community is based on the taken-for-granted and unreflected culture of the same application era. The culture becomes then "a collective phenomenon, which is shared with people who live or lived within the same social environment, which is where it was learned; culture consists of the unwritten rules of the social

---

[4] L. Fleck has introduced many ideas before T. Kuhn [5]: "... What Kuhn named "paradigm" offers a periphrastic rendering or oblique translation of Fleck's Denkstil/Denkkollektiv, a derivation that may also account for the lability of the term "paradigm". This was due not to Kuhn's unwillingness to credit Fleck but rather to the cold war political circumstances surrounding ...".

[5] We prefer the notion of community over collective since it is less laden.

game; it is the collective programming of the mind that separates the member of one group or category of people from others." [25] As noted in the preface to [16], the members of a thought community such as a modelling community are "not free to be scientifically active otherwise than on the basis of his stylized socialisation. His thought must start with the propositions created by the collective."

The evolution of the thought style is based on classical evolution principles (selection, mutation, drift), enrichment, adaption, diffusion and dissemination, recombination, innovation, mobility and migration, and consumption and production as well as the Baldwin effect for view(point) evolution.

Second-order cybernetics [82, 84] adds to these principles the tolerant change of the origin based on the changes derived from a model. It includes how the community of practice observes and constructs their "reality". The community of practice considers within the domain of what is described and analyzed. The goal of a model is describable by a ternary relation on current state, acceptable future state, and the community. Within a school of thought, modellers use similar methods and share the same assumptions. They might be biased to these disregard those preferred by other schools of thought. However, they are not mutually unintelligible or incommensurable.

1.5 The Usefulness of a General Notion of Model

The work [78] offered a rational reconstruction of the notion of model. Definition 1 introduces an explicit and clear the notion of model, the reasons behind modelling, modelling decisions, and modelling practices. This work follows the positivist approach by [53, 54] and sees modelling as an a priori engineering discipline. Additionally, models have also an explanatory value. Models and modelling methods have their own obstinacy. They widen the understanding, theory, and engineering within an area and restrict at the same time.

We realised that models may be artifacts or mental things. Models are however going to be used in a certain way. This way can be stereotyped as a scenario. Models function within such scenarios. This functioning explains then what is the purpose and the goal of the model itself. Taking this turn, models are then *instruments* as an element of technology.

1.6 (Conceptual) Modelling is Biased Modelling

Modelling does not start from scratch. Rather it uses previous experience, approaches developed so far, commonalities agreed within a community of practice, disciplinary and other context, and also a consensus within an application. Therefore, it is not 'greenfield' work. It might be 'brownfield' work for migration or modernisation project, e.g. evolution, migration, and reconstruction projects. The distinction between greenfield modelling and brownfield modelling is typically observed in such projects. Modernisation requires to understand the 'philosophy' behind the model and the system which are under revision. The old model has already its style. The style must be properly understood before the modernisation project starts.

The state-of-the-art is however more ticklish for most modelling approaches. Model development does also not start from scratch in the case of greenfield projects. Model developers inject into the model their own experience, their own philosophy, their own understanding of a model, their own work culture, and their own biases without making an explicit reference of that. The background is implicit in the model.

A third problem of modelling is the implicit understanding of the function of the model. The function is governed by the utility and use somebody wants to gain from the model. A typical example is the modelling method that is used in Mathematics as well as in other disciplines. The mathematical modelling method (e.g. [9, 23, 51, 55, 83]) starts with an investigation, reformulation, understanding, idealisation, and description of the problem situation. It continues with development of some reality model, with a formulation of the problem situation, and a deep understanding of the task. The third phase aims at a model development within the chosen mathematical language. Problem solution is the issue in the fourth phase. The firth phase is the problem verification against the reality model. The validation of the solution may result in accepting the solution. The mathematical model must thus preserve the problem situation (*invariance*). At the same time, the solution must be *faithful*.

A fourth problem is the modelling practice. The modelling method, the modelling environment, the modelling style, and the modelling language are not taken into consideration. Instead, they are taken for granted. Modern Computer Engineering starts with some language without knowing the restrictions and biases of this language. A similar situation can be observed for monographs and teaching, e.g. on conceptual modelling[6].

So, models have explicit components from one side and hidden implicit components from the other side. A scientific approach to modelling, to model activities, and to models must answer the question *whether the implicit part of a model can be made explicit, can be thoroughly handled, and can provide a better understanding and a higher utility of a model.*

1.7 Outline of the Paper

In all four cases above, modelling is laden by its grounding, its basis, its specific application scenarios, its community of practice, and its context. This laddeness

---

[6] Even worse: There is nowadays no commonly accepted notion of conceptual model, i.e. 40 years after the notion of a 'conceptual database model' has been coined for database modelling. Many modelling languages much later got their theoretical underpinning if at all. Strengths are well communicated. Limitations, weaknesses, opportunities, and threats of these languages are often unknown.

can be understood by the deep model or the 'logos' underneath the normal[7] model.

The main issue of this paper is the development of an *understanding of the matrix of (conceptual) models*. Matrices[8] consist of deep models and some modelling scenario setting. Normal models are governed by such modelling matrices. A matrix is "something within or from which something else originates, develops, or takes from" [1]. The matrix is assumed to be correct for normal models. Normal modelling involves showing how systems and their models can be fitted into the elements the matrix provides. Most of this work is detail-oriented. So, the matrix governs the modelling process. A failure to solve a modelling task reflects on the modellers' skills, and not on the legitimacy of the setting. We start in Section 2 with an investigation whether modelling is mainly normal modelling that latently uses an underlying matrix. As a case study we use conceptual modelling in Section 3. Based on this experience, we develop a general approach to modelling matrices in Section 4.

2 Normal Modelling

Normal modelling is the day-to-day business of modelling in most areas and also in Computer Science and Computer Engineering. The matrix introduced in the sequel can be seen as a pervasive, disciplinary and well-accepted framework in which modellers perceive and develop their model. Normal modelling is concerned with model development and model deployment for the given application task and nothing else beyond that.

2.1 Modelling is Often Stereotyped

The *modelling process* is often stereotyped [77]. It reuses experience gained in the community of practice and especially by the modellers. The *modelling method* does not start modelling from scratch. It starts with some insights into an application domain, with some settings that are commonly accepted in the thought community, with approaches that have shown to be successful in the past, and with a *modelling agenda* that is either based on explicit or implicit modelling rules or at least based on some methodological rules for

---

[7] The word 'normal' has different meanings [1]. It seems that Computer Science and Logics prefer 'normal' as conforming or constituting a norm or standard or level or type or social norm. We prefer the meaning as being appropriately average or within certain limits or occurring naturally or being characterised by average development. Being 'normal' also means to be in accordance to accepted consensus or rules or laws.

[8] The word matrix has several meanings beside the one in Mathematics as a mathematical matrix. We don't use the mathematical notion. Instead, the non-mathematical meaning (e.g. engineering, psychology, etc.) of the notion of matrix is used. German language uses two different words: 'Matrize' and 'Matrix' that separate the meaning in a better form.

specific kinds of models. Agenda setting restricts potential utilisation scenarios of models. It thus results in a clarification of the model functions and thus also purpose and goal. It is additionally governed by puzzles and expectations and especially by the origins with the selected concept space. This approach assures modellers that each model is adequate and dependable, and provides standards for evaluating its adequateness and dependability. It uses a definitional frame, is based on certain modelling situations - or better on certain *modelling scenarios* - that determine the *agenda* of the modelling process, and can be started from scratch or with an *initial model*, e.g. a generic model.

The *definitional frame* defines the setting of the modelling process, i.e. (1) its *priming and orientation* that is governed by the context (application domain or discipline, school of thought, time, space, granularity, scope) and the grounding (paradigms, postulates, restrictions, theories, culture, foundations, conventions, authorities), (2) the *actors* (which form its community of practice) with their roles, responsibilities, and obligations, and (3) the *language* (as a carrier) and *basics* (assumptions, concepts, practices, language as carrier, thought community and thought style, methodology, pattern, routines, commonsense). Another part of the definitional frame is defined as specific adequacy and dependability criteria which are applied to a model.

2.2 Modelling is Mainly Normal Modelling

We will realise in the sequel that such kind of stereotyped modelling uses two models behind a model. It defines a *macro-model* for the modelling process itself, i.e the way how the model and especially the surface model is going to be developed. It is also the basis for the *deep model* that directs the modelling process and the surface or normal model. The deep model can be understood as the common basis for a number of models. Education on conceptual modelling starts, for instance, directly with the deep model. In this case, the deep model has to be accepted and is thus hidden and latent.

Normal modelling is similar to normal science. It is based on some kind of consensus about how modelling should be done. It is governed by some implicit knowledge, by commonsense, and - more generally - by consensus behind that we call in the sequel '*matrix*'. It is thus puzzle-solving. Normal modelling is what modellers do most of the time. A typical puzzle-solving task is the development of a conceptual schema for a given application within a given business context for a given community of practice, and within a given system orientation. The next puzzle is then solved by the next conceptual schema. Textbooks and education mainly develop puzzle-solving skills.

In normal modelling, the modelling theories, modelling tools, modelling attitudes, and modelling assumptions comprise the modelling matrix. They are kept fixed, permitting the cumulative generation of puzzle-solutions The modelling matrix undergoes revision whenever the underlying technology, the context, or the application are changing. If the consensus on modelling is lacking then competing schools of thought possess differing procedures, theories, even practices. Normal modelling proceeds on the basis of perceived similarity to exemplars.

Education and also edification is governed more by examples than by rules or methods. E.g. the field of conceptual modelling is mainly taught on the basis of examples; even more: nowadays on the basis of toy examples. In daily practice, models should be used and understood. Therefore, we need a notion that is as simple as possible in the given scenario and given situation. At the same time, we should not loose the specific agreements we have made for models. Models must be effective, efficient, user-friendly, economic, and well-organised. Otherwise, nobody can properly use the conclusions and results that have been generated by the help of models. Sometimes, models may mis-orientate, condition, biase or persuade [72] users in their understanding and must be corrected after paradigmatic revision and synthesis.

The orientation on normal models has also its pitfalls. For instance, cardinality constraints [68] have mainly be developed for relational technology of the early 90ies or 80ies. At his time, the mapping of these constraints was a deep research issue. Nowadays, object-oriented database system technology allows a far more sophisticated handling of constraints. Maintenance can be deferred (eager or lazy integrity enforcement). Time management allows handling more optimal timepoints for consistence maintenance. Consistency can be supported at the row level. Integrity constraints can be maintained at the application level. Integrity can be made through views. Finally, flexible strategies may be used, besides the no-action and rollback approach, e.g. on the basis of triggers or stored procedures. Therefore, we may generalise cardinality constraints to conditional cardinality constraints [66].

2.3 Normal Models are Governed by Their Modelling Matrix

Normal modelling accepts one notion of model as normal. It just happens in a broad set of presupposed, unquestioned assumptions that govern among other things the sort of models to be developed, how these models are investigated and deployed, and how these models are interpreted. If the matrix would be question then modelling becomes difficult if not too time consuming. The matrix guides and instructs. Normal modelling is perfectly good modelling as long as the tasks are solved by the models that have been developed. So, modellers can be 'blinded' by the success although they are close-minded. The consensus provides a good means for collaboration and minor modernisation. The modelling tasks are focussed on the task spectrum that is preferred at present. So, the matrix got its kind of faith and trust. This kind of 'brainwashing' or indoctrination is the basis for teaching.

2.4 Normal Modelling Develops its Boundaries

All matrices have their limit, restrictions and even pitfalls. It might happen that the model does not solve the task in a proper form or that the model is not sufficient or that models develop their obstinacy or that models result in anomalies. Problematic tasks are not counterexamples to normal modelling. In this case, the matrix must be revised since it is not adequate anymore.

The first resolution step is the introduction of new elements to the current matrix. E.g. the entity-relationship modelling language has been heavily extended by about 50 constructs in the 80ies and early 90ies [68] until it has been detected that these extensions will not become coherent. At the same time, work-arounds have been built for overcoming limitations. Currently object-relational technology is available. It seems that ER modelling might somehow suffice with model creation. However, classical theories are not sufficient anymore. The confidence into the ER approach to modelling weakens nowadays. With the advent of object-centred modelling and the supporting XML technology one might ask whether conceptual modelling can be based on the ER matrix. Since data collections might also evolve in their structuring, the class-oriented technique must be nowadays revised what has already been discovered within the research on conceptual modelling for 'big' data. Before changing the matrix we must, however, develop a proper understanding of it. We will turn to this problem in the next section.

The transformation of models become a bottleneck whenever the matrices of the given model and of its transformation do not match properly. Such impedance mismatches have already widely been discussed for object-oriented programming in imperative environments. A similar observation is valid too for conceptual modelling based on sets and physical modelling based on multi-sets and references (see, for instance, SQL with specific referential integrity and multi-set handling).

3 A Case Study: Information Systems Modelling

3.1 Historical Matrices

The matrix is an essential component of the identity within a community of practice or within a scientific community. It identifies puzzles to be solved, governs expectation, assures modellers that each puzzle fulfills its purpose, and provides standards for evaluating.

According to [1], a matrix is "something within or from which something else originates, develops, or takes from". Normal models will be understood as models crammed with the modelling matrix. The existence of such modelling matrices makes modelling simpler and supports parsimony. It is a kind of complex laddeness of a model [77]. What modellers develop depends, in pertinent part, on what they already believe or expect. Developing is less passive, less receptive than many had thought. Modelling is dependent on the chosen modelling matrix.

Let us consider one example where we observe surprisingly many postulates, paradigms, theories, assumptions, accepted practices, bindings to a school of thought, context, and commonsense. The matrix is well-accepted but not explicitly explained in textbooks and research papers. The *entity-relationship modelling language* became popular in the late 70ies as a means for documenting logical relational schemata and for visualising the association among types[9]. The entity-relationship modelling language became now some kind of standard despite the unknown and not explicitly given matrix underlying this language. It uses a *Global-As-Design* approach where the schema reflects all viewpoints. Local viewpoints are derivable and somehow reflectable. The default semantics (and sometimes the only one to be considered) is *set semantics* for collections of objects for a type. The *reference semantics* for relationship types is hidden and not properly understandable during schema development but used in transformation. *Explicit existence* postulates that any object must exist before there can be a reference to it, i.e. rigid separation of creation and use. The model assumes a *closed-world view* and *unique names*. It is based on a well understood *name space* or glossary or ontology. *Salami-slice* representation uses homogenous, decomposed types (potentially with complex attributes) with incremental type construction. It is *type-centric*. According to tradition of logic-based computer science, *semantics follows syntax*, i.e. the definition of semantics can be given if the syntax that is used is already defined. The user perspective is cut out, i.e. we base the model on *neglected pragmatics*. *Functionality representation* is deferred without consideration of the performance impact to the schema. *Separation into syntax and semantics* allows defining semantics on top of the syntax. Explicit semantics is based on constraints. *Paradigms, postulates, assumptions* of database technology and database support are assumed due to the three main quality criteria (performance, performance, performance). *Basic data types* are hidden until mapping to facilities provided by DBMS typing systems or to logical models. *Visualisation* is represented by one holistic diagram that displays the entire syntax and semantics. One satyr or misbelief is representation of associations among types on the basis of binary types despite the valence of normal verbs in natural languages. In general, *binarisation* is possible by introduction of abstract artificial types and by relating the new type to each of the association components [10]. *The advent of data cubes has shown that an explicit* co-handling of views empowers database technology. *Star and snowflake schemata* introduced for data warehouses are nothing else than view in the basis of high-order relationship types [68].

---

[9] In his oral presentation of his keynote speech at ER96, C. Bachman [6] claimed that the new modelling languages have been introduced as a reaction of the inflexibility and due to the insufficiency of his network modelling language for representation of relational schemata. His claim has been the basis for the development of a new approach under his supervisorship [15] (reprinted in [18]).

[10] RDF representation has chosen this way on the price of the maintenance and retrieval nightmare. A better binarisation has been used by MIMER resp. RAPID on the basis of sixth normal form storage (called nowadays one-column representation).

3.2 Puzzle-Solving in Information Systems Modelling

In the sequel, we observe that the matrix forms the implicit and tacit knowledge behind modelling, i.e. the second component of the rigor cycle in design science research. The matrix generates a concensus about how modelling should be done. This consensus distinguishes modelling from other scientific or engineering endeavours.

Modern application with dynamic structuring of objects such as big data collections cannot be properly represented by the static structuring which is one of the silver bullet assumptions of DBMS since it provides optimisation facilities that brought the victory of relational technology over network or hierarchical technology.

Puzzle-solving left open a good number of problems for future research [73]. One of the lacunas is the NULL marker problem [29, 57]. It becomes a bottleneck whenever aggregation functions are going to be applied [40]. The representation of NULL-polluted types by a collection of NULL-free subtypes is computationally infeasible. Schema-wide constraint maintenance is another big problem at present.

Education in this area has been built on success stories and proceeds on the basis of perceived similarity to success cases. At present, information system modelling still modelling in the small. Modelling in the large or modelling in the world must be based on different matrices; which ones is not clear yet. Puzzle-solving allows to transfer experience gained for one problem to another class of problems and to evaluate and appreciate solutions of other (e.g. reference models or generic models). Design science research is oriented on cumulative addition of new knowledge in terms of the application of the modelling and designing method.

3.3 Limitations and Pitfalls of Conceptual Modelling

Like any language, the entity-relationship approach is not covering all issues. It is not cognitive complete since it represents only 2 of 6 cognitive categories[11] (container, link). Pitfalls of this approach are similar to the 88 pitfalls of object-oriented programming [85]. Salami-slice tactics is often not appropriate. Things in the application domain are however multi-facetted. For instance, a human is represented via a Person type that is separated from the Student type etc. The ER language is still based on the one-schema-one-diagram approach. Schemata are typically flat. Applications are however structured. For instance, [33] presented a three-dimensional structure of schemata with the application dimension, the volatile workflow data change recording dimension, and the metadata dimension. Additionally one might think of the user involvement dimension within a schema.

---

[11] As shown in [71], the extended ER language HERM [68] covers 5 of 6 where the last sixth category center-periphery can be represented on the basis of HERM views.

The ER modelling language has nowadays also been aging. Object-relational DBMS support features that should be representable at the conceptual level due to their utility. Modern technology provides user defined types, identification trees for components of relationship types and subtypes with overwriting by new surrogate types, flexible view-oriented handling of integrity constraints, indexing mechanism, maintenance of data blocks etc. These features are not representable in the classical ER modelling language but are useful for conceptualisation.

The modelling process is far more dogmatic than understood in cookbooks or textbooks. It is somehow ritual or routine-based in education as well in practice. Models are mainly shallow models. They represent a part of the application. The specifics of constraints are not well applied. For instance, cardinality constraints specify the extremal minimal/maximal cases. Users however concentrate on the normal situation.

3.4 Views: The Overlooked Element of Conceptual Modelling

One of the limitations of the ER matrix is neglecting user views and viewpoints we consider now. The misunderstanding of view and viewpoints causes a small crisis in understanding database technology and modelling. The crisis has manifested in the development of data warehouses, data marts, star and snowflake schemata. Star and snowflake schemata are nothing else than conceptual or business user views on the database. Currently the approach falls into a lot of difficulties and resulted in the development of a 'novel' technology[12].

The three-layer architecture of information systems is a commonly accepted and widely taught conception. It is however neither true nor useful. It is only a starting point for understanding how a system might work. In reality, user viewpoints come with the application or business user level. They are represented by user schemata. Then these viewpoint schemata are integrated into the conceptual schema in the 'global-as-design' and 'local-as-view' approach. In order to represent the viewpoints we should use view schemata at the conceptual level what is however not consensus [79]. The logical level turns all the user view schemata to view definitions with the loss of the association between the relational views due to the limitations of relational technology. So far, this

---

[12] A similar trillion $ mistake is now the evolution of big data. The hype will shred a lot of resources until a new consensus occurs. In order to understand we remember the discussions about the relational data model in panels at ER92, ER93, ER94, and ER95. The new consensus was relatively undeveloped. It was not able to represent all modelling situations the network or hierarchical data models could. Later it was realised that the different modelling styles could not had been judged on a common scale. All three approaches have some shared habits and ways of seeing things. Proponents of these different approaches tended to talk past each other. Dogmatism and idiosyncrasy function in a complex social arrangement [36] such as conferences and journals.

is the current state-of-the-art. It is nothing else than an anomaly since this schematology repeatedly resisted solution.

It could be improved with the approach developed in [27]. [79] defines a conceptual model to consist of a conceptual schema and of a collection of conceptual views that are associated (in most cases tightly by a mapping facility) to the conceptual schema. A conceptual schema is then mapped to a collection of logical views.

A database schema could not be anymore seen as an integrated, holistic schema with the same level of detail. Instead, we are able to represent a number of viewpoints at different abstraction level, with different foci and scopus, with different aging and currency, with supporting mechanisms depending on currency requirements, etc. So, the database structure model forms some kind of 'web' [31] instead of one schema with derived views. Viewpoints represent structures in whatever order is best for human comprehension and thus expressing it in a stream of consciousness order.

4 The Modelling Matrix

4.1 Deep Models and Scenarios Form the Modelling Matrix

T. Kuhn [37] widely used the notion of paradigm in a variety of forms and explanations. Essentially his notions can be understood as a disciplinary matrix [41], i.e. a symbolic generalisation, a meta-model, and collection of sample cases. Based on the observations on stereotyped modelling we may distinguish four initialisation phases:

(i) orientation on modelling scenarios and used macro-models for development with derivation of the function (and thus purpose and goal) a model has;
(ii) acceptance of the grounding, of a language and of the general concept space;
(iii) setting of a deep model as the hidden, latent model or acceptance of such for some context and a community of practice;
(iv) acquisition of origins for modelling.

Definition 2 *The* deep model *consists of the grounding for modelling (paradigms, postulates, restrictions, theories, culture, foundations, conventions, authorities), the outer directives (context and community of practice), and basis (assumptions, general concept space, practices, language as carrier, thought community and thought style, methodology, pattern, routines, commonsense) of modelling.*

The deep model thus uses a collection of undisputable elements of the background as grounding and additionally a disputable and adjustable basis which is commonly accepted in the given context by the community of practice. It is typically used for many normal models but not explicitly stated whenever a normal model has been stated. The deep model is far more dogmatic than often understood. It is some kind of model 'logos' behind the normal model.

At the same time, the deep model is a rich source of knowledge [41, 43] that is already provided by the deep model, i.e. the deep model carries the knowledge and beliefs as well as the culture of the community of practice. It supports communication within the community of practice that accepts the deep model as common ground and has already agreed on the judgements made for the deep model. This common background also includes a common ontology. The deep model provides an identity within this community for the shared 'correct' opinion. The normal model becomes an epistemic instrument that is based on the common ground.

Definition 3 *The* modelling matrix *consists of the deep model and the modelling scenarios. The agenda is derived from the modelling scenario and the utilisation scenarios.*

So within a model development process, the modelling scenario and the deep model serve as a part of the definitional frame. They define also the capacity and potential of a model whenever it is utilised. The normal model can be deployed in a specific form as long as the scenarios and the deep model are not changed. For instance, database structure modelling on the basis of the entity-relationship approach has an ordinary interpretation for all developed schemata.

Different matrices solve different problems. It might happen that a normal model with one matrix does not make sense if the matrix is changed. A typical case is co-modelling that is modelling on the basis of the entity-relationship modelling language for structures and on the basis of BPMN diagrams for processes [80].

Models typically represent a number of origins. It is often the case that these origins use a common application-specific concept space, e.g. an application ontology with its lexicology and lexicography [71]. The application-specific concept space is annotated by a namespace.

A modelling matrix may be enhanced by generic models or reference models. Generic models are abstractions of a set of models that represent similar solutions. They are later tailored to suit the particular purpose and function. A generic generally represents origins under interest, provides means to establish adequacy and dependability of the model, and establishes focus and scope of the model. A reference model is used as a blueprint for a fully fledged model and provides a general solution in an application area..

4.2 Adequacy and Dependability Governed by the Modelling Matrix

The modelling matrix allows deriving specific pattern for specification of adequacy and dependability of a model. The general notion of model in Definition 1 defines adequacy based on an analogy property, a focus property, and purposefulness. Dependability is based on a justification and a quality certificate.

Justification is given by an empirical corroboration according to modelling objectives, by rational coherence and conformity stated through conformity formulas or statements, by falsifiability, and by stability and plasticity within a collection of origins. Quality can be defined by characteristics that state the internal and the external quality as well as the quality in use. The certificate is the result of an evaluation of an accepted bundle of quality characteristics through some evaluation procedure.

It seems that the statement of adequacy and dependability is a heavy and sumptuous procedure. In reality it is far more simpler due to the existence of a modelling matrix. The entity-relationship modelling matrix uses, for instance, a homomorphism or infomorphism [30] mapping property for analogy, a focus that is already determined by the situation or perception models or other origins, the purpose of full representation within the language setting, an empirical corroboration due to the mapping from the situation or perception model or other origins, the conformity that is already inherent in the modelling matrix, falsifiability via validation of the model against the origins, and stability against origins as a general class of situation and perception models or other origins. A similar definitional frame can be observed for many model kinds in Computer Science and Engineering.

As already discussed in Section 3.1, the matrix of entity-relationship modelling is quite comprehensive. We are explicitly using this matrix in dependence on the scenario. For instance, in system construction scenarios: closed-world schemata, Salami slice schemata, methods for simple transformation; preparedness for direct incorporation; hierarchical schemata; separation of syntax and semantics; tools with well-defined semantics; viewpoint derivation; componentisation and modularisation; integrity constraint formulation support; methods for integration, variation. In communication scenarios we orient the matrix to: viewpoint and flavour representation; flexible usage (full logical independence); variable name space representation; methods for reasoning, understanding, presentation, exploration; methods for explanation, check, appraise, experience

This orientation governs the well-formedness criteria such as[13]:

- unambiguous esp. for transformation,
- easy to read,
- aspect-separated, e.g. by colouring different parts,
- naming styles, e.g. either singular or plural,
- higher normal forms,
- optional structure routed to the subtype,
- freeness of semantical cycles,
- distinguishability of attributes, e.g. unique name assumption,
- meaningful names, avoidance of auxiliary verbs, e.g. 'has', 'is', ...,
- non-empty classes, and
- flag avoidance.

---

[13] For details we refer to classical database design books [45, 7, 62, 67, 68]

Central characteristics for well-formed schema are: closed world and unique name assumptions; concept enhancement and well-defined name space; no sharpening or contrasting; well-founded logics; layering of functionality, views, and interaction.

The adequacy of eER[14] schemata is based on the following properties for origins $\mathfrak{A}$ and for the scenario $\mathfrak{S}$:

1. $\mathfrak{A}$-analogous: structural analogy (homomorphic, but not qualitative, functional) resulting in structural alignment; metaphysical, epistemological and heuristic adequacy;
2. $\mathfrak{A}$-reduced (or $\mathfrak{A}$-focused): compactness, no repetition, high-level descriptive abstraction; conceptual minimal;
3. $\mathfrak{S}$-purposeful: either for construction of another representation (thus with construction hints and tactics; with simple transformation; normalised, simple integrity enforcement) or for communication with the (business) user (thus with different viewpoints and flavours; simple viewpoints; cognitive complete).

The focus of eER Schemata is based on the following characteristics:

– Separation into kernel object types, dependent types, and properties: Kernel objects have their own relative existence independence.
– Kernel object types and typical/central types become entity types; properties may be complex and are typically mapped to (complex) attribute types; hierarchies are separated and then represented by generalisation/specialisation hierarchies; relationship types are either application association types, user-relating types, meta-associations, or workflow hocks. This is similar to good practices for E/R/A/C mappings.
– All derivable constructs are represented otherwise. Irrelevant, specific elements are avoided.
– The schema concentrates on important, relevant and typical elements.
– The schema must be as simple as possible, avoid unnecessary abstractions, provide a precise meaning for each type, reduce any complexity on Salami slice techniques, and should combine similar elements.
– Rigid incremental schema.

The purposefulness of eER Schemata is given by the following orientation in dependence on the purpose:

Orientation model: corporate overview as a context, data as a source and sink, "environment" model;
Communication model: external schemata depending on the context;
Conceptual model: things of significance, concepts, assertions; semantic model for the business language ("divergent"); architectural model (general categories, "convergent", platform independent);

---

[14] Extended ER language: eER; an example is the Higher-Order ER Modelling language (HERM).

Realisation model: based on technology and platform; internal (logical) schemata (platform-specific: relational, XML, ...) (with technological twists), physical schema (storage, (vendor-specific); and
Documentation model of ground structures used in a given application system.

Scenarios typically combine a number of functions for the ER models. So we might use several schemata and especially view(point) schemata as a model suite [69].

Dependability of eER schemata and models is defined by:

1. Justification is based on embedding the model into the understanding of the application domain, i.e. through an external corroboration. The internal corroboration is based on the language. The origins determine items of the model. So, no additional acquisition and elicitation is needed. The model conforms to standards accepted in CoP such as: Salami slice tactics; correctness; restriction to essential business items; approved, closed world schemata; partially evolution prone, partial flexibility; simple diagramming with overlay diagrams.

   The model should be cognitive complete based on an appropriate representation of things of interest in real world with some ordering (e.g. hierarchies (up-down, front-back)) and additionally based on other cognitive dimensions (container, part-whole, link, centre-periphery, source-path-goal) [71]. The model is a deputy of relationships of interest in the real word with some ordering additionally based on other cognitive dimensions. We might use additional characteristics of interest for both sides [32, 50, 17].

2. Sufficiency is defined by an evaluation form and by characteristics for internal and external quality and quality of use. Typical criteria are [68]: completeness, naturalness, minimality, system independence, flexibility, self-explanation, ease of reading and using of firm quality and evaluation. We mainly use quality in use characteristics without any error tolerance. Additionally, we assume (a) avoidance of redundancy (or at least restriction to necessary (controlled)) , (b) avoidance imposed implementation restrictions, (c) internal and external characteristics for the usage of the model as blueprint without requirement for completeness of constraint sets, (d) natural keys, (e) avoidance of mega-attributes, and (f) complete confidence in all model components.

We notice, that most of the adequacy and dependability characteristics are assumed to be given with any eER schema or model. They are not mentioned but assumed. So, they are a part of the matrix.

A similar definitional frame can be defined for BPMN and other workflow diagrams.

Specific definitional frames are used for adequacy and dependability statements for models

· which provide specific extensions as an *amplification* which are not observed in the origins,

· which are *distortions* and are used for improving the origins (e.g. the physical world) or for inclusion of visions of better reality, e.g. for construction via transformation or in *Galilean models*, and
· which are *idealisations* through abstraction from origins by scoping the model to the ideal state of affairs.

Therefore, the modelling matrix allows reducing and simplifying the statement whether a model is adequate and dependable. The reduction also stems from the definitional frame that is already used for the deep model.

4.3 Development of the Normal Model and the Matrix

Education and practice in modelling typically starts with acceptance of a matrix. Whether this matrix is adequate or not is not questioned. So, we can use the modelling matrix and define the specific model in dependence on a function or purpose. Let us now consider, revise and extend model notions in [79].

Definition 4 *The* normal conceptual eER database structure model for communication and negotiation *comprises the database schema, reflects viewpoints and perspectives of different involved parties and their perception models. The matrix for communication scenario implicitly links to (namespaces or) concept fields of parties which are partially used. It defines adequacy and dependability based on the association of the perception models to viewpoints and of the viewpoints with the schema. A partial communication model does not use a schema and does not associate viewpoints to schema elements.*

As already observed in [79], normal models used for *communication and negotiation* follow additional principles: Viewpoints and specific semantics of users are explicitly given. The normal model is completely logically independent from the platform for realisation. The name space is rather flexible. The normal model is functioning and effective if methods for reasoning, understanding, presentation, exploration, explanation, validation, appraisal and experimenting are attached.

Definition 5 *The* normal conceptual eER database structure model for conceptualisation *consists of a collection of views for support of business users. The* deep model *is based on a mapping for schema elements that associates potential elements of the normal model to the common concept field and the perception models of business users. They may be extended by a skeleton that combine business user viewpoints or by a global schema which integrates these viewpoints. The matrix uses a strict adequacy and dependability. It is based on a context-driven conceptualisation of the application domain.*

Conceptualisation is based on one or more concept or conception spaces of business users. Semantics is typically rather flexible. The normal model and the viewpoint rather reflect the normal cases and do not extend these cases to the extremal cases, e.g. for (cardinality) constraints.

The deep model and the normal model for description can be defined in a similar way. They are representations, refinements and amplifications [70, 75] of situation or reality models and therefore refinements and extensions of the communication model.

Definition 6 *The normal conceptual database structure model for description comprises the database schema, and a collection of views for support of business users. The model reflects a collection of a commonly accepted reality models that reflects perception or situation models with explicit association to views, and a shallow declaration of model adequacy and dependability. The deep model and the matrix are driven by the description scenario and completely bound to the understanding in the application area and to technology, methodology and theory which is commonly agreed within the community of practice.*

The descriptive normal model reflects the origins and abstracts from reality by scoping the model to the normally considered state of affairs. The deep model also provides an idealisation.

Prescriptive models that are used for system construction are filled with anticipation of the envisioned system. They deliberately diverge from reality in order to simplify salient properties of interest, transforming them into artifacts that are easier to work with.

Definition 7 *The normal conceptual database structure model for prescription comprises the database schema and a collection of views for both support of business users and system operating. It is based on a deep model that provides a number of a realisation templates according to the platform capabilities. The matrix declaration of model uses strict adequacy and dependability.*

The matrix also defines directives (or pragmas) [2] and transformation parameters [68] . The deep model also consists of general descriptions or templates for realisation style and tactics, for configuration parameters (coding, services, policies, handlers), for generic operations, for hints for realisation of the database, for performance expectations, for constraint enforcement policies, and for support features for the system realisation.

These notions of normal models, deep models, and matrices specialise general notions like those given in the introduction or the notion by W. Steinmüller (*"A model is information: on something (content, meaning), created by someone (sender), for somebody (receiver), for some purpose (usage context).*)[65] or B. Mahr (*" A model is always at the same time a model 'of something' and a model 'for something'. Its function is to 'carry' some 'cargo' from its 'matrix' to its 'applicate'."*)[44] or F. Matthes and J. Schmidt (*"A relational database model on the basis of the approach by E.F. Codd describes semantics of declarations and statements within a database specification language and thus corresponds to an abstract model of a programming language with its static and dynamic semantics which can be specified through formal type and evaluation rules."*)[47].

Modelling is often 'brownfield' work. The model exists already and has been developed based on another matrix. Consider, for instance the schemata in [20, 24, 46, 61, 60, 63]. The schemata follow a certain matrix, e.g. in this case IDEF. Therefore, typical applications combine a number of matrices.

5 Conclusion

5.1 Model ≡ Normal Model ⋈ Deep Model

A model can thus be understood as a normal model combined with some deep model similar to the visible (or surface) and invisible parts of an iceberg. The deep model is an essential part of the modelling matrix. The matrix forms a relatively stable component for a larger collection of models and can be thus neglected for these models. The matrix is considered to be valid and does not thus need a justification. This observation led us to the conclusion that modelling is mainly normal modelling.

The matrix and especially the underlying deep models standardise the accepted and practised approach to normal models. It directs perception with factual and mental processing of observations. The matrix is not absolute. It depends on various context, on thought communities, and on utilisation scenarios. It typically also includes ortho-normalised forms of utterances within a language environment.

The deep model and the modelling matrix must be given in a form that allows to extract them whenever this becomes important. Models carry their deep model in the inner part. We may now try to measure the impact that a deep model has on the model. Figure 3 displays a typical case for the share that each kind of model has to satisfaction of properties of a given model. In the case of conceptual database models, this share determines the attention that a model property has to satisfaction. The share and thus the contribution explains why most researchers pay attention to the normal model case and do not consider deep models. Quality became an issue only recently. Such shares also explain why classical notions of model[15] have been mainly paying attention to adequacy properties and neglected dependability as well as method development. The property of being well-formed and quality properties are determined by the origins. Their share is higher due to this determination. The shares also explain why normal models are considered to be throwaway inspiration models during software system development and are not continuously redeveloped whenever the software system or the database system evolve.

5.2 The Model Matrix as the Stabile Ground of Normal Models

We observed that a model contains its matrix (or a number of its matrices). The matrix is neither questioned nor a matter of redefinition in a modelling process. It is taken for granted. A special case are 'brownfield' models which have a legacy matrix and a current matrix and which may consist of a model

---

[15] H. Stachowiak [64] defines three characteristic properties of models: the *mapping* property (have an original) [as a far more restrictive form of analogy], *truncation* property (the model lacks some of the ascriptions made to the original) [as a more specific form of focus], and *pragmatic* property (the model use is only justified for particular model users, tools of investigation, and period of time).

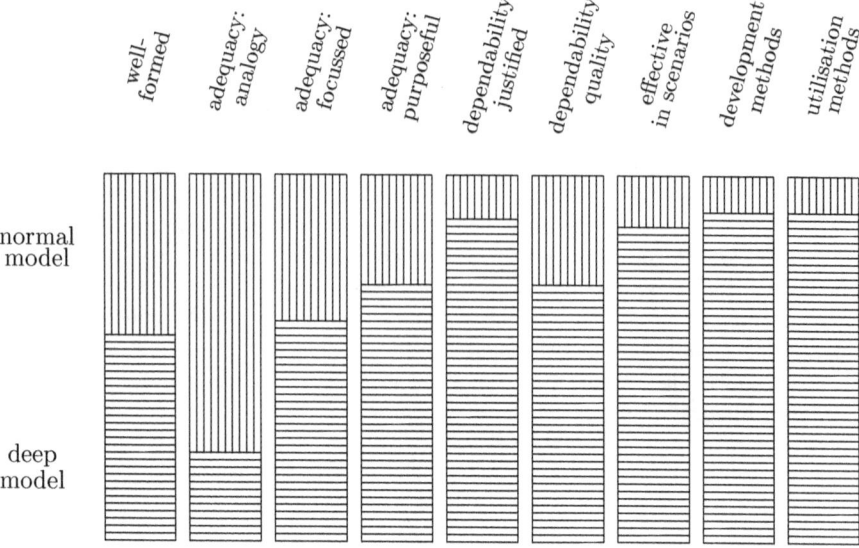

Fig. 3. Share and contribution of deep models and of normal models to satisfaction of properties; a typical case for conceptual database models

suite of mutual models for each of the matrices. The deep model also provides a deep and latent structure, the rationale behind, and first principles. Normal models are also called lumped models [88]. They deliver an understanding of the surface structure.

A matrix may evolve as well due to its limitations, revisions of dependability and adequacy required for an application, misconceptions, or missing elements. In our area, we observe changes of the deep model only for cases when technology entirely changes, e.g. the transfer from network or hierarchical modelling languages to the relational ones. The relational environment has changed however as well. So far, it is at its best an evolution step for matrices if at all. Database structure modelling has not changed for more than two decades although technology has changed a lot. Matrix evolution is also caused by changes in the scenarios.

Matrices are relatively stable. They are foremost an object of consensus within a thought community. They govern education and also applications. They generate a consensus how normal models should be developed. The main ingredient of a matrix is the deep model that is considered to be correct within the given application field and the utilisation scenarios. Normal models are under continuous change also due to rational and empirical evaluation or due to quality problems, e.g. validity & completeness, reliability & coherence, and conformity & correspondence. Therefore, normal model evolution is mainly based on a stabile ground, i.e. a stabile matrix. The matrix tests modellers more than modellers test the matrix. A failure to develop a normal model reflects on the skills of the modeller, not on the legacy of the matrix. Only when the matrix losses its grip in the thought community (e.g. nowadays observable for big data modelling) and the matrix repeatedly resisted development of a normal model then the matrix is going to be changed. The modeller community lost confidence in the given matrix. Normal modelling involves showing how applications from the application field can be fitted into the deep model. Therefore, this work is detail-oriented.

One advantage of such stabile grounds is the potential for accumulation and maturation of normal model development and utilisation. It enables knowledge elicitation and acquisition used in design science [87]. It is then part of the rigor cycle.

A simple form of matrix evolution is the combination of scenarios into a coherent set of scenarios. This combination or adduction allows combining the matrices into holistic ones. The deep models are then typically model suites. A specific form of matrix evolution is consolidation of the matrix, for instance, by development of supporting theories and by maturing methodologies. In this case, the normal models can still be used in the same form.

5.3 Model Notions for Normal Models

We may now elaborate the notions in the introduction. It is a nice exercise to see how these notions match our understanding of normal models and their matrices. For instance, the last one of those considers system construction and documentation scenarios with a classical IT system conception space. The first one is governed by the business problem model and by problem-solving scenarios within a mathematical modelling method. So, let us consider two additional examples.

An Example for Database IT Practice. [52] considers mediator/communication scenarios. The model is used for "the representation of some aspects" of the situation model, "enables clearer communication" about the situation model, and "serves as a blueprint to shape and to construct the proposed structures" in the situation model. So, a *normal* "data model is a device that
 — helps the users or stakeholders understand clearly the database system that is being implemented based on the information requirements of an organization, and
 — enables the database practitioners to implement the database system exactly conforming to the information requirements."

This notion of model is determined by the given two scenarios, by the deep model of database models, by the community of business users and database developers, by data engineering and DBMS as its context.

Models for Domain resp. Software Engineering. An application domain is a universe of discourse, an area of human activity or an area of science. Domain engineering is understood as modelling: "a careful description of the domain as it is, void of any reference to possibly desired new software, including requirements to new software". [12] "By a domain theory we understand a formal model of a domain such that properties of the domain can be stated and formally verified - claiming that these properties are properties of the domain being modelled." "A domain model is thus a description of a sufficient number of domain entities, domain functions, domain events and domain behaviours - so formulated and detailed that one is able to answer most relevant questions about the domain."

The deep model is partially explicitly given as a domain theory. The implicit part is, for instance, the notion of application domain, the focus to description and the functions of the model, the underlying mathematical theory, the modelling language (entities, ...), and the way of associating. All this forms the matrix of the "domain model".

A similar observation to the ones above can be made for classical software engineering, e.g. [34, 77].

5.4 Modelling from Art to Science

Modelling is still considered to be an art. It will become a science in the future in the understanding of [10]. Moving paths are thus: from practices to principles, from skilled performance to fundamental recurrences, from action to explanation, from invention to discovery, from synthesis to analysis, and from construction to dissection. Modelling as science is organised to understand, exploit and cope with an application. It encompasses natural and artificial aspects of the application. It codifies the body of knowledge mainly on the basis of deep models and matrices. It will have a commitment to normal models for discovery and validation. Models will thus become reproducible. Modelling is enhanced by falsifiability, testing, validation and verification. Modelling as a science has the ability to make reliable predictions, some of which might be surprising. Modelling might also be based on other techniques than presented in this paper. All models in [78] have their matrix. So, modelling based on normal models with their matrix will still be one of the main forms of modelling culture.

Acknowledgement. We are thankful to our reviewers and their suggestions which led to a substantial improvement of the paper.

References

1. Websters ninth new collegiate dictionary, 1991.
2. ISO/IEC JTC 1/SC 22. Information technology – Programming languages – C. ISO/IEC 9899:2011, Stage: 60.60, 2011.
3. D. Abts and W. Mülder. *Grundkurs Wirtschaftsinformatik : Eine kompakte und praxisorientierte Einführung.* Vieweg, 2004.

4. P. Alpar. *Computergestützte interaktive Methodenauswahl.* PhD thesis, Frankfurt Main Univ., 1980.
5. B. E. Babich. From Fleck's Denkstil to Kuhn's paradigm: conceptual schemes and incommensurability. *International Studies in the Philosophy of Science*, 17(1):75–92, 2003.
6. C. W. Bachman. Impact of object oriented thinking on ER modeling. LNCS 1157, pages 1–4. Springer, Berlin, 1996.
7. C. Batini, S. Ceri, and S. Navathe. *Conceptual database design (an entity-relationship approach)*. Benjamin/Cummings, Redwood City, 1992.
8. J. Becker and R. Schütte. *Handelsinformationssysteme : Domänenorientierte Einführung in die Wirtschaftsinformatik.* Moderne Industrie, 2004.
9. R. Berghammer and B. Thalheim. *Wissenschaft und Kunst der Modellierung: Modelle, Modellieren, Modellierung,* chapter Methodenbasierte mathematische Modellierung mit Relationenalgebren, pages 67–106. De Gryuter, Boston, 2015.
10. M. Bichler, U. Frank, D. Avison, J. Malaurent, P. Fettke, D. Hovorka, J. Krämer, D. Schnurr, B. Müller, L. Suhl, and B Thalheim. Theories in business and information systems engineering. *Business & Information Systems Engineering*, pages 1–29, 2016.
11. A. Bird. *Thomas Kuhn.* Routledge, 2014.
12. D. Bjørner. *Domain engineering,* volume 4 of *COE Research Monographs.* Japan Advanced Institute of Science and Technolgy Press, Ishikawa, 2009.
13. E. Brann. *The logos of Heraclitus.* Paul Dry Books, 2011.
14. S. Chadarevian and N. Hopwood, editors. *Models - The third dimension of science.* Stanford University Press, Stanford, California, 2004.
15. P. P. Chen. The entity-relationship model: Toward a unified view of data. *ACM TODS*, 1(1):9–36, 1976.
16. R.S. Cohen and T. Schnelle, editors. *Cognition and fact: Materials on Ludwik Fleck,* volume 87 of *Boston Studies in the Philosophy of Science.* Dortrecht, Reidel, 1986.
17. F. de Saussure. *Cours de Linguistique Générale.* Payot, 1995.
18. D. Embley and B. Thalheim, editors. *The Handbook of Conceptual Modeling: Its Usage and Its Challenges.* Springer, 2011.
19. L. Fleck. *Denkstile und Tatsachen, edited by S. Werner and C. Zittel.* Surkamp, 2011.
20. M. Fowler. *Refactoring.* Addison-Wesley, Boston, 2005.
21. P. Godfrey-Smith. Models and fictions in science. *Philosophical studies,* 143(1):101–116, 2009.
22. J. Gray. eScience: A transformed scientific method. Technical report, Talk given Jan 11, 2007. Edited by T. Hey, S. Tansley, and K. Tolle. http://research.microsoft.com/en-us/um/people/gray/talks/NRC-CSTB_eScience.ppt, Microsoft Research Publications, 2007.
23. G. Greefrath, G. Kaiser, W. Blum, and R. Borromeo Ferri. *Mathematisches Modellieren für Schule und Hochschule,* chapter Mathematisches Modellieren - Eine Einführung in theoretische und didaktische Hintergründe, pages 11–37. Springer, 2013.
24. D. C. Hay. *Data model pattern: Conventions of thought.* Dorset House, New York, 1995.
25. G. Hofstede and G.J. Hofstede. *Cultures and Organizations: Software of the Mind: Intercultural Cooperation and Its Importance for Survival.* McGraw-Hill, New York, 2004.

26. T. Illenseer. *Wissenschaft und Kunst der Modellierung: Modelle, Modellieren, Modellierung*, chapter Astrophysikalische Modellbildung am Beispiel aktiver galaktischer Kerne, pages 251–280. De Gryuter, Boston, 2015.
27. H. Jaakkola and B. Thalheim. Architecture-driven modelling methodologies. In *Information Modelling and Knowledge Bases*, volume XXII, pages 97–116. IOS Press, 2011.
28. R. Kaschek. *Konzeptionelle Modellierung*. PhD thesis, University Klagenfurt, 2003. Habilitationsschrift.
29. H.-J. Klein. Null values in relational databases and sure information answers. In L. Bertossi et al., editor, *Semantics in Databases*, pages 102–121, Dagstuhl Castle, December 2002. Springer-Verlag. LNCS 2582.
30. M. Klettke and B. Thalheim. Evolution and migration of information systems. In *The Handbook of Conceptual Modeling: Its Usage and Its Challenges*, chapter 12, pages 381–420. Springer, Berlin, 2011.
31. D. E. Knuth. *Literate Programming*. Number 27 in CSLI Lecture Notes. Center for the Study of Language and Information at Stanford/ California, 1992.
32. B. Kralemann and C. Lattmann. Models as icons: modeling models in the semiotic framework of peirce's theory of signs. *Synthese*, 190(16):3397–3420, 2013.
33. F. Kramer and B. Thalheim. Component-based development of a metadata datadictionary. In *BIS*, volume 176 of *Lecture Notes in Business Information Processing*, pages 110–121. Springer, 2014.
34. J. Krogstie. *Model-based development and evolution of information systems*. Springer, 2012.
35. T. Kuhn. *The Structure of Scientific Revolutions*. University of Chicago Press, Chicago, Illinois, 1962.
36. T. Kuhn. Objectivity, value judgment, and theory choice. *Arguing About Science*, pages 74–86, 1977.
37. T. Kuhn. *Die Entstehung des Neuen*. Surkamp, Frankfurt a. M., 1978.
38. I. Lakatos. Falsification and the methodology of scientific research programmes. *Criticism and the Growth of KnowledgeÄ_Ä, ed. I. Lakatos and A. Musgrave (London: Cambridge University Press, 1974)*, 2014.
39. A.V. Lebedev. *The Logos Heraklits - A recomnstruction of thoughts and words; full commented texts of fragments (in Russian)*. Nauka, 2014.
40. H.-J. Lenz and B. Thalheim. OLAP databases and aggregation functions. In *13th SSDBM 2001*, pages 91 – 100, 2001.
41. B. Mahr. *Das Wissen im Modell*. Technische Universität, Fakultät IV Berlin, 2004.
42. B. Mahr. Intentionality and modeling of conception. In *Judgements and Propositions. Logical, Linguistic, and Cognitive Issues*. Logos, 2010.
43. B. Mahr. *Mathesis & Graphé: Leonhard Euler und die Entfaltung der Wissenssysteme*, chapter Denken in Modellen, pages 85–100. Akademie-Verlag, 2010.
44. B. Mahr. Modelle und ihre Befragbarkeit - Grundlagen einer allgemeinen Modelltheorie. *Erwägen-Wissen-Ethik (EWE)*, Vol. 26, Issue 3:329–342, 2015.
45. H. Mannila and K.-J. Räihä. *The design of relational databases*. Addison-Wesley, Wokingham, England, 1992.
46. D. Marco and M. Jennings. *Universal meta data models*. Wiley Publ. Inc., 2004.
47. F. Matthes and J. W. Schmidt. *Datenbankhandbuch*, chapter Datenbankmodelle und Datenbanksprachen, pages 1–154. Springer Berlin, 1987.
48. M.S. Morgan and M. Morrison, editors. *Models as mediators*. Cambridge Press, 1999.

49. P.P. Nayak. *Automated modeling of physical systems.* Springer, LNCS 1003, Berlin, 1995.
50. C.S. Peirce. What is a sign? In Peirce Edition Project, editor, *The essential Peirce: selected philosophical writings*, volume 2, pages 4 – 10. Indiana University Press, Bloomington, Indiana, 1998.
51. G. Polya. *How to solve it: A new aspect of mathematical method.* Princeton University Press, Princeton, 1945.
52. P. Ponniah. *Data modeling fundamentla.* John Wiley & Sons, Hoboken, 2007.
53. K. R. Popper. *Logik der Forschung.* J.C.B. Mohr (Paul Siebeck), Tübingen, 10th. edition, 1994.
54. K. R. Popper. Subjektive oder objektive Erkenntnis? In David Miller, editor, *Karl Popper Lesebuch*, pages 40–59. J.C.B. Mohr (Paul Siebeck), Tübingen, 1994/1995.
55. A. Rutherford. *Mathematical Modelling Techniques.* Dover publications, 1995.
56. J.E. Safra, I. Yeshua, and et. al. *Encyclopædia Britannica.* Merriam-Webster, 2003.
57. K.-D. Schewe and B. Thalheim. NULL value algebras and logics. In *Information Modelling and Knowledge Bases*, volume XXII, pages 354–367. IOS Press, 2011.
58. K.-D. Schewe and B. Thalheim. Co-design of web information systems. Texts & Monographs in Symbolic Computation, pages 293–332, Wien, 2015. Springer.
59. B. Scholz-Reiter. *Konzeption eines rechnergesttzten Werkzeugs zur Analyse und Modellierung integrierter Informations- und Kommunikationssysteme in Produktionsunternehmen.* PhD thesis, TU Berlin, 1990.
60. L. Silverston. *The data model resource book. Revised edition*, volume 2. Wiley, 2001.
61. L. Silverston, W. H. Inmon, and K. Graziano. *The data model resource book.* John Wiley & Sons, New York, 1997.
62. G. Simsion. *Data modeling essentials - Analysis, design and innovation.* Van Nonstrand Reinhold, New York, 1994.
63. G. Simsion and G.C. Witt. *Data modeling essentials.* Morgan Kaufmann, San Francisco, 2005.
64. H. Stachowiak. Modell. In Helmut Seiffert and Gerard Radnitzky, editors, *Handlexikon zur Wissenschaftstheorie*, pages 219–222. Deutscher Taschenbuch Verlag GmbH & Co. KG, München, 1992.
65. W. Steinmüller. *Informationstechnologie und Gesellschaft: Einführung in die Angewandte Informatik.* Wissenschaftliche Buchgesellschaft, Darmstadt, 1993.
66. V. Storey and B. Thalheim. Conceptual modeling: Enhancement through semiotics. In *Proc. ER'17*, LNCS, 10650, page forthcoming, Cham, 2017. Springer.
67. T. J. Teorey. *Database modeling and design: The entity-relationship approach.* Morgan Kaufmann, San Mateo, 1989.
68. B. Thalheim. *Entity-relationship modeling – Foundations of database technology.* Springer, Berlin, 2000.
69. B. Thalheim. *The Conceptual Framework to Multi-Layered Database Modelling based on Model Suites*, volume 206 of *Frontiers in Artificial Intelligence and Applications*, pages 116–134. IOS Press, 2010.
70. B. Thalheim. Towards a theory of conceptual modelling. *Journal of Universal Computer Science*, 16(20):3102–3137, 2010. http://www.jucs.org/jucs_16_20/towards_a_theory_of.
71. B. Thalheim. Syntax, semantics and pragmatics of conceptual modelling. In *NLDB*, volume 7337 of *Lecture Notes in Computer Science*, pages 1–12. Springer, 2012.

72. B. Thalheim. The conception of the model. In *BIS*, volume 157 of *Lecture Notes in Business Information Processing*, pages 113–124. Springer, 2013.
73. B. Thalheim. Open problems of information systems research and technology. In *Invited Keynote, BIR'2013. LNBIB 158*, pages 10–18. Springer, 2013.
74. B. Thalheim. The conceptual model ≡ an adequate and dependable artifact enhanced by concepts. In *Information Modelling and Knowledge Bases*, volume XXV of *Frontiers in Artificial Intelligence and Applications, 260*, pages 241–254. IOS Press, 2014.
75. B. Thalheim. Models, to model, and modelling - Towards a theory of conceptual models and modelling - Towards a notion of the model. Collection of recent papers, http://www.is.informatik.uni-kiel.de/∼thalheim/indexkollektionen.htm, 2014.
76. B. Thalheim. Conceptual modeling foundations: The notion of a model in conceptual modeling. In *Encyclopedia of Database Systems*. Springer US, 2017.
77. B. Thalheim. General and specific model notions. In *Proc. ADBIS'17*, LNCS 10509, pages 13–27, Cham, 2017. Springer.
78. B. Thalheim and I. Nissen, editors. *Wissenschaft und Kunst der Modellierung: Modelle, Modellieren, Modellierung*. De Gruyter, Boston, 2015.
79. B. Thalheim and M. Tropmann-Frick. The conception of the conceptual database model. In *ER 2015*, LNCS 9381, pages 603–611, Berlin, 2015. Springer.
80. B. Thalheim and M. Tropmann-Frick. Models and their capability. In C. Beierle, G. Brewka, and M. Thimm, editors, *Computational Models of Rationality*, volume 29 of *College Publications Series*, pages 34–56. College Publications, 2016.
81. O. Thomas. Das Modellverständnis in der Wirtschaftsinformatik: Historie, Literaturanalyse und Begriffsexplikation. Technical Report Heft 184, Institut für Wirtschaftsinformatik, DFKI, Saarbrücken, Mai 2005.
82. S. A. Umpleby. Second-order cybernetics as a fundamental revolution in science. *Constructivist Foundations*, 11(3):455–465, 2016.
83. C. von Dresky, I. Gasser, C. P. Ortlieb, and S Günzel. *Mathematische Modellierung: Eine Einführung in zwölf Fallstudien*. Vieweg, 2009.
84. H. von Foerster, editor. *Cybernetics of cybernetics: Or, the control of control and the communication of communication*. Future Systems, 1995.
85. B. F. Webster. *Pitfalls of object-oriented development: a guide for the wary and entusiastic*. M&T books, New York, 1995.
86. S. Wenzel. Referenzmodell für die Simulation in Produktion und Logistik. *ASIM Nachrichten*, 4(3):13–17, 2000.
87. R. Wieringa. *Design science methodology for information systems and software engineering*. Springer, Heidelberg, 2014.
88. B. P. Zeigler, T. G. Kim, and H. Praehofer. *Theory of Modeling and Simulation*. Elsevier Academic Press, 2000.

Part II
Databases

Extremal Combinatorics of SQL Keys

Sven Hartmann[1], Markus Kirchberg[2], Henning Koehler[3], Uwe Leck[4], and Sebastian Link[5]

[1] Department of Informatics, Clausthal University of Technology, Germany
[2] The National University of Singapore, Singapore
[3] School of Engineering & Advanced Technology, Massey University, New Zealand
[4] Department of Mathematics, University of Flensburg, Germany
[5] Department of Computer Science, University of Auckland, New Zealand

Abstract. Keys form the most fundamental class of constraints in database practice. A key is a set of column names that is satisfied by a table if no two distinct rows of the table have matching values on all the columns that belong to the key. Some fundamental combinatorial problems ask how many minimal keys can possibly coexist, and which families of minimal keys attain this cardinality. Without missing data, Sperner's theorem has been used to establish a complete solution to these problems. In the real world of industry-standard compliant SQL databases, missing data is the norm. Defining a possible world of an SQL table as a table obtained by independently replacing each missing data occurrence by some actual data value, one arrives naturally at the notions of possible and certain keys. Indeed, a key is possible (certain) for a given SQL table if the key holds in some (all) possible worlds. SQL also permits users to specify column names as NOT NULL, meaning that missing data must not occur in that column. We establish a complete solution to the fundamental combinatorial problems applied to possible and certain SQL keys under NOT NULL constraints. In fact, the problems are solved for any given upper bound on the number of column names that keys may have. The results provide insight into maintenance costs and optimization opportunities that SQL databases may entail.

Keywords: Database; Key; Missing data; Sperner theory; SQL

1 Motivation

Keys have always been a core enabler for data management. They are fundamental for understanding the structure and semantics of data. Given a collection of entities, a key is a set of attributes whose values uniquely identify an entity in the collection. For example, a key for a relational table is a set of columns such that no two different rows have matching values in each of the key columns. Keys are essential for many other data models, including semantic models [44,57], object models [27,47], XML [23], RDF [36], and uncertain data [5, 28]. They help in many classical areas of data management, including data modeling, database design, indexing, and query optimization. Knowledge about keys enables us to i) uniquely reference entities across data repositories, ii) minimize data redundancy at schema design time to process updates efficiently at run time, iii) provide better selectivity estimates in cost-based query optimization, iv) provide a query optimizer with new access paths that can lead to substantial speedups in query processing, v) allow the database administrator to improve the efficiency of data access via physical design techniques such as data partitioning or the creation of indexes and materialized views, and vi) provide new insights into application data. Modern applications raise the importance of keys further. They facilitate the data integration process, help with the detection of duplicates and anomalies, provide guidance in repairing data, and return consistent answers to queries over dirty data. The discovery of keys is one of the core activities in data profiling.

Keys, and other classes of integrity constraints, have been extensively studied in Codd's relational model of data [7]. Here, a relation is a finite set of tuples in which no data is missing. In real world databases, i.e. those compliant with the industry standard SQL, missing data is the norm. SQL accommodates missing data by permitting occurrences of the distinguished null marker, which we denote by $\bot$. The main reason is that information is stored in a rigid table format and one cannot expect all data to be available at the time of acquisition. Not acquiring such data at all would incur loss of information, insight, and business value. In general, an SQL table is a multiset of partial rows, where partial means that the row may feature occurrences of $\bot$. Relations are idealized SQL tables in which no duplicate rows and no null markers occur.

While many interpretations for the occurrences of the null marker exist, it is often argued that SQL has adopted the interpretation "value exists, but unknown", compliant with Codd's original proposal. This interpretation leads naturally to a possible world semantics, in which a possible world of an SQL table is the result of independently replacing each occurrence of $\bot$ by some value from the domain of the column in which $\bot$ occurs. This approach suggests two semantics of keys. A *possible key* $p \langle X \rangle$ is satisfied by an SQL table I if and only if there is some possible world of I in which the key X is satisfied. A *certain key* $c \langle X \rangle$ is satisfied by an SQL table I if and only if the key X is satisfied in every possible world of I. In particular, certain keys make it possible to store incomplete rows that can still be identified uniquely by the values in the key columns. For example, the snapshot I in Table 1 satisfies the possible key $p \langle journal \rangle$ as witnessed by the possible world in which $\bot$ is replaced by 'PLOS One 2000', but violates the certain key $c \langle journal \rangle$ as witnessed by the possible world in which $\bot$ is replaced by 'Nucleic Acids 1997'. Moreover, I satisfies the certain keys $c \langle title, journal \rangle$ and $c \langle author, journal \rangle$.

In practice, data engineers can benefit from knowing how complex the maintenance of their database can grow. For example, it is useful to know how large a non-redundant family of SQL keys over a table with n column names can

title	author	journal
The uRNA database	Zwieb C	Nucleic Acids 1997
The uRNA database	Zwieb C	Nucleic Acids 1996
Genome wide detect.	Ragh. R	⊥

Table 1. Snippet I of the RFAM data set

title	author	journal
The uRNA database	Zwieb C	Nucleic Acids 1997
Genome wide detect.	Ragh. R	⊥
GenBank	⊥	Nucleic Acids 2008
The uRNA database	Danot O	Nucleic Acids 2015

Table 2. Snippet I' that satisfies three possible and three certain keys

be, and which of these families attain such extreme cardinality. Here, non-redundancy of the family Σ means that none of the keys $\sigma \in \Sigma$ is implied by the remaining keys $\Sigma \backslash \{\sigma\}$ in the family. That is, for every $\sigma \in \Sigma$ there is some table that satisfies all the keys in $\Sigma \backslash \{\sigma\}$ but violates σ. Non-redundancy is an important property in practice, guaranteeing that no resources are wasted in redundantly validating the satisfaction of any keys when a database is updated. That is, the maximal cardinality of non-redundant families of SQL keys also represents the worst possible number of SQL keys that must necessarily be validated whenever updates occur. More positively, this number can also be interpreted as the best possible number of SQL keys from which query optimizations may result.

In the relational model of data, in which all column names are by default NOT NULL, a family of keys is non-redundant if and only if it has the Sperner property (i.e. forms an anti-chain with respect to set inclusion). Consequently, the maximum cardinality of a non-redundant family is $\binom{n}{\lfloor n/2 \rfloor}$, which is attained by the family of $n/2$-ary subsets of the given n-ary schema when n is even, and by the family of $(n+1)/2$-ary subsets and the family of $(n-1)/2$-ary subsets of the given n-ary schema when n is odd.

The situation is rather different in the context of SQL, that is, for possible and certain keys in the presence of arbitrary NOT NULL constraints. For example, the snippet I' in Table 2 satisfies the non-redundant family that contains the two possible keys $p \langle author \rangle$, $p \langle journal \rangle$ and the two certain keys $c \langle title, author \rangle$, and $c \langle title, journal \rangle$.

The example illustrates two major differences to the idealized special case of the relational model: i) In SQL, the non-redundancy of a family of possible and certain keys is no longer characterized by simply satisfying the Sperner property, and ii) The extremal bounds from Sperner's theorem do no longer apply.

Contribution. Motivated by the dominance of the industry standard SQL in practice and motivated by the non-applicability of results from the idealized special case of the relational model within SQL, we investigate Sperner-type properties of SQL keys. Our first contribution is a Sperner-like characterization of non-redundant families of possible and certain keys in the presence of NOT NULL constraints. Based on this characterization, we establish a complete solution to the problems what cardinality non-redundant families of possible and certain keys in the presence of NOT NULL constraints can have, and which families attain this cardinality. We solve these problems for any fixed upper bound that one may put on the number of column names that may comprise a key.

Organization. We summarize some related work on the extremal combinatorics of database constraints in Section 2, which provides further motivation for our research. We particularly emphasize contributions by Klaus-Dieter in this area. We give preliminary definition of the data model in Section 3. Section 4 establishes the Sperner-like characterization of non-redundant families of SQL keys. The main combinatorial problem and its unique solution are reported in Section 5. Two lemmata are proven in Section 6 that prepare the proof of the main theorem in Section 7. We conclude in Section 8 where we also list some potential future work.

2 Selected Related Work

Integrity constraints enforce the semantics of application domains in database systems. Klaus-Dieter has been one of the thought leaders in developing formal frameworks for maintaining and enforcing integrity in various data models [42, 46, 49], and showing the limitations of rule trigger systems [48]. Integrity constraints form one of the cornerstones of database technology [52]. Besides domain and referential integrity, entity integrity is one of the three inherent integrity rules proposed by Codd [8]. These three types of constraints are the only ones, amongst around 100 classes of studied constraints [52], that enjoy built-in support by SQL database systems.

In the relational model, combinatorial properties of keys have attracted much attention [10, 11, 16, 13, 14, 44, 50, 54–56]. Surveys include [12, 26] and the book [52] contains various results on the topic. Since we consider missing data, our results are different. The relational model is the idealized special case of SQL in which all column names are specified to be NOT NULL.

Some papers have investigated combinatorial problems for keys over data models with missing data. Here, different meaningful notions of a key co-exist. Codd argued that each table must have one distinguished key (called the primary key) whose column names are all NOT NULL. Families of keys with this property, called Codd families, were studied in [21]. Codd families are different from possible and certain keys, in particular their non-redundancy and combinatorial properties. For example, the snapshot I from Table 1 satisfies no Codd key since every subset of column names features either some null marker or cannot separate some pair of distinct rows. However, I satisfies some possible and certain keys such as $p\langle journal\rangle$ and $c\langle title, journal\rangle$, respectively. Thalheim [53], Levene and Loizou [37] studied the notion of a key set. A relation satisfies a key set if, for each pair of distinct rows, there is some key in the key set on which the two rows have no missing data and are distinct. A certain key is equivalent to a key set consisting of all the singleton subsets of the column names that belong to a key, e.g., $c\langle title, journal\rangle$ corresponds to the key set $\{\{title\}, \{journal\}\}$. The extremal combinatorics of key sets enjoys the same bounds of keys from relational databases [53, 52] and is fundamentally different from our results.

Klaus-Dieter investigated Sperner families in complex-value data models which utilize constructors for tuples, lists, sets, multisets, optionality and disjoint unions. In these models, he studied keys that are equivalent with certain ideals, permitting an ordering between sets of keys. He then established a sufficient conditions for a Sperner family of such ideals to be a system of minimal keys, and determined lower and upper bounds for the size of the smallest Armstrong instance [44]. In follow-up work [45], Klaus-Dieter showed that under some mild technical assumptions, keys in these models can be identified with closed sets of subattributes, given a particular closure operator, and minimal keys correspond to closed sets that are minimal under set-wise containment. The existence of Armstrong databases for given minimal key systems is investigated. It is for this particular line of work that we hope Klaus-Dieter will enjoy our results that apply extremal combinatorics to a natural notion of keys in the industry standard SQL.

3 Preliminary Definitions

We begin with basic terminology. Let $\mathfrak{A} = \{A_1, A_2, \ldots\}$ be a (countably) infinite set of distinct symbols, called *attributes*. Attributes represent column names of tables. A *table schema* is a finite non-empty subset T of $\mathfrak{A}$. Each attribute A of a table schema T is associated with an infinite domain $dom(A)$ which represents the possible values that can occur in column A. In order to encompass incomplete information the domain of each attribute contains the null marker, denoted by $\bot$. The interpretation of $\bot$ is to mean "value unknown at present". We stress that the null marker is not a domain value. In fact, it is a purely syntactic convenience that we include the null marker in the domain of each attribute. For attribute sets X and Y we may write XY for their set union $X \cup Y$. If $X = \{A_1, \ldots, A_m\}$, then we may write $A_1 \cdots A_m$ for X. In particular, we may write A to represent the singleton $\{A\}$. A *tuple* over T is a function $t : T \to \bigcup_{A \in T} dom(A)$ with $t(A) \in dom(A)$ for all $A \in X$. For $X \subseteq T$ let $t[X]$ denote the restriction of the tuple t over T to X. We say that a tuple t is X-*total* if $t[A] \neq \bot$ for all $A \in X$. A tuple t over T is said to be a *total tuple* if it is T-total. A *table I* over T is a finite multiset of tuples over T. A table I over T is a *total table* if every tuple $t \in I$ is total. Let t, t' be tuples over T. We define *possible/certain similarity* of t, t' on $X \subseteq T$ as follows:

$$t[X] \sim_p t'[X] :\Leftrightarrow \forall A \in X. (t[A] = t'[A] \vee t[A] = \bot \vee t'[A] = \bot)$$
$$t[X] \sim_c t'[X] :\Leftrightarrow \forall A \in X. (t[A] = t'[A] \neq \bot) \ .$$

Possible and certain similarity become identical for tuples that are X-total. In such "classical" cases we denote similarity by $t[X] \sim t'[X]$.

A *null-free subschema* (NFS) over the table schema T is an expression T_S where $T_S \subseteq T$. The NFS T_S over T is satisfied by a table I over T if and only if I is T_S-total. SQL allows the specification of attributes as NOT NULL, so the set of attributes declared NOT NULL forms an NFS over the underlying table schema. For convenience we sometimes refer to the pair (T, T_S) as table schema.

We say that $X \subseteq T$ is a *key* for the total table I over T, denoted by $I \vdash X$, if there are no two different tuples $t, t' \in I$ that agree on X. We will now define two different notions of keys over general tables, using a possible world semantics. Given a table I on T, a *possible world* of I is obtained by independently replacing every occurrence of $\bot$ in I with a domain value. We say that $X \subseteq T$ is a *possible/certain key* for I, denoted by $p\langle X \rangle$ and $c\langle X \rangle$ respectively, if the following hold:

$$I \vdash p\langle X \rangle :\Leftrightarrow X \text{ is a key for some possible world of } I$$
$$I \vdash c\langle X \rangle :\Leftrightarrow X \text{ is a key for every possible world of } I \ .$$

The following result establishes a syntactic characterization for the satisfaction of possible and certain keys.

Theorem 1. *$X \subseteq T$ is a possible (certain) key for I iff no distinct tuples in I are certainly (possibly) similar on X.*

Proof. (possible key $\Rightarrow$ not certainly similar) Let X be a possible key for I and $t, t' \in I$ be distinct. Then there exists a possible word $\rho(I)$ of I for which X is a key. Denote by $\rho(t), \rho(t')$ the copies of t, t' in $\rho(I)$. Since X is a key for $\rho(I)$, there exists $A \in X$ with $\rho(t)[A] \neq \rho(t')[A]$. Since only $\bot$ values get replaced, this means that $t[A] \neq t'[A]$ or $t[A] = \bot$ or $t'[A] = \bot$ holds. In either case t, t' are not certainly similar.

(possible key $\Leftarrow$ not certainly similar) Let no two distinct tuples in I be certainly similar on X. We construct a possible world $\rho(I)$ of I by replacing all $\bot$ values occurring in I with mutally distinct domain values that did not occur in I previously[6]. Let again $\rho(t), \rho(t') \in \rho(I)$ be two distinct copies of $t, t' \in I$. Since t, t' are not certainly similar on X, there exists $A \in X$ with $t[A] \neq t'[A]$ or $t[A] = \bot$ or $t'[A] = \bot$. As we have chosen our replacement values to be

---

[6] Such a replacement exists since domains are infinite and tables finite.

unique in $\rho(I)$, we have $\rho(t)[A] \neq \rho(t')[A]$ in all three cases. Thus $\rho(t), \rho(t')$ are not similar on X, so X is a possible key for I.

(not certain key $\Rightarrow$ possibly similar) Let X not be a certain key for I. Then there exists a possible world $\rho(I)$ of I and distinct tuples $t, t' \in I$ such that $\rho(t)[X] \sim \rho(t')[X]$. Thus for every $A \in X$ we have $\rho(t)[A] = \rho(t')[A]$ and hence $t[A] = t'[A]$ or $t[A] = \bot$ or $t'[A] = \bot$, i.e., $t[X] \sim_p t'[X]$.

(not certain key $\Leftarrow$ possibly similar) Let $t, t' \in I$ with $t[X] \sim_p t'[X]$. Then we can construct a possible world $\rho(I)$ of I which replaces $\bot$ values on t, t' as follows: i) If $t[A] = t'[A] = \bot$ let $\rho(t)[A] = \rho(t')[A]$ be arbitrary, ii) If $t[A] = \bot \wedge t'[A] \neq \bot$ let $\rho(t)[A] = t'[A]$, and iii) If $t[A] \neq \bot \wedge t'[A] = \bot$ let $\rho(t')[A] = t[A]$. In each case we have $\rho(t)[A] = \rho(t')[A]$ and hence $\rho(t)[X] \sim \rho(t')[X]$. Thus X is not a certain key for I. □

For a set Σ of constraints over table schema T we say that a table I over T *satisfies* Σ if I satisfies every $\sigma \in \Sigma$. If for some $\sigma \in \Sigma$ the table I does not satisfy σ we say that I *violates* σ (and violates Σ). A table I over (T, T_S) is a table I over T that satisfies T_S. A table I over (T, T_S, Σ) is a table I over (T, T_S) that satisfies Σ. For a set $\Sigma \cup \{\varphi\}$ of constraints over (T, T_S) we say that Σ *implies* φ, denoted by $\Sigma \models \varphi$, if and only if every table over (T, T_S) that satisfies Σ also satisfies φ. The *implication problem* for a class $\mathcal{C}$ of constraints is to decide, for an arbitrary (T, T_S) and an arbitrary set $\Sigma \cup \{\varphi\}$ of constraints in $\mathcal{C}$, whether Σ implies φ. For possible and certain keys the implication problem can be characterized as follows.

Theorem 2. *Let Σ be a set of possible and certain keys.*

1. *Σ implies $c\langle X \rangle$ iff $c\langle Y \rangle \in \Sigma$ for some $Y \subseteq X$ or $p\langle Z \rangle \in \Sigma$ for some $Z \subseteq X \cap T_S$.*
2. *Σ implies $p\langle X \rangle$ iff $c\langle Y \rangle \in \Sigma$ or $p\langle Y \rangle \in \Sigma$ for some $Y \subseteq X$.*

Proof. The "if" directions follow from Theorem 1. We show the "only if" direction next.

1. Let $c\langle Y \rangle \notin \Sigma$ for every $Y \subseteq X$ and $p\langle Z \rangle \notin \Sigma$ for every $Z \subseteq X \cap T_S$. Consider a table $I = \{t, t'\}$ on (T, T_S) such that i) $t = (0, \ldots, 0)$, ii) $t'[X \cap T_S] = (0, \ldots, 0)$, iii) $t'[X \setminus T_S] = (\bot, \ldots, \bot)$, and iv) $t'[T \setminus X] = (1, \ldots, 1)$. Now consider Theorem 1. Since t, t' possibly agree on X only, the only certain keys $c\langle Y \rangle$ violated by I are those with $Y \subseteq X$. Hence no certain keys in Σ are violated by I. Since t, t' certainly agree on $X \cap T_S$ only, the only possible keys $p\langle Z \rangle$ violated by I are those with $Z \subseteq X \cap T_S$. Hence no possible keys in Σ are violated by I. Hence I respects Σ but violates $c\langle X \rangle$, so Σ does not imply $c\langle X \rangle$.
2. Analogous with $t'[X] = (0, \ldots, 0)$. □

A family Σ of constraints over table schema (T, T_S) is said to be *non-redundant* if and only if for all $\sigma \in \Sigma$, $\Sigma \setminus \{\sigma\}$ does not imply σ. Our first goal is a Sperner-like characterization for non-redundant families of SQL keys.

4 Non-redundant Families of SQL Keys

In this section we write $[n] := \{1, \ldots, n\}$ instead of $T = \{A_1, \ldots, A_n\}$, and A instead of T_S. For $X \subseteq [n]$, we use the notations $X^{(i)} := \{Y \subseteq X : |Y| = i\}$ and $X^{(\leq i)} := \{Y \subseteq X : |Y| \leq i\}$. A family $\mathcal{A} \subseteq [n]^{(\leq n)}$ is an *antichain* (or *Sperner family*) if $X \not\subseteq Y$ for all distinct $X, Y \in \mathcal{A}$. For a set Σ of possible and certain keys over $[n]$, define $\mathcal{F} := \{X \mid c\langle X \rangle \in \Sigma\}$ and $\mathcal{G} := \{X \mid p\langle X \rangle \in \Sigma\}$.

Theorem 3. *Σ is non-redundant if and only if all of the following conditions are satisfied (1) $\mathcal{F}$ is an antichain, (2) $\mathcal{G}$ is an antichain, (3) $\forall F \in \mathcal{F}, G \in \mathcal{G}: F \not\subseteq G$, and (4) $\forall F \in \mathcal{F}, G \in \mathcal{G}: G \not\subseteq F \cap A$.*

Proof. By Theorem 2, Σ is non-redundant iff (1') $\neg \exists c\langle X \rangle, c\langle Y \rangle \in \Sigma$ such that $Y \subseteq X$, (2') $\neg \exists p\langle X \rangle, p\langle Y \rangle \in \Sigma$ such that $Y \subseteq X$, (3') $\neg \exists p\langle X \rangle, c\langle Y \rangle \in \Sigma$ such that $Y \subseteq X$, and (4') $\neg \exists c\langle X \rangle, p\langle Y \rangle \in \Sigma$ such that $Y \subseteq X \cap A$ hold. Conditions (1')–(4') are equivalent to conditions (1)–(4). □

5 Problem and Solution

The problem we study reads as follows. Given non-negative integers n, k with $k \leq n$ and a set $A \subseteq [n]$, find all pairs $(\mathcal{F}, \mathcal{G}) \in [n]^{(\leq k)} \times [n]^{(\leq k)}$ that satisfy conditions (1)–(4) and maximize $|\mathcal{F} \cup \mathcal{G}|$. Note that $\mathcal{F}$ and $\mathcal{G}$ must be disjoint because of (3). In what follows, a complete solution is given.

First, we briefly discuss the case when $A = [n]$. In this case, (1)–(4) are satisfied if and only if $\mathcal{F} \dot\cup \mathcal{G} \subseteq [n]^{(\leq k)}$ is an antichain. If $k > n/2$, then by Sperner's Theorem [51] $|\mathcal{F} \cup \mathcal{G}|$ attains its maximum if and only if $\mathcal{F}$ and $\mathcal{G}$ form a partition of $[n]^{(\lfloor n/2 \rfloor)}$ or of $[n]^{(\lceil n/2 \rceil)}$, respectively. If $k \leq n/2$, then $|\mathcal{F} \cup \mathcal{G}|$ is maximized if and only if $\mathcal{F}$ and $\mathcal{G}$ form a partition of $[n]^{(k)}$. This is well-known and follows from the original proof of Sperner's Theorem [51]. In the sequel, we will assume that $0 \leq |A| < n$. Note that the bound (1) below does not hold when $A = [n]$ for even n and $k > n/2$.

Theorem 4. *Let n, a, k be non-negative integers with $n \geq 2$, $a < n$ and $k \leq n$, and let $A \subseteq [n]$ with $|A| = a$. Furthermore, let $\mathcal{F}, \mathcal{G} \subseteq [n]^{(\leq k)}$ be antichains such that $F \not\subseteq G$ and $G \not\subseteq F \cap A$ for all $F \in \mathcal{F}$ and $G \in \mathcal{G}$. Then*

$$|\mathcal{F} \cup \mathcal{G}| \leq \binom{n+1}{m} - \binom{a}{m-1}, \qquad (1)$$

where $m := \min\{k, \lfloor n/2 \rfloor + 1\}$. This bound is best possible, and equality is attained if and only if

(i) $\mathcal{F} \dot\cup \mathcal{G} = [n]^{(m)} \cup ([n]^{(m-1)} \setminus A^{(m-1)})$,
where $[n]^{(m)} \setminus A^{(m)} \subseteq \mathcal{F}$ and $[n]^{(m-1)} \setminus A^{(m-1)} \subseteq \mathcal{G}$, or

(ii) n is even, $k > n/2$, and $a \leq n/2 - 2$ or $a = n - 1$,
$$\mathcal{F} \dot{\cup} \mathcal{G} = [n]^{(m-1)} \cup ([n]^{(m-2)} \setminus A^{(m-2)}),$$
where $[n]^{(m-1)} \setminus A^{(m-1)} \subseteq \mathcal{F}$ and $[n]^{(m-2)} \setminus A^{(m-2)} \subseteq \mathcal{G}$.

Note that $[n]^{(m)}$ and $[n]^{(m-1)}$ are the two largest disjoint anti-chains over $[n]^{(<=k)}$. To prevent $F \subseteq G$, F must contain the larger sets. To prevent $G \subseteq F \cup A$ we drop small subsets of A from G. The remaining subsets of A may be attributed to either F or G as they do not conflict with others. This reflects the equivalence of certain/possible keys over NOT NULL attributes. While case (i) always presents a non-redundant family of maximum cardinality, case (ii) occurs because for even n and large k, an additional pair of largest disjoint anti-chains exists: $[n]^{(m-1)}$ and $[n]^{(m-2)}$. The difference between the resulting sets lies in the size of $A^{(m-1)}$ and $A^{(m-2)}$, leading to the conditions for a. The trivial case $n = 1$ is excluded in the above theorem to avoid certain technicalities in its proof.

Example 1. The maximum cardinality of a non-redundant family of possible and certain keys over table schema $(T = \{A, B, C, D\}, T_S = \{A, B\})$, or equivalently, $n = 4$ and $A = \{1, 2\}$, is nine. This cardinality is attained by the set Σ consisting of

$$c \langle A,B,C \rangle, \, c \langle A,B,D \rangle, \, c \langle A,C,D \rangle, \, c \langle B,C,D \rangle, \, p \langle A,C \rangle, \, p \langle B,C \rangle, \, p \langle A,D \rangle,$$
$$p \langle B,D \rangle, \, c \langle B,D \rangle.$$

The powerset lattice for this example is shown in Figure 1, where the top marked circles represent the certain keys and the bottom marked circles represent the possible keys above.

6 Main Lemmata

The two lemmas below will be applied in the proof of Theorem 4. To prove the lemmas, we proceed similar to the original proof of Sperner's Theorem [51]. For $A \subsetneq [n]$ and $\mathcal{K} \subseteq [n+1]^{(i)}$, we will use the following notations:

$$\Delta\mathcal{K} := \{X \in [n+1]^{(i-1)} : X \subset K \text{ for some } K \in \mathcal{K}\},$$
$$\tilde{\Delta}\mathcal{K} := \{X \in \Delta\mathcal{K} : n+1 \notin X \text{ or } X \not\subseteq A \cup \{n+1\}\},$$
$$\nabla\mathcal{K} := \{X \in [n+1]^{(i+1)} : X \supset K \text{ for some } K \in \mathcal{K}\},$$
$$\tilde{\nabla}\mathcal{K} := \{X \in \nabla\mathcal{K} : n+1 \notin X \text{ or } X \not\subseteq A \cup \{n+1\}\}.$$

In the proof of Lemma 1 we will use the Normalized Matching Inequality which can be stated as follows.

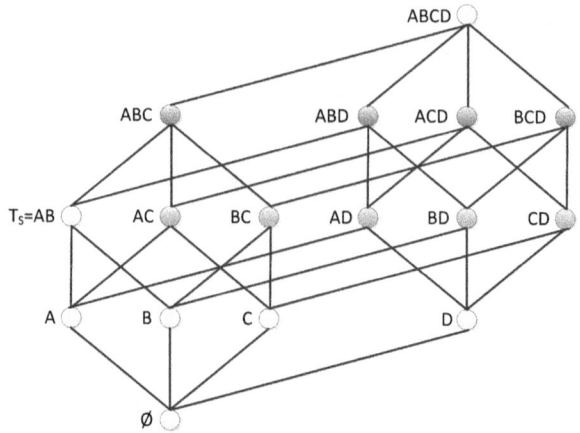

Fig. 1. Maximum Non-redundant Family of Keys

Normalized Matching Inequality (cf. [17] or [51]).
Let n, i be integers with $0 \leq i \leq n$.
(a) If $\mathcal{F} \subseteq [n+1]^{(i)}$, then $|\nabla \mathcal{F}| \geq \frac{n-i+1}{i+1}|\mathcal{F}|$.
(b) If $\mathcal{G} \subseteq [n+1]^{(i+1)}$, then $|\Delta \mathcal{G}| \geq \frac{i+1}{n-i+1}|\mathcal{G}|$.

Lemma 1. *Let n, a, A be as in Theorem 4, and let $i \leq n+1$ be a nonnegative integer. Furthermore, let $\emptyset \neq \mathcal{K} \subseteq [n+1]^{(i)}$ such that there is no $K \in \mathcal{K}$ with $n+1 \in K \subseteq A \cup \{n+1\}$.*

(a) *If $i \leq n/2$, then $|\tilde{\nabla}\mathcal{K}| \geq |\mathcal{K}|$.*
   Equality is attained if and only if n is even, $i = n/2$, $a \leq n/2 - 2$ or $a = n - 1$, and $\mathcal{K} = [n+1]^{(n/2)} \setminus \left(A^{(n/2-1)} \times \{n+1\}\right)$.
(b) *If $i \geq (n+3)/2$, then $|\tilde{\Delta}\mathcal{K}| > |\mathcal{K}|$.*

Proof. (a) Assume that $i \leq n/2$. Consider the bipartite graph G on the vertex set $V(G) = \mathcal{K} \cup \tilde{\nabla}\mathcal{K}$ and with edge set

$$E(G) = \{(X, Y) : X \in \mathcal{K}, Y \in \tilde{\nabla}\mathcal{K}, X \subset Y\}.$$

Let $\mathcal{A} := \{X \in \mathcal{K} : X \subseteq A\}$. Now $X \in \mathcal{K}$ has degree $n - i$ in G if $X \in \mathcal{A}$ and degree $n - i + 1$ otherwise. Hence,

$$|E(G)| = (n - i + 1)(|\mathcal{K}| - |\mathcal{A}|) + (n - i)|\mathcal{A}| = (n - i + 1)|\mathcal{K}| - |\mathcal{A}|. \quad (2)$$

As an immediate consequence of (2) we obtain

$$|E(G)| \geq (n - i)|\mathcal{K}|. \quad (3)$$

On the other hand, as every $Y \in \tilde{\nabla}\mathcal{K}$ is adjacent to at most $i+1$ elements of $\mathcal{K}$, we have
$$|E(G)| \leq (i+1)|\tilde{\nabla}\mathcal{K}|. \tag{4}$$

By (3) and (4), we have
$$(n-i)|\mathcal{K}| \leq (i+1)|\tilde{\nabla}\mathcal{K}|. \tag{5}$$

If $i < (n-1)/2$, then (5) implies $|\tilde{\nabla}\mathcal{K}| > |\mathcal{K}|$.

Assume that $i = (n-1)/2$, where n is odd. In this case, (5) reads $|\tilde{\nabla}\mathcal{K}| \geq |\mathcal{K}|$, and we need to show that equality can not hold. Assume that $|\tilde{\nabla}\mathcal{K}| = |\mathcal{K}|$. For this to be the case, equality must also hold in (3) and in (4). Equality in (3) implies that $\mathcal{K} = \mathcal{A}$. For equality in (4) it is necessary that every $Y \in \tilde{\nabla}\mathcal{K}$ is adjacent to exactly $(n+1)/2$ elements of $\mathcal{K}$, i.e., that $\Delta(\tilde{\nabla}\mathcal{K}) \subseteq \mathcal{K}$. Consider some $x \in K \in \mathcal{K}$ and $y \in [n]\setminus A$ (which exists as $a < n$). The set $K \cup \{y\}\setminus\{x\}$ is in $\Delta(\tilde{\nabla}\mathcal{K})$. On the other hand, it is not in $\mathcal{K}$ because of $\mathcal{K} = \mathcal{A}$, a contradiction.

For the remainder of the proof of (a), assume that $i = n/2$, where n is even. In this case, equation (2) becomes
$$|E(G)| = \left(\frac{n}{2}+1\right)|\mathcal{K}| - |\mathcal{A}|. \tag{6}$$

We will show the following strengthening of (4):
$$|E(G)| \leq \left(\frac{n}{2}+1\right)|\tilde{\nabla}\mathcal{K}| - |\mathcal{A}|. \tag{7}$$

Together with (6), this implies the claim $|\tilde{\nabla}\mathcal{K}| \geq |\mathcal{K}|$.

To prove (7), we will show that at least $|\mathcal{A}|$ elements of $\tilde{\nabla}\mathcal{K}$ have degree at most $n/2$ in G. As $a < n$, there is an $y \in [n] \setminus A$. Let $\mathcal{A}' := \mathcal{A} \times \{y\}$. Clearly, $\mathcal{A}'$ is a subset of $\tilde{\nabla}\mathcal{K}$, and $|\mathcal{A}'| = |\mathcal{A}|$. Consider the collection $\mathcal{B} \times \{y\}$ of those elements of $\mathcal{A}'$ that have degree $n/2+1$ in G. The remaining $|\mathcal{A}|-|\mathcal{B}|$ elements of $\mathcal{A}'$ have degree at most $n/2$ in G. By the choice of $\mathcal{B}$ we have $\Delta(\mathcal{B}\times\{y\}) \subseteq \mathcal{K}$ which implies
$$\Delta\mathcal{B} \times \{y, n+1\} \subseteq \tilde{\nabla}\mathcal{K} \setminus \mathcal{A}'. \tag{8}$$

As $\Delta\mathcal{B} \times \{n+1\}$ is a subset of $A^{(n/2-1)} \times \{n+1\}$, it does not contain any element of $\mathcal{K}$. That is, all $|\Delta\mathcal{B}|$ elements of $\Delta\mathcal{B}\times\{y, n+1\}$ have degree at most $n/2$ in G. In total, we have shown the number of elements of $\tilde{\nabla}\mathcal{K}$ that have degree at most $n/2$ in G to be at least $|\mathcal{A}| - |\mathcal{B}| + |\Delta\mathcal{B}|$. Since $\mathcal{B} \subseteq A^{(n/2)}$ and $a < n$, the normalized matching inequality gives $|\Delta\mathcal{B}| \geq |\mathcal{B}|$. This concludes the proof of (7).

It is easy to verify that $|\tilde{\nabla}\mathcal{K}| = |\mathcal{K}|$ holds for $\mathcal{K}$ as given in Lemma 1(a). It remains to show that $|\tilde{\nabla}\mathcal{K}| = |\mathcal{K}|$ implies that $\mathcal{K}$ is as in Lemma 1(a). If $n = 2$, then it is straightforward to check that $|\tilde{\nabla}\mathcal{K}| = |\mathcal{K}|$ only holds for A being a singleton subset of $\{1,2\}$ and $\mathcal{K} = \{\{1\},\{2\}\}$. Assume that $n > 2$ and $|\tilde{\nabla}\mathcal{K}| = |\mathcal{K}|$. Clearly, equality must hold in (6) and (7) then. By the above discussion, a necessary condition for equality in (7) is $|\Delta \mathcal{B}| = |\mathcal{B}|$.

Case 1. Assume that $\mathcal{B} = \emptyset$. Then, by the definition of $\mathcal{B}$ and for equality in (7), all elements of $\mathcal{A}'$ have degree $n/2$ in G, the remaining elements of $\tilde{\nabla}\mathcal{K}$ degree $n/2+1$. If $\mathcal{A} \neq \emptyset$, then consider some $Z \in \mathcal{A}$. The set $Z \cup \{y\}$ is in $\tilde{\nabla}\mathcal{K}$, and, as it has degree $n/2$ in G and as $n > 2$, there is a $z \in Z$ such that $Z \cup \{y\} \setminus \{z\} \in \mathcal{K}$. Now $Z' := Z \cup \{y, n+1\} \setminus \{z\}$ is in $\tilde{\nabla}\mathcal{K} \setminus \mathcal{A}'$. As $Z \cup \{n+1\} \setminus \{z\} \notin \mathcal{K}$, the set Z' has degree at most $n/2$ in G, a contradiction. Hence, $\mathcal{A} = \emptyset$. It follows that $\tilde{\nabla}\mathcal{K} = \nabla \mathcal{K}$. By the normalized matching inequality, we have $|\nabla \mathcal{K}| \geq |\mathcal{K}|$, and it is well-known (see [17], for instance) that equality only holds for $\mathcal{K} = [n+1]^{(n/2)}$, i.e., when $a \leq n/2 - 2$ and $\mathcal{K}$ as claimed.

Case 2. Assume that $\mathcal{B} \neq \emptyset$. Using the normalized matching inequality again, we obtain that $|\Delta \mathcal{B}| = |\mathcal{B}|$ is only possible when $a = n - 1$ and $\mathcal{B} = \mathcal{A} = A^{(n/2)}$. Now (8) implies

$$\nabla(\mathcal{K} \dot{\cup} (A^{(n/2-1)} \times \{n+1\})) = \tilde{\nabla}\mathcal{K} \dot{\cup} (A^{(n/2)} \times \{n+1\}),$$

and the normalized matching inequality gives

$$|\mathcal{K}| + \binom{n-1}{n/2-1} \leq |\tilde{\nabla}\mathcal{K}| + \binom{n-1}{n/2}$$

(which is equivalent to $|\mathcal{K}| \leq |\tilde{\nabla}\mathcal{K}|$), where equality holds if and only if

$$\mathcal{K} = [n+1]^{(n/2)} \setminus \left(A^{(n/2)} \times \{n+1\}\right).$$

(b) The proof of Lemma 1(b) is analogous to the proof of Lemma 1(a) for $i \leq (n-1)/2$. □

Lemma 2. *Let n, a, k, A, m be as in Theorem 4. Furthermore, let $\mathcal{H} \subseteq [n+1]^{(\leq k)}$ be an antichain such that there is no $H \in \mathcal{H}$ with $n+1 \in H \subseteq A \cup \{n+1\}$. Then*

$$|\mathcal{H}| \leq \binom{n+1}{m} - \binom{a}{m-1}, \qquad (9)$$

and equality holds if and only if

(i) $\mathcal{H} = [n+1]^{(m)} \setminus \left(A^{(m-1)} \times \{n+1\}\right)$, or
(ii) $\mathcal{H} = [n+1]^{(m-1)} \setminus \left(A^{(m-2)} \times \{n+1\}\right)$,
    where n is even, $k > n/2$, and $a \leq n/2 - 2$ or $a = n - 1$.

Proof. Among all antichains $\mathcal{H}$ as in the theorem, consider one of maximum cardinality. Let $\ell := \min\{|H| : H \in \mathcal{H}\}$ and $u := \max\{|H| : H \in \mathcal{H}\}$, and $\mathcal{H}_i := \{X \in \mathcal{H} : |X| = i\}$ for $\ell \leq i \leq u$.

Assume that $u \geq (n+3)/2$. It is straight forward to verify that $\mathcal{H}' := (\mathcal{H} \setminus \mathcal{H}_u) \cup \tilde{\Delta}\mathcal{H}_u$ is an antichain that also satisfies the conditions in the theorem.

By Lemma 1(b), we have $|\mathcal{H}'| > |\mathcal{H}|$, a contradiction to the maximality of $|\mathcal{H}|$. Hence,
$$u \leq (n+2)/2. \tag{10}$$

Next, assume that $\ell < \min\{n/2, k\}$. Now $\mathcal{H}'' := (\mathcal{H} \setminus \mathcal{H}_\ell) \cup \tilde{\nabla}\mathcal{H}_\ell$ satisfies the conditions in the theorem. By Lemma 1(a), we have $|\mathcal{H}''| > |\mathcal{H}|$, a contradiction. Consequently,
$$\ell \geq \min\{n/2, k\}. \tag{11}$$

If n is odd or $k \leq n/2$, then, by (10) and (11), $\mathcal{H} \subseteq [n+1]^{(m)}$. By the maximality of $|\mathcal{H}|$, it follows that $\mathcal{H}$ must be the antichain given in (i) for which equality in (9) is obviously attained.

Assume that n is even and that $k > n/2$. (10) and (11) imply $\mathcal{H} = \mathcal{H}_{n/2} \cup \mathcal{H}_{n/2+1}$. By the maximality of $|\mathcal{H}|$, we have $|\tilde{\nabla}\mathcal{H}_{n/2}| \leq |\mathcal{H}_{n/2}|$. According to Lemma 1(a), this is only possible if $\mathcal{H}_{n/2} = \emptyset$ or if $\mathcal{H}_{n/2} = [n+1]^{(n/2)} \setminus (A^{(n/2-1)} \times \{n+1\})$, where $a \leq n/2 - 2$ or $a = n - 1$. In the first case, $\mathcal{H}$ must be as in (i). In the latter case, $\mathcal{H}_{n/2+1}$ must be empty because $\mathcal{H}$ is an antichain, and $\mathcal{H}$ is as in (ii). Finally, it is straightforward to verify that for $\mathcal{H}$ as in (ii) equality is attained in (9). $\square$

7 Proof of Theorem 4

Let $\mathcal{G}^* := \{G \in \mathcal{G} : G \subset F \text{ for some } F \in \mathcal{F}\}$. Then
$$\mathcal{H} := \mathcal{F} \cup (\mathcal{G} \setminus \mathcal{G}^*) \cup \{G \cup \{n+1\} : G \in \mathcal{G}^*\}$$
has the same cardinality as $\mathcal{F} \cup \mathcal{G}$ and satisfies the conditions in Lemma 2. Now (9) implies the bound (1).

It is easy to verify that any $\mathcal{F}$ and $\mathcal{G}$ as in (i) or (ii) satisfy the conditions in Theorem 4, and that equality in (1) is attained for such families. It remains to show that these are the only optimal choices of $\mathcal{F}$ and $\mathcal{G}$.

For equality to be attained in (1), we must have equality in (9), i.e., $\mathcal{H}$ must be as in Lemma 2 (i) or (ii).

If $\mathcal{H}$ is as in Lemma 2 (i), then the definition of $\mathcal{H}$ implies that
$$\mathcal{F} \cup \mathcal{G} = [n]^{(m)} \cup ([n]^{(m-1)} \setminus A^{(m-1)})$$
and $[n]^{(m-1)} \setminus A^{(m-1)} = \mathcal{G}^* \subseteq \mathcal{G}$. As $\mathcal{G}$ is an antichain, $[n]^{(m)} \setminus A^{(m)} \subseteq \mathcal{F}$, and $\mathcal{F}$ and $\mathcal{G}$ are as in (i).

Similarly, if $\mathcal{H}$ is as in Lemma 2 (ii), then $\mathcal{F}$ and $\mathcal{G}$ are as in (ii). $\square$

8 Conclusion and Future Work

We have established the first Sperner-like theorems for important classes of keys in industry-compliant SQL databases, which subsume well-known results from relational databases as a special case. More results on possible and certain keys can be found in [29, 33, 34]. For future work, it would be interesting to investigate combinatorial problems for more expressive classes of SQL constraints, such as functional dependencies [24, 31], or other types of constraints, such as contextual keys [58], possibilistic keys and functional dependencies [28, 30, 40, 41], or even for systems of combinations with classical keys [3], functional dependencies [15], independence atoms [19, 35, 43], multivalued dependencies [1, 2, 4, 18, 39, 38, 22], and inclusion dependencies [6, 9, 20, 25, 32].

Acknowledgement. The authors would like to congratulate Klaus-Dieter on his 60th anniversary. They would like to take the opportunity to acknowledge his leadership over the years, and express gratitude for his influence on the authors' career. Thank you and the best wishes for your future, Klaus-Dieter.

References

1. Biskup, J.: On the complementation rule for multivalued dependencies in database relations. Acta Inf. 10, 297–305 (1978)
2. Biskup, J.: Inferences of multivalued dependencies in fixed and undetermined universes. Theor. Comput. Sci. 10, 93–105 (1980)
3. Biskup, J.: Some remarks on relational database schemes having few minimal keys. In: Düsterhöft, A., Klettke, M., Schewe, K. (eds.) Conceptual Modelling and Its Theoretical Foundations - Essays Dedicated to Bernhard Thalheim on the Occasion of His 60th Birthday. Lecture Notes in Computer Science, vol. 7260, pp. 19–28. Springer (2012)
4. Biskup, J., Link, S.: Appropriate inferences of data dependencies in relational databases. Ann. Math. Artif. Intell. 63(3-4), 213–255 (2011)
5. Brown, P., Link, S.: Probabilistic keys. IEEE Trans. Knowl. Data Eng. 29(3), 670–682 (2017)
6. Casanova, M.A., Fagin, R., Papadimitriou, C.H.: Inclusion dependencies and their interaction with functional dependencies. J. Comput. Syst. Sci. 28(1), 29–59 (1984)
7. Codd, E.F.: A relational model of data for large shared data banks. Commun. ACM 13(6), 377–387 (1970)
8. Codd, E.F.: The Relational Model for Database Management, Version 2. Addison-Wesley (1990)
9. Cosmadakis, S.S., Kanellakis, P.C., Vardi, M.Y.: Polynomial-time implication problems for unary inclusion dependencies. J. ACM 37(1), 15–46 (1990)
10. Demetrovics, J.: On the number of candidate keys. Inf. Process. Lett. 7(6), 266–269 (1978)
11. Demetrovics, J.: On the equivalence of candidate keys with sperner systems. Acta Cybern. 4(3), 247–252 (1979)
12. Demetrovics, J., Katona, G.O.H.: Extremal combinatorial problems of database models. In: Biskup, J., Demetrovics, J., Paredaens, J., Thalheim, B. (eds.) MFDBS 87, 1st Symposium on Mathematical Fundamentals of Database Systems, Dresden, GDR, January 19-23, 1987, Proceedings. Lecture Notes in Computer Science, vol. 305, pp. 99–127. Springer (1987)

13. Demetrovics, J., Katona, G.O.H., Miklós, D., Seleznjev, O., Thalheim, B.: The average length of keys and functional dependencies in (random) databases. In: Gottlob, G., Vardi, M.Y. (eds.) Database Theory - ICDT'95, 5th International Conference, Prague, Czech Republic, January 11-13, 1995, Proceedings. Lecture Notes in Computer Science, vol. 893, pp. 266–279. Springer (1995)
14. Demetrovics, J., Katona, G.O.H., Miklós, D., Seleznjev, O., Thalheim, B.: Asymptotic properties of keys and functional dependencies in random databases. Theor. Comput. Sci. 190(2), 151–166 (1998)
15. Demetrovics, J., Katona, G.O.H., Miklós, D., Thalheim, B.: On the number of independent functional dependencies. In: Dix, J., Hegner, S.J. (eds.) Foundations of Information and Knowledge Systems, 4th International Symposium, FoIKS 2006, Budapest, Hungary, February 14-17, 2006, Proceedings. Lecture Notes in Computer Science, vol. 3861, pp. 83–91. Springer (2006)
16. Demetrovics, J., Thi, V.D.: Some problems concerning keys for relation schemes and relations in the relational datamodel. Inf. Process. Lett. 46(4), 179–184 (1993)
17. Engel, K.: Sperner Theory. Cambridge Uni Press (1997)
18. Fagin, R.: Multivalued dependencies and a new normal form for relational databases. ACM Trans. Database Syst. 2(3), 262–278 (1977)
19. Hannula, M., Kontinen, J., Link, S.: On the finite and general implication problems of independence atoms and keys. J. Comput. Syst. Sci. 82(5), 856–877 (2016)
20. Hannula, M., Kontinen, J., Link, S.: On the interaction of inclusion dependencies with independence atoms. In: Eiter, T., Sands, D. (eds.) LPAR-21, 21st International Conference on Logic for Programming, Artificial Intelligence and Reasoning, Maun, Botswana, May 7-12, 2017. EPiC Series in Computing, vol. 46, pp. 212–226. EasyChair (2017)
21. Hartmann, S., Leck, U., Link, S.: On Codd families of keys over incomplete relations. Comput. J. 54(7), 1166–1180 (2011)
22. Hartmann, S., Link, S.: On a problem of Fagin concerning multivalued dependencies in relational databases. Theor. Comput. Sci. 353(1-3), 53–62 (2006)
23. Hartmann, S., Link, S.: Efficient reasoning about a robust XML key fragment. ACM Trans. Database Syst. 34(2), 10:1–10:33 (2009)
24. Hartmann, S., Link, S.: The implication problem of data dependencies over SQL table definitions: Axiomatic, algorithmic and logical characterizations. ACM Trans. Database Syst. 37(2), 13:1–13:40 (2012)
25. Kanellakis, P.C.: Elements of relational database theory. In: Handbook of Theoretical Computer Science, Volume B: Formal Models and Sematics (B), pp. 1073–1156 (1990)
26. Katona, G.O.H.: Combinatorial and algebraic results for database relations. In: Biskup, J., Hull, R. (eds.) Database Theory - ICDT'92, 4th International Conference, Berlin, Germany, October 14-16, 1992, Proceedings. Lecture Notes in Computer Science, vol. 646, pp. 1–20. Springer (1992)
27. Khizder, V.L., Weddell, G.E.: Reasoning about uniqueness constraints in object relational databases. IEEE Trans. Knowl. Data Eng. 15(5), 1295–1306 (2003)
28. Köhler, H., Leck, U., Link, S., Prade, H.: Logical foundations of possibilistic keys. In: Logics in Artificial Intelligence - 14th European Conference, JELIA 2014, Funchal, Madeira, Portugal, September 24-26, 2014. Proceedings. pp. 181–195 (2014)
29. Köhler, H., Leck, U., Link, S., Zhou, X.: Possible and certain keys for SQL. VLDB J. 25(4), 571–596 (2016)

30. Köhler, H., Link, S.: Qualitative cleaning of uncertain data. In: Mukhopadhyay, S., Zhai, C., Bertino, E., Crestani, F., Mostafa, J., Tang, J., Si, L., Zhou, X., Chang, Y., Li, Y., Sondhi, P. (eds.) Proceedings of the 25th ACM International Conference on Information and Knowledge Management, CIKM 2016, Indianapolis, IN, USA, October 24-28, 2016. pp. 2269–2274. ACM (2016)
31. Köhler, H., Link, S.: SQL schema design: Foundations, normal forms, and normalization. In: Proceedings of the 2016 International Conference on Management of Data, SIGMOD Conference 2016, San Francisco, CA, USA, June 26 - July 01, 2016. pp. 267–279 (2016)
32. Köhler, H., Link, S.: Inclusion dependencies and their interaction with functional dependencies in SQL. J. Comput. Syst. Sci. 85, 104–131 (2017)
33. Köhler, H., Link, S., Zhou, X.: Possible and certain SQL keys. PVLDB 8(11), 1118–1129 (2015)
34. Köhler, H., Link, S., Zhou, X.: Discovering meaningful certain keys from incomplete and inconsistent relations. IEEE Data Eng. Bull. 39(2), 21–37 (2016)
35. Kontinen, J., Link, S., Väänänen, J.A.: Independence in database relations. In: Libkin, L., Kohlenbach, U., de Queiroz, R.J.G.B. (eds.) Logic, Language, Information, and Computation - 20th International Workshop, WoLLIC 2013, Darmstadt, Germany, August 20-23, 2013. Proceedings. Lecture Notes in Computer Science, vol. 8071, pp. 179–193. Springer (2013)
36. Lausen, G.: Relational databases in RDF: Keys and foreign keys. In: SWDB-ODBIS. pp. 43–56 (2007)
37. Levene, M., Loizou, G.: A generalisation of entity and referential integrity. ITA 35(2), 113–127 (2001)
38. Link, S.: Charting the completeness frontier of inference systems for multivalued dependencies. Acta Inf. 45(7-8), 565–591 (2008)
39. Link, S.: Characterisations of multivalued dependency implication over undetermined universes. J. Comput. Syst. Sci. 78(4), 1026–1044 (2012)
40. Link, S., Prade, H.: Possibilistic functional dependencies and their relationship to possibility theory. IEEE Trans. Fuzzy Systems 24(3), 757–763 (2016)
41. Link, S., Prade, H.: Relational database schema design for uncertain data. In: Mukhopadhyay, S., Zhai, C., Bertino, E., Crestani, F., Mostafa, J., Tang, J., Si, L., Zhou, X., Chang, Y., Li, Y., Sondhi, P. (eds.) Proceedings of the 25th ACM International Conference on Information and Knowledge Management, CIKM 2016, Indianapolis, IN, USA, October 24-28, 2016. pp. 1211–1220. ACM (2016)
42. Link, S., Schewe, K.: An arithmetic theory of consistency enforcement. Acta Cybern. 15(3), 379–416 (2002)
43. Paredaens, J.: The interaction of integrity constraints in an information system. J. Comput. Syst. Sci. 20(3), 310–329 (1980)
44. Sali, A., Schewe, K.: Counter-free keys and functional dependencies in higher-order datamodels. Fundam. Inform. 70(3), 277–301 (2006)
45. Sali, A., Schewe, K.: Keys and Armstrong databases in trees with restructuring. Acta Cybern. 18(3), 529–556 (2008)
46. Schewe, K.: Consistency enforcement in Entity-Relationship and object-oriented models. Data Knowl. Eng. 28(1), 121–140 (1998)
47. Schewe, K., Schmidt, J.W., Wetzel, I.: Identification, genericity and consistency in object-oriented databases. In: Database Theory - ICDT'92, 4th International Conference, Berlin, Germany, October 14-16, 1992, Proceedings. pp. 341–356 (1992)

48. Schewe, K., Thalheim, B.: Limitations of rule triggering systems for integrity maintenance in the context of transition specifications. Acta Cybern. 13(3), 277–304 (1998)
49. Schewe, K., Thalheim, B.: Towards a theory of consistency enforcement. Acta Inf. 36(2), 97–141 (1999)
50. Selesnjew, O., Thalheim, B.: On the numbers of shortest keys in relational databases on non-uniform domains. Acta Cybern. 8(3), 267–271 (1988)
51. Sperner, E.: Ein Satz über Untermengen einer endlichen Menge. Math. Z. 27, 544–548 (1928)
52. Thalheim, B.: Dependencies in relational databases. Teubner (1991)
53. Thalheim, B.: On semantic issues connected with keys in relational databases permitting null values. Elektronische Informationsverarbeitung und Kybernetik 25(1/2), 11–20 (1989)
54. Thalheim, B.: The number of keys in relational and nested relational databases. Discrete Applied Mathematics 40(2), 265–282 (1992)
55. Tichler, K.: Minimum matrix representation of some key system. Discrete Applied Mathematics 117(1-3), 267–277 (2002)
56. Tichler, K.: Extremal theorems for databases. Ann. Math. Artif. Intell. 40(1-2), 165–182 (2004)
57. Toman, D., Weddell, G.E.: On keys and functional dependencies as first-class citizens in description logics. J. Autom. Reasoning 40(2-3), 117–132 (2008)
58. Wei, Z., Link, S., Liu, J.: Contextual keys. In: Mayr, H.C., Guizzardi, G., Ma, H., Pastor, O. (eds.) Conceptual Modeling - 36th International Conference, ER 2017, Valencia, Spain, November 6-9, 2017, Proceedings. Lecture Notes in Computer Science, vol. 10650, pp. 266–279. Springer (2017)

Polynomially Bounded Valuations in Higher-Order Logics over Relational Databases[*]

Flavio Ferrarotti[1], Loredana Tec[2], and José María Turull-Torres[3]

[1] Software Competence Center Hagenberg, Austria
Flavio.Ferrarotti@scch.at,
[2] RISC Software GmbH, Hagenberg, Austria
Loredana.Tec@risc-software.at,
[3] Depto. de Ingeniería e Investigaciones Tecnológicas Universidad Nacional de La Matanza, Argentina and Massey University, New Zealand
J.M.Turull@massey.ac.nz

Abstract. There are many examples of queries to relational database instances that can be expressed by simple and elegant third-order logic (TO) formulae. For many of those queries the expressive power of TO is not required, but the equivalent second-order logic (SO) formulae can be very complicated or unintuitive. From the point of view of the study of highly expressive query languages is then relevant to identify fragments of TO (and, in general, of higher-order logics of order ≥ 3) which *do have* an SO equivalent formula. In this article we investigate this precise problem as follows. Firstly, we define a general *schema* of existential third-order ($\exists$TO) formulae and show that all $\exists$TO-formulae of this schema can be translated into equivalent SO-formulae. We give examples which show that this is a very usual, intuitive, and convenient schema in the expression of database queries. Secondly, aiming to characterize the fragment of TO which can be translated to SO, we define TO^P as the restriction of TO to valuations of polynomially bounded cardinality. We show constructively that TO^P collapses to SO. Moreover, we define a similar restriction for every higher-order logic of order $i \geq 4$ (HO^i) and show that they also collapse to SO.

1 Introduction

In the framework of computable queries to relational databases and Finite Model Theory where queries define (second-order) relations on the input database

---

[*] Work supported by **Austrian Science Fund (FWF):[I2420-N31]**. Project: *Higher-Order Logics and Structures*. Initiated during a project sponsored visit of Prof. José María Turull-Torres. The research reported in this paper has been partly supported by the Austrian Ministry for Transport, Innovation and Technology, the Federal Ministry of Science, Research and Economy, and the Province of Upper Austria in the frame of the COMET center SCCH.

instance or finite relational structure, there are many examples of properties (queries) that can be expressed by simple and elegant third-order logic (TO) formulae [4]. Let us consider three such properties:

a) Consider the property *hypercube graph* (see [8]). An *n-hypercube graph* $\mathbf{Q}_n$, also called an *n*-cube, is an undirected graph whose vertices are binary *n*-tuples. Two vertices of $\mathbf{Q}_n$ are adjacent iff they differ in exactly one bit. Note that we can build an $(n+1)$-cube $\mathbf{Q}_{n+1}$ starting with two isomorphic copies of an *n*-cube $\mathbf{Q}_n$ and adding edges between corresponding vertices. Using this fact, we can define in TO the class of *hypercube graphs*, by saying that there is a sequence of graphs (i.e., a *third-order linear digraph*, where every TO node is an undirected (second-order) graph) which starts with the graph K_2, ends with a graph which is equal to the input graph, and such that every graph G_2 in the sequence results from finding two total, injective functions f_1, f_2 from the previous graph G_1, so that f_1 and f_2 induce in G_2 two isomorphic copies of G_1, the images of those functions define a partition in the vertex set of G_2, and there is an edge in G_2 between the images $f_1(x)$ and $f_2(x)$ of every node x in G_1.

b) Another definition of *hypercube graphs* that yields a simple (TO) formula is the following. We say that there is a proper non empty subset V' of the vertex set V of the input graph G, and a (third-order) bijective function f from the vertex set of G to the power set of V', such that for every pair of nodes x and y in G, there is an edge between them iff $f(x)$ can be obtained from $f(y)$ by adding or removing a single element (note that V' is necessarily of size $\log_2 |V|$).

c) As another example consider the *Formula-Value query*, i.e., given a propositional formula with constants in $\{F, T\}$ decide whether it is true. We can express it with a simple and intuitive TO formula by saying that there is a sequence of word models (which represent formulae) which starts with the input formula, ends with the formula T, and such that every formula φ_2 in the sequence results from the application to the previous formula φ_1 of one of the operations of conjunction, disjunction, or negation which is ready to be evaluated (i.e., like in $(T \wedge F)$), or elimination of a pair of redundant parenthesis (i.e., like in $((T))$) to exactly one sub formula of φ_1.

Actually, the expressive power of TO *is not required* to characterize hypercube graphs, since they can be recognized in nondeterministic polynomial time, and by Fagin's theorem [7], existential second-order logic ($\exists$SO) captures NP. Thus, there are formulae in $\exists$SO which can express this property. Nevertheless, to define the class of hypercube graphs in $\exists$SO, and even in full second-order logic (SO), is certainly more challenging than to define it in TO (see the two

---

[4] TO extends second-order logic with third-order quantifiers which bind third-order relation variables; those variables are valuated with sets of tuples of (second-order) relations.

strategies for hypercube graphs in [10]). Also, we *don't need* TO to express the Formula-Value query, since it is in DLOGSPACE [2], and hence can be expressed in ∃SO, since DLOGSPACE ⊆ P ⊆ NP = ∃SO.

On the other hand, if we consider the query SATQBF of satisfiable quantified Boolean formulae, we can express it in existential third-order logic (∃TO), since the problem is PSPACE-complete, and it is a well-known fact that ∃TO is powerful enough as to characterize every problem in PSPACE, since it captures NTIME($2^{n^{O(1)}}$) (see [13]). Note that as PSPACE can be captured by SO extended with a transitive closure operator, and furthermore this logic is widely conjectured to be strictly more expressive than the standard SO, the existence of an SO characterization of this problem is unlikely.

Then, it would be very interesting to *distinguish* in some way the TO formulae which *do have* an SO equivalent formula, such as in the first three examples above, from the TO formulae which (most likely) do not, as in SATQBF. In the general case, it would mean that for those queries in the first class we can take advantage of the much higher expressibility and simplicity of TO, and be able to express a query in a more simple and intuitive way, *though still formal*, but *without* having to pay the price of a higher complexity to evaluate the corresponding formulae. Note that by the results in [7] and [13], ∃SO = NTIME($n^{O(1)}$), while ∃TO = NTIME($2^{n^{O(1)}}$).

In addition, there are well known problems such as hypercube graph and satisfiability of quantified Boolean formulae with k alternating blocks of quantifiers (SATQBF$_k$) which, despite of the fact that can be characterized in ∃SO and in the prenex fragment Σ_k^1 of SO with k alternating blocks of quantifiers[5], respectively, do not appear to have a straightforward characterization in SO, even if we consider full SO. In [10] we gave detailed formulae for those properties, and the two sentences turned out to be complex and several pages long.

From an applied perspective, this indicates that it makes sense to investigate higher-order logics and structures in the context of database query languages. Even though most of the queries commonly used in the industry are in P, the use of higher-order quantifiers can potentially simplify the way in which many of those queries are expressed. Think for instance of PERT charts, which are extensively used in Software Engineering in the context of planning and scheduling tasks of project management. Formally, these charts correspond to graphs with edges representing tasks or activities that need to be done, while nodes represent events or milestones. In the case of planning and scheduling the many interrelated tasks in a large and complex project, where for instance, a node represents a PERT chart itself, the encoding of higher-order relations of order ≥ 3 into SO relations as studied in this paper could be exploited as a normal form to store such type of complex higher-order objects into a standard relational database. Furthermore, for querying such a complex PERT chart, it

---

[5] Note that SATQBF$_k$ is complete for the level Σ_k^p of the polynomial-time hierarchy and $\Sigma_k^1 = \Sigma_k^p$, for all $k \geq 1$ (see [17]).

becomes necessary to perform a so-called "zooming" in order to retrieve e.g. a sub activity used in the node of a higher level activity or simply to determine if or which sub activities can be done in parallel. All these kind of queries can be naturally expressed using the higher-order logics studied in this work. The translation of higher-order queries of order ≥ 3 to SO logic as proposed in this paper could then be a first fundamental step to synthesise the resulting SO queries into corresponding efficient queries over the (normalized) relational databases, much in the same style as [14].

Taking into account these considerations, is then relevant to identify ways to isolate the "good behaving" fragments of TO (and, in general, HO^i, for $i \geq 3$) formulae. In this line we define in Section 3 a general *schema* of $\exists TO$ formulae which generalizes the examples (a) and (c) above, and we give a constructive proof of the fact that all $\exists TO$ sub formulae of that schema can be translated into an equivalent SO formula. The schema is essentially the expression of an iteration of polynomial length, unfolded as a sequence of relational structures which represents a computation or derivation in the sense of Complexity and Computability Theories, by explicitly stating which operations are the ones that can be involved in the construction of a given structure in the sequence, when applied to the previous one. This is a very usual, intuitive, and convenient schema in the expression of properties (see Section 3 for further examples).

Then, in Section 4 aiming to formally characterize the fragment of TO which can be translated to SO, we define a restriction of TO, which we denote TO^P, for *polynomial TO*, and we give a constructive proof of the fact that it collapses to SO. We conjecture that TO^P is the exact characterization of the class of TO formulae that can be translated to SO, but we do not think that it can be proved, since most likely the problem is undecidable as it would be equivalent to the problem of deciding whether a given TO formula is equivalent to a SO formula. We define TO^P as the fragment of TO where valuations can assign to TO relation variables only TO relations whose cardinalities are bounded by a polynomial that depends on the quantifier. Note that the example (b) above doesn't seem to be expressible by a TO formula of the *well-behaved* schema described above, but is clearly expressible in $\exists TO^P$. Nevertheless, in the final part of this section we argue that the general schema of $\exists TO$ formulae proposed in Section 3 is, from the perspective of database query languages, still relevant. Regarding expressive power, we also discuss briefly (in the light of our result) the case of SO extended with the deterministic inflationary fixed-point (IFP) quantifier, where the variable which is bounded by the IFP quantifier is a third-order variable. If this SO+IFP logic is restricted to fixed points with a polynomially bounded number of stages, then it also collapses to SO. Moreover, we define a hierarchy of prenex TO^P formulae and show its exact correspondence with the polynomial time hierarchy.

Finally, in Section 5, we define a similar restriction of all higher-order logics HO^i, for each $i \geq 4$, which we denote $HO^{i,P}$, for *polynomial HO^i*. Then we give a constructive proof of the fact that for all $i \geq 4$, $HO^{i,P}$ collapses to

SO. Roughly, HO^i is first order logic extended with quantifiers of any order $2 \leq j \leq i$, which in turn bind j-th order relation variables. For $j \geq 3$, j-th order relation variables are valuated with sets of tuples of $(j-1)$-th order relations. A second-order relation is a relation in the usual sense (i.e., fixing $r \geq 1$, a set of r-tuples of elements from the domain of a structure), and for $j \geq 3$, a j-th order relation is a set of tuples of $(j-1)$-th order relations, of some fixed width. For $\mathrm{HO}^{i,P}$ we use a different strategy which we give in detail for the case of the collapse of $\mathrm{HO}^{4,P}$ to SO. That strategy can be generalized in a straightforward way to all orders $i \geq 4$. To define $\mathrm{HO}^{i,P}$ we do it inductively, starting with $\mathrm{HO}^{4,P}$, stating that $\mathrm{HO}^{i+1,P}$ is an extension of $\mathrm{HO}^{i,P}$, where the $(i+1)$-th order quantifiers restrict the cardinality (i.e., the number of tuples of (i)-th relations) of the valuating $(i+1)$-th order relations to be bounded by a polynomial that depends on the quantifier.

The main results in this paper already appeared in [9]. However, due to space limitations only brief sketches of the corresponding proofs were included in that work. Here we provide the full detailed proofs of those results and discuss their significance from the perspective of highly expressive query languages for relational databases.

2 Preliminaries

We assume the reader is familiar with the basic concepts and the framework of Finite Model Theory [6]. We only consider signatures, or vocabularies, which are purely *relational*, and for simplicity we do not allow constant symbols. We use the classical Tarski's semantics, except that in the context of finite model theory, *only finite* structures or interpretations are considered. Thus our structures will always be *finite relational structures*, or equivalent, *relational database instances*. If **I** is a structure of vocabulary σ, or σ-structure, we denote its domain by $dom(\mathbf{I})$ or I, which is a finite set containing all elements of the structure. Recall that a *valuation* is a function which assigns to every variable in the logic, an element of a given structure. By $\varphi(x_1, \ldots, x_r)$ we denote a formula of some logic whose free variables are exactly $\{x_1, \ldots, x_r\}$. If $\varphi(x_1, \ldots, x_r)$ is a formula of vocabulary σ, **I** is a σ-structure, and $\bar{a} = (a_1, \ldots, a_r)$ is an r-tuple over I, we use the notation $\mathbf{I}, v \models \varphi(x_1, \ldots, x_r)[\bar{a}]$ to denote that φ is satisfied by the structure **I** under the valuation v and that $v(x_i) = a_i$ for $1 \leq i \leq r$. In turn, the notation $\mathbf{I} \models \varphi(x_1, \ldots, x_r)[\bar{a}]$ denotes that φ is satisfied by the structure **I** under *all* valuations v such that $v(x_i) = a_i$ for $1 \leq i \leq r$.

With HO^i, for any $i \geq 2$, we denote i-th order logic which extends first order logic with quantifiers of any order $2 \leq j \leq i$, which in turn bind j-th order relation variables. In particular, HO^2 denotes second-order logic (SO) as usually studied in the context of finite model theory (see [6,15] for a formal definition), and HO^3 denotes third-order logic (TO). second-order variables of arity r are valuated with r-ary relations in the usual sense (i.e., a set of r-tuples of elements from the domain of a structure). For $j \geq 3$, j-th order relation variables are valuated with sets of tuples of $(j-1)$-th order relations according to their relation types. A *third-order relation type of width* w is a w-tuple $\tau = (r_1, \ldots, r_w)$ where $w, r_1, \ldots, r_w \geq 1$, and $r_1, \ldots, r_w$ are arities of (second-order) relations. For $i \geq 4$, an *i-th order relation type of width* w is a w-tuple $\tau = (\rho_1, \ldots, \rho_w)$ where $w \geq 1$ and $\rho_1, \ldots, \rho_w$ are $(i-1)$-th order relation types. A *second-order relation* is a relation in the usual sense. A *third-order relation* of type $\tau = (r_1, \ldots, r_w)$ is a set of tuples of (second-order) relations of arities $r_1, \ldots, r_w$, respectively. For $i \geq 4$, an *i-th order relation* of type $\tau = (\rho_1, \ldots, \rho_w)$ is a set of tuples of $(i-1)$-th order relations, of types $\rho_1, \ldots, \rho_w$, respectively. A more formal definition, but also more cumbersome of the type of higher-order relations in this context can be found among others in [8]. We use uppercase calligraphic letters $\mathcal{X}, \mathcal{Y}, \mathcal{Z}, \ldots$ to denote i-th order variables of order $i \geq 3$, uppercase letters $X, Y, Z, \ldots$ to denote second-order variables, and lower case letters $x, y, z, \ldots$ to denote first order variables. With $\mathcal{X}^{i,\tau}$ we denote an i-th order variable of type τ. If $\mathcal{X}^{i,\tau}$ is a third-order variable, we tend to omit the superindices. We sometimes use X^r to denote that X is a second-order variable or arity r. For any $i \geq 3$, we define the notion of *satisfaction* in HO^i as follows: $\mathbf{A}, val \models \exists \mathcal{X}^{i,(\rho_1,\ldots,\rho_w)}(\varphi(\mathcal{X}))$, where $\mathcal{X}$ is an i-th order relation variable and φ is a well-formed formula, iff there is at least one i-th order relation $\mathcal{R}$ of type $\tau = (\rho_1, \ldots, \rho_w)$ in A, such that $\mathbf{A}, val' \models \varphi(\mathcal{X})$, where val and val' are $\mathcal{R}$-equivalent valuations on $\mathbf{A}$, and $val'(\mathcal{X}) = \mathcal{R}$.

Since we use graphs throughout the paper, let us recall in this context that an *undirected graph* is a finite relational structure $\mathbf{G}$ of vocabulary $\sigma = \{E\}$, satisfying $\varphi_1 \equiv \forall xy(E(x,y) \rightarrow E(y,x))$ and $\varphi_2 \equiv \forall x(\neg E(x,x))$. If we do not require $\mathbf{G}$ to satisfy both φ_1 and φ_2, we talk about *directed graph* or *digraph*. We denote by V the domain of the structure $\mathbf{G}$, i.e., the set of vertices of the graph $\mathbf{G}$.

3 A General Schema of TO Formulae

We define next a *general schema* of $\exists$TO formulae which consists of existentially quantifying a third-order linear digraph of polynomial length (i.e., a sequence of structures that represents a computation) by explicitly stating which operations are the ones that can be involved in the construction of a given structure in the sequence, when applied to the previous one. The schema is as follows:

$$\exists \mathcal{C}^{\bar{s}} \exists \mathcal{O}^{\bar{s}\bar{s}} \big(\text{TotalOrder}(\mathcal{C}, \mathcal{O}) \wedge$$
$$\forall G_1 \forall G_2 \big((\text{First}(G_1) \rightarrow \alpha_{\text{First}}(G_1)) \wedge (\text{Last}(G_1) \rightarrow \alpha_{\text{Last}}(G_1)) \wedge \quad (1)$$
$$((\mathcal{C}(G_1) \wedge \mathcal{C}(G2) \wedge \text{Pred}(G1, G2)) \rightarrow \varphi(G1, G2))\big)\big),$$

where
- the relational structures in $\mathcal{C}$ have type $\bar{s} = (i_1, \ldots, i_s)$ with $i_j \geq 1$ for $j = 1, \ldots, s$.

- TotalOrder($\mathcal{C}, \mathcal{O}$), First(G_1), Last(G_1) and Pred($G1, G2$) denote fixed SO formulae which express that $\mathcal{O}$ is a total order over $\mathcal{C}$, G_1 is the first relational structure in $\mathcal{O}$, G_1 is the last relational structure in $\mathcal{O}$ and G_1 is the immediate predecessor of G_2 in $\mathcal{O}$, respectively.
- $\alpha_{\text{First}}(G_1)$ and $\alpha_{\text{Last}}(G_1)$ denote arbitrary SO formulae which define, respectively, the properties that the first and last structure in $\mathcal{O}$ should satisfy.
- $\varphi(G1, G2)$ denote an arbitrary SO formula that expresses how we get G_2 out of G_1, i.e., which operations can be used to obtain G_2 from G_1.

This is a very usual, intuitive, and convenient schema in the expression of natural properties, as confirmed by the examples in [10] and by those discussed through this paper.

Example 1. *Consider the problem of deciding whether a graph is a hypercube, described as Example (a) in the introduction. This can be expressed in TO following the schema (1) by letting:*

- α_{First} *express that "the first graph in the order $\mathcal{O}$ is K_2",*
- α_{Last} *express that "the last graph in the order $\mathcal{O}$ is the input graph", and*
- φ *express that "G_2 can be built from two isomorphic copies of G_1 by adding edges between the corresponding vertices".*

φ can be expressed in the following way: Every graph G_2 in the sequence results from finding two total, injective functions f_1, f_2 from the previous graph G_1, so that f_1 and f_2 induce in G_2 two isomorphic copies of G_1, the images of those functions define a partition in the vertex set of G_2, and there is an edge in G_2 between the images $f_1(x)$ and $f_2(x)$ of every node x in G_1 (see also the formulae A4.1–A4.5 in [10], pp. 5 for details).

Example 2. *Consider the Formula-Value query, described as Example (c) in the Introduction. Every propositional Boolean formula ϕ can be viewed as a word model[6] G_ϕ of vocabulary $\pi = \{\leq, P_(, P_), P_\wedge, P_\vee, P_\neg, P_F, P_T\}$. The Formula-Value query can then be expressed in TO by following the general schema (1), letting:*

- α_{First} *express that "the first word model (formula) in $\mathcal{O}$ is the input formula",*
- α_{Last} *express that "the last word model (formula) in $\mathcal{O}$ is the formula T", and*
- φ *express that "G_{ϕ_2} is obtained by applying to G_{ϕ_1} one of the operations of conjunction, disjunction or negation or elimination of a pair of redundant parenthesis, which is ready to be evaluated to exactly one sub formula of G_{ϕ_1}".*

---

[6] For $u = a_1 \ldots a_n \in A^+$, a *word model* for u is a structure of the form $(B, <, (P_a)_{a \in A})$ where $|B| = length(u)$, $<$ is a linear order of B, and P_a corresponds to the positions in u carrying an a (see [6] among others).

Interesting additional examples are a set of very relevant relationships between pairs of undirected graphs $\langle G, H \rangle$ defined as orderings of special sorts, which can be expressed following such a schema, by defining a set of possible operations that can be applied repeatedly to H, until a graph which is isomorphic to G is obtained. Furthermore, in all these cases the length of the sequence is at most linear. Namely, the following properties of such kind can be expressed by ∃TO formulae that follow the schema described above:

1. $G \leq_{immersion} H$ (i.e., G being an *immersion* in H, see [1,5,12]),
2. $G \leq_{top} H$ (i.e., G being *topologically embedded*, or *topologically contained* in H, see [1,5,12]),
3. $G \leq_{minor} H$ (i.e., G being a *minor* of H, see [11,5]),
4. $G \leq_{induced-minor} H$ (i.e., G being an *induced minor* of H, see [5]).

The operations on graphs that are used to define those orderings are: (E) delete an edge, (V) delete a vertex, (C) contract an edge, (T) degree 2 contraction, or subdivision removal, and (L) lift an edge. In particular the set of allowable operations for each of those orderings are: $\{E, V, L\}$ for $\leq_{immersion}$, $\{E, V, C\}$ for $\leq_{minor}$, $\{E, V, T\}$ for $\leq_{top}$, and $\{V, C\}$ for $\leq_{induced-minor}$ (see [5]).

Another example of the use of a polynomially bounded sequence of structures is the classical definition of *planarity* in undirected graphs, where one way of stating the classical definition of Kuratowski, which is due to Wagner (1937) and which makes use of one of the orderings mentioned above is the following: a graph is planar if and only if it contains neither K_5 nor $K_{3,3}$ as a minor [3].

A Translation of TO formulae of the General Schema to SO

We show next that if Ψ is a TO formula of the general schema (1) and there is a polynomial p such that in all the valuations that satisfy Ψ the cardinality of the third-order relation assigned to $\mathcal{C}$ is bounded by p in the size of the input structure, then we can translate Ψ into an equivalent SO formula Ψ'.

To simplify the presentation, we first consider the case of graphs and assume that there must be at least one node in each graph in $\mathcal{C}$. Let us denote by d and t the degree of the polynomials that bound the number of graphs that can appear in any valuation of $\mathcal{C}$ which satisfies Ψ and the size of each graph in $\mathcal{C}$, respectively, both in terms of the size n of the (input) structure in which Ψ is evaluated.

Our strategy consists on encoding the TO relation $\mathcal{C}$ as a pair of second-order relations C and E_C of arities $d+t$ and $2(d+t)$, respectively. Notice that every formula of the schema (1) stipulates that $\mathcal{O}$ is a linear order of the graphs in $\mathcal{C}$ which represents the stages (or steps) of a computation. Consequently the number of stages needed is bounded by n^d. Since in turn each stage has a bound on the number of elements it adds or changes (at most n^t), we have to consider a set of $(d+t)$-tuples. That is, each graph has at most n^t nodes and we have to allow for sequences of at most n^d stages, where each stage has at most n^t nodes. Regarding the TO relation $\mathcal{O}$, we replace it by a pair of second-order relations ST and E_{ST}, in this case of arities d and $2d$, respectively.

The encoding into second-order relations is completed by a relation $R \subseteq ST \times C$ which is left total, where C is the union of the domains of all the structures in the sequence. Every node in ST represents one stage, and through the forest R defines a subset of nodes, which is the vertex set of a sub graph (not necessarily connected) of the whole graph (C, E_C). We use $C|_{R(\bar{x})}$, $E_C|_{R(\bar{x})}$ to denote the restriction of C and E_C, respectively, to $R(\bar{x})$, i.e., $C|_{R(\bar{x})} = \{\bar{y} \mid C(\bar{y}) \land R(\bar{x}, \bar{y})\}$ and $E_C|_{R(\bar{x})} = \{(\bar{v}, \bar{w}) \in E_C \mid R(\bar{x}, \bar{v}) \land R(\bar{x}, \bar{w})\}$. The sub graph of (C, E_C) which corresponds to the stage $ST(\bar{x})$ is denoted as $(C|_{R(\bar{x})}, E_C|_{R(\bar{x})})$.

Then the translation to SO of TO formulae of the schema (1) for the case of (non-empty) arbitrary graphs can be done as follows:

$$\exists C^{d+t} E_C^{2(d+t)} ST^d E_{ST}^{2d} R^{2d+t}\big($$
$$\text{Linear}(ST, E_{ST}) \land R \subseteq ST \times C \land \text{LeftTotal}(R) \land$$
$$\forall \bar{x} \forall \bar{y} ((\text{First}(\bar{x}) \to \hat{\alpha}_{\text{First}}) \land (\text{Last}(\bar{x}) \to \hat{\alpha}_{\text{Last}}) \land \qquad (2)$$
$$((ST(\bar{x}) \land ST(\bar{y}) \land \text{Pred}(\bar{x}, \bar{y})) \to$$
$$\hat{\varphi}((C|_{R(\bar{x})}, E_C|_{R(\bar{x})}), (C|_{R(\bar{y})}, E_C|_{R(\bar{y})}))))\big),$$

where

- Linear(ST, E_{ST}), First$(\bar{x})$, Last$(\bar{x})$ and Pred$(\bar{x}, \bar{y}))$ denote SO formulae which express that (ST, E_{ST}) is a linear digraph, $\bar{x}$ is the first node in (ST, E_{ST}), x is the last node in (ST, E_{ST}) and $\bar{x}$ is the immediate predecessor of $\bar{y}$ in (ST, E_{ST}), respectively.
- $R \subseteq ST \times C$ and LeftTotal(R) are shorthands for $\forall \bar{x} \bar{y}(R(\bar{x}, \bar{y}) \to (ST(\bar{x}) \land C(\bar{y})))$ and $\forall \bar{x}(ST(\bar{x}) \to \exists \bar{y}(R(\bar{x}, \bar{y})))$, respectively.
- $\hat{\alpha}_{\text{First}}$ and $\hat{\alpha}_{\text{Last}}$ are SO formulae built from α_{First} and α_{Last}, respectively, by modifying them to talk about the graph described by $\bar{x}$ through $ST(\bar{x})$, E_{ST} and R.
- $\hat{\varphi}$ is an SO formula built from φ by modifying it to talk about the graphs described by $\bar{x}$ and $\bar{y}$ through $ST(\bar{x})$, $ST(\bar{y})$, E_{ST} and R.

Example 3. *Take the TO formula described in Example 2 for expressing the Formula-Value query. In this case $t = 1$, since the size of the input formulae is equal to the size n of the word model presenting it, and the whole evaluation process takes up to n steps. Its translation to SO using the strategy described in this section then results in a SO formula of the schema (2) where (C, E_C) encodes a graph whose nodes are binary tuples and whose edges are quadruples, and (ST, E_{ST}) encodes a linear digraph of length at most n. In turn, R encodes a ternary relation such that $(x, y, z) \in R$ iff $(y, z) \in C|_{R(\bar{x})}$. Figure 1 depicts* one of the valuations for the key SO variables which satisfies the resulting SO query when it is evaluated over a word model with domain $\{1, 2, \ldots, 8\}$ that encodes the formula $(T \wedge (\neg F))$.

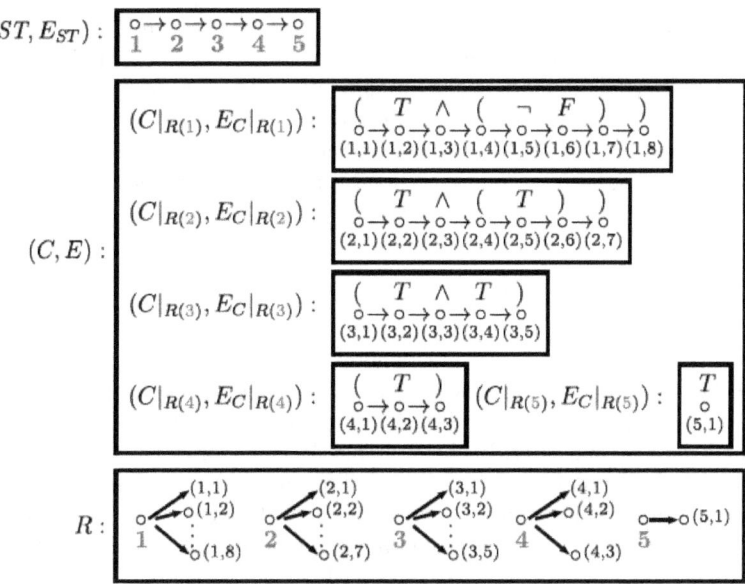

Fig. 1. Formula-Value Query Example.

For the case of relations of arbitrary arity, say S of arity $r \geq 1$, we simply need to consider E_C as an r-ary relation (denoted E_C^S). Thus $E_C^S|_{R(\bar{x})} = \{(\bar{v}_1, \ldots, \bar{v}_r) \in E_C^S : R(\bar{x}, \bar{v}_1) \wedge \ldots \wedge R(\bar{x}, \bar{v}_r)\}$. If we have a tuple of relations, say $l \geq 1$ relations of arities $r_1, \ldots, r_l \geq 1$, respectively, then we have to consider similarly $E_{C_1}^{S_1}, \ldots, E_{C_l}^{S_l}$.

Thus, we get the following (recall that given a relational structure $\mathbf{I}$, with $I^{\bar{s}}$ we denote the set of all TO relations (or relational structures) of type (signature) $\bar{s}$ in the set $I = dom(\mathbf{I})$).

Theorem 1. Let $\Psi \equiv \exists \mathcal{C}^{\bar{s}} \mathcal{O}^{\bar{s}\bar{s}} \psi(\mathcal{C}, \mathcal{O})$ be a TO formula of some relational vocabulary σ, of the form (1). There is a translation of Ψ to an equivalent SO formula if the following holds:

a. The sub formulae α_{First}, α_{Last} and φ are SO formulae.
b. There is a positive integer d such that for every σ-structure $\mathbf{I}$, every TO relation $\mathcal{R}$ in $I^{\bar{s}}$, and every valuation v with $v(\mathcal{C}) = \mathcal{R}$, if $\mathbf{I}, v \models \exists \mathcal{O}^{\bar{s}\bar{s}} \psi(\mathcal{C})$, then $|\mathcal{R}| \leq |dom(\mathbf{I})|^d$.

Remark. If we restrict α_{First}, α_{Last} and φ to $\exists SO$ formulae, then Theorem 1 can be seen as a direct consequence of Fagin's famous theorem [7] which states that $\exists SO$ captures NP. Note that every property definable by a TO formula of the form (1) such that α_{First}, α_{Last} and φ are $\exists SO$ formulae and property (b.) in Theorem 1 holds, can be checked in NP exactly as it happens for every property definable in SO (it suffices to additionally guess a polynomial-sized valuation for the two existentially quantified TO variable). Then, by Fagin's theorem, we get that every property definable by such kind of TO formulae can also be defined in $\exists SO$. Nevertheless, the constructive approach that we follow in this paper has the advantage of providing an actual translation to SO which is clear and intuitive, as well as new insight into the problem, in particular if we look at it from the perspective of database query languages.

4 The Fragment TO^P of Third-Order Logic

Now we define a restriction of TO, denoted TO^P, standing for *polynomial third-order logic*. In TO^P the cardinality of the TO relations which can be assigned by a valuation to a TO variable will be bounded by the degree of a polynomial that depends on the quantifier. In the alphabet of TO^P, for every positive integer d we have a third-order quantifier $\exists^{P,d}$ and for every third-order type $\bar{r}$, we have countably many third-order variable symbols $\mathcal{X}^{d,\bar{r}}$. Here, we will usually avoid the superindex d for clarity. A *valuation* in a structure $\mathbf{A}$ in this setting assigns to each TO^P variable $\mathcal{X}^{d,\bar{r}}$ a TO relation $\mathcal{R}$ in $A^{\bar{r}}$, such that $|\mathcal{R}| \leq |dom(\mathbf{A})|^d$. As usual in Finite Model Theory, given that we study logics as a means to express queries to relational structures (which unless they are Boolean, they define a SO relation in each structure of the corresponding signature) we do not allow free SO or TO variables in TO^P. The TO^P quantifier $\exists^{P,d}$ has the following semantics: let $\mathbf{A}$ be a structure; then $\mathbf{A} \models \exists^{P,d}\mathcal{X}^{d,\bar{r}}\varphi(\mathcal{X})$ iff there is a TO relation $\mathcal{R}^{\bar{r}}$ of type $\bar{r}$, such that $\mathbf{A} \models \varphi(\mathcal{X})[\mathcal{R}]$ and $|\mathcal{R}| \leq |dom(\mathbf{A})|^d$. Therefore we have

$$\mathbf{A}, v \models \exists^{P,d}\mathcal{X}^{d,(r_1,\ldots,r_s)}(\varphi(\mathcal{X}))[\mathcal{R}^{(r_1,\ldots,r_s)}] \text{ iff} \tag{3}$$
$$\mathbf{A} \models \varphi[\mathcal{R}/\mathcal{X}] \text{ and } |\mathcal{R}| \leq dom(\mathbf{A})^d,$$

where $\mathbf{A}$ is a structure, $d \geq 1$ is the degree of the polynomial, φ is a TO^P formula of the same signature as $\mathbf{A}$, $\mathcal{X}^{d,(r_1,\ldots,r_s)}$ is a free TO^P variable in φ, and v is a valuation which assigns the TO relation $\mathcal{R}^{(r_1,\ldots,r_s)}$, of the same type, to the variable $\mathcal{X}$, i.e., $v(\mathcal{X}) = \mathcal{R}$.

4.1 A Translation of TO^P formulae to SO

We show next that the above formula (3), and in general every TO^P formula, can be translated to an equivalent SO formula. Note that in classical third-order

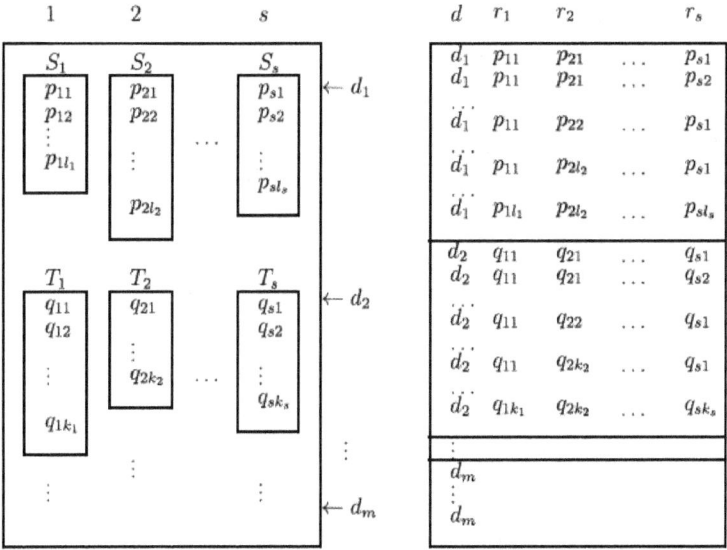

Fig. 2. Encoding of TO relation $\mathcal{R}^{(r_1,\ldots,r_s)}$ to SO relation $R_\mathcal{R}^{d+r_1+\ldots+r_s}$.

logic, a TO relation $\mathcal{R}^{(r_1,\ldots,r_s)}$ in a structure $\mathbf{A}$ satisfies $\mathcal{R} \subseteq \mathcal{P}(dom(\mathbf{A})^{r_1}) \times \ldots \times \mathcal{P}(dom(\mathbf{A})^{r_s})$. Hence, $|\mathcal{R}| \leq 2^{|dom(\mathbf{A})|^{O(1)}}$.

To encode the polynomially bounded TO relations that can be assigned to a TO relation variable in TO^P we use SO relations as follows. Let $\mathcal{R}^{(r_1,\ldots,r_s)}$ be a TO relation of type $(r_1,\ldots,r_s)$ as above, and let $d \geq 1$ be the degree of the bounding polynomial such that $|\mathcal{R}| \leq |dom(\mathbf{A})|^d$, for some structure $\mathbf{A}$. We use an SO relation $R_\mathcal{R}^{d+r_1+\ldots+r_s}$, of arity $(d+r_1+\ldots+r_s)$ to encode $\mathcal{R}^{(r_1,\ldots,r_s)}$, where we use d-tuples from $dom(\mathbf{A})^d$ as identifiers of tuples of SO relations in $\mathcal{R}$, so that whenever a tuple $(a_1,\ldots,a_d,a_{d+1},\ldots,a_{d+r_1},\ldots,a_{d+r_1+\ldots+r_{s-1}+1},\ldots,a_{d+r_1+\ldots+r_s}) \in R_\mathcal{R}$ it means that there is a tuple of SO relations in $\mathcal{R}$ identified by the sub-tuple $(a_1,\ldots,a_d)$, which consists of s SO relations $S_1^{r_1},\ldots,S_s^{r_s}$, of arities $r_1,\ldots,r_s$, respectively, such that $(a_{d+1},\ldots,a_{d+r_1}) \in S_1,\ldots,$ $(a_{d+r_1+\ldots+r_{s-1}+1},\ldots,a_{d+r_1+\ldots+r_s}) \in S_s$. This encoding is depicted in Figure 2, where we used the following notation: $(S_1^{r_1}, S_2^{r_2},\ldots,S_s^{r_s})$, $(T_1^{r_1}, T_2^{r_2},\ldots,T_s^{r_s})$ are SO relations in $\mathcal{R}$ of cardinality $l_i = |S_i^{r_i}|$, respectively $k_i = |T_i^{r_i}|$, for $1 \leq i \leq s$. The SO relations $S_i^{r_i}$, $T_i^{r_i}$ contain in turn r_i-tuples p_{ij} with $1 \leq i \leq s$ and $1 \leq j \leq l_i$ and respectively r_i-tuples q_{ij}, with $1 \leq i \leq s$ and $1 \leq j \leq k_i$.

Atomic Formulae

Let $\alpha \equiv \mathcal{X}^{d,(r_1,\ldots,r_s)}(X_1^{r_1},\ldots,X_s^{r_s})$, with $d, s \geq 1$, $r_1,\ldots,r_s \geq 1$, and where $\mathcal{X}$ is a TO^P relation variable of type $(r_1,\ldots,r_s)$.

First, for a better understanding of the translation, let's assume that there are *no empty SO relations* in the tuples of SO relations in the TO relations that can be assigned by a valuation to the TO^P relation variable $\mathcal{X}$. Then, considering the encoding of polynomially bounded TO relations described above, every TO^P relation variable $\mathcal{X}^{d,(r_1,\ldots,r_s)}$ of type $(r_1,\ldots,r_s)$, which has been quantified by a quantifier $\exists^{P,d}$, with $d \geq 1$, can be encoded in an SO relation variable $X_{\mathcal{X},ne}^{d+r_1+\ldots+r_s}$ of arity $(d+r_1+\ldots+r_s)$ (with the sub-index ne in $X_{\mathcal{X},ne}$ we denote the restriction assumed above). Accordingly, *in this specific case* the SO formula $\hat{\alpha}_{ne}$ equivalent to α is as follows:

$$\hat{\alpha}_{ne} \equiv \exists z_1 \ldots z_d \ldots z_{d+r_1+\ldots+r_s}(\text{``}X_1 \neq \emptyset \wedge \ldots \wedge X_s \neq \emptyset\text{''} \wedge$$
$$X_{\mathcal{X},ne}(z_1,\ldots,z_d,\ldots,z_{d+r_1+\ldots+r_s}) \wedge \forall v_1 \ldots v_{d+r_1+\ldots+r_s}([v_1 = z_1 \wedge \ldots \wedge$$
$$v_d = z_d \wedge X_{\mathcal{X},ne}(v_1,\ldots,v_d,\ldots,v_{d+r_1+\ldots+r_s})] \to [X_1(v_{d+1},\ldots,v_{d+r_1}) \wedge$$
$$\ldots \wedge X_s(v_{d+r_1+\ldots+r_{s-1}+1},\ldots,v_{d+r_1+\ldots+r_s})]) \wedge \forall v_{d+1} \ldots v_{d+r_1+\ldots+r_s}($$
$$[X_1(v_{d+1},\ldots,v_{d+r_1}) \wedge \ldots \wedge X_s(v_{d+r_1+\ldots+r_{s-1}+1},\ldots,v_{d+r_1+\ldots+r_s})] \to \exists v_1$$
$$\ldots v_d(v_1 = z_1 \wedge \ldots \wedge v_d = z_d \wedge X_{\mathcal{X},ne}(v_1,\ldots,v_d,\ldots,v_{d+r_1+\ldots+r_s})))).$$

However, in the general case we *do have to* consider the cases where either some, all or none of the SO relations that form a given tuple in a TO relation, are empty. Then, instead of encoding a TO relation $\mathcal{R}^{(r_1,\ldots,r_s)}$ with a single SO relation $R_{\mathcal{R}}^{d+r_1+\ldots+r_s}$, we use several SO relations to encode it. In fact we use exactly 2^s SO relations, since there are 2^s possible patterns of empty and non empty relations in a tuple of s SO relations. We denote by $\omega = (i_1,\ldots,i_{|\omega|})$ such a pattern of *empty* relations, with $1 \leq i_1 < i_2 < \ldots < i_{|\omega|} \leq s$ being the indices of the components in an s-tuple of SO relations that are empty. Correspondingly, we denote by $\bar{\omega} = (j_1,\ldots,j_{|\bar{\omega}|})$ the corresponding pattern of *non empty* relations, with $1 \leq j_1 < j_2 < \ldots < j_{|\bar{\omega}|} \leq s$ being the indices of the components in an s-tuple of SO relations, that are non empty. By abuse of notation, we will denote as $\{\omega\}$ and $\{\bar{\omega}\}$ the sets of indices in ω and $\bar{\omega}$, respectively. Note that the case considered above, where all the components of an s-tuple of SO relations are non empty, is one particular value of those patterns, namely $\{\omega\} = \emptyset$ and $\bar{\omega} = (1,\ldots,s)$.

In the formula $\hat{\alpha}$ for the general case, we need to refer to tuples of varying length, since for each pattern ω, the arity of the SO relation $X_{\mathcal{X},e,\omega}$ which contains the tuples of SO relations in $\mathcal{X}$ (to be precise, in the TO relation assigned to the TO^P variable $\mathcal{X}$ by a given valuation) whose components with empty relations follow the pattern ω, depends on ω. For that matter we use the following notation (recall that by $\bar{s}\bar{s}'$ we mean the concatenation of the sequences $\bar{s}$ and $\bar{s}'$): $\bar{f}_{\bar{\omega}} = \bar{f}_{j_1} \ldots \bar{f}_{j_{|\bar{\omega}|}}$, where $\bar{f}_{j_1} = (f_{j_1 1},\ldots,f_{j_1 r_{j_1}}),\ldots,\bar{f}_{j_{|\bar{\omega}|}} = (f_{j_{|\bar{\omega}|}1},\ldots,f_{j_{|\bar{\omega}|}r_{j_{|\bar{\omega}|}}})$. We also use $\bar{f}'_{\bar{\omega}}$ with the same meaning.

Then, the SO translation of the TO^P atomic formula

$$\alpha \equiv \mathcal{X}^{d,(r_1,\ldots,r_s)}(X_1^{r_1},\ldots,X_s^{r_s}),$$

in the general case, is the following SO formula $\hat{\alpha}$:

$$\hat{\alpha} \equiv \bigvee_{w \in \Omega} \Big([\text{``}(X_{i_1} = \emptyset \wedge \ldots \wedge X_{i_{|w|}} = \emptyset)\text{''} \wedge \text{``}(X_{j_1} \neq \emptyset \wedge \ldots \wedge X_{j_{|\bar{w}|}} \neq \emptyset)\text{''}] \wedge$$

$$(\exists v_1 \ldots v_d \bar{f}_{\bar{w}}(X_{\mathcal{X},e,w}(v_1,\ldots,v_d,\bar{f}_{\bar{w}}) \wedge \forall u_1 \ldots u_d \bar{f}'_{\bar{w}}[[u_1 = v_1 \wedge \ldots \wedge u_d = v_d \wedge$$

$$X_{\mathcal{X},e,w}(u_1,\ldots,u_d,\bar{f}'_{\bar{w}})] \to [\bigwedge_{l \in \{j_1,\ldots,j_{|\bar{w}|}\}} X_l(f'_{l1},\ldots,f'_{lr_l})]] \wedge$$

$$\forall \bar{f}'_{\bar{w}}[[\bigwedge_{l \in \{j_1,\ldots,j_{|\bar{w}|}\}} X_l(f'_{l1},\ldots,f'_{lr_l})] \to \exists u_1 \ldots u_d(u_1 = v_1 \wedge \ldots \wedge u_d = v_d \wedge$$

$$X_{\mathcal{X},e,w}(u_1,\ldots,u_d,\bar{f}'_{\bar{w}}))]]) \Big),$$

where $\Omega = \{w \mid w = (i_1,\ldots,i_{|w|}); 1 \leq i_1 < i_2 < \ldots < i_{|w|} \leq s; 0 \leq |w| \leq s; \bar{w} = (j_1,\ldots,j_{|\bar{w}|}); \{\bar{w}\} \cup \{w\} = \{1,\ldots,s\}; \{\bar{w}\} \cap \{w\} = \emptyset\}$.

The Existential Case

Now, let $\alpha \equiv \exists^{P,d} \mathcal{X}^{d,(r_1,\ldots,r_s)}(\varphi)$, with $d, s \geq 1$, $r_1,\ldots,r_s \geq 1$, and where $\mathcal{X}$ is a TO^P relation variable of type $(r_1,\ldots,r_s)$. Note that no d-tuple can be in more than one of the different SO relations that encode a given polynomially bounded TO relation. The SO translation of the TO^P formula α in this case, is the following SO formula $\hat{\alpha}$:

$$\hat{\alpha} \equiv \{\exists X_{\mathcal{X},e,w}^{d+|\bar{f}_{\bar{w}}|}\}_{w \in \Omega} (\forall z_1 \ldots z_d [\bigwedge_{\substack{w=(i_1,\ldots,i_{|w|}) \\ 1 \leq i_1 < i_2 \ldots < i_{|w|} \leq s \\ 1 \leq |w| \leq s}} \forall \bar{f}_{\bar{w}}[X_{\mathcal{X},e,w}(z_1,\ldots,z_d,\bar{f}_{\bar{w}}) \to$$

$$(\bigwedge_{\substack{w'=(i'_1,\ldots,i'_{|w'|}) \\ 1 \leq i'_1 < i'_2 \ldots < i'_{|w'|} \leq s \\ 1 \leq |w'| \leq s; \, w' \neq w}} \forall \bar{f}'_{\bar{w}'}(\neg X_{\mathcal{X},e,w'}(z_1,\ldots,z_d,\bar{f}'_{\bar{w}'})))]]) \wedge \hat{\varphi},$$

where $\Omega = \{w \mid w = (i_1,\ldots,i_{|w|}); 1 \leq i_1 < i_2 < \ldots < i_{|w|} \leq s; 0 \leq |w| \leq s; \bar{w} = (j_1,\ldots,j_{|\bar{w}|}); \{\bar{w}\} \cup \{w\} = \{1,\ldots,s\}; \{\bar{w}\} \cap \{w\} = \emptyset\}$, and $\hat{\varphi}$ is the SO formula equivalent to the TO^P formula φ, obtained by applying inductively the translations described above.

The cases for the translation on logical connectives are trivial.

4.2 Some Considerations on the Expressive Power of TO^P

The following result is an immediate consequence of the translation presented in the previous section.

Theorem 2. TO^P *collapses to SO. That is, for every formula in* TO^P *there is an equivalent SO formula.*

The schema of TO formulae introduced in Section 3 is a special case of TO^P formulae, and hence besides the SO translation given in Subsection 3, the TO formulae that follow that schema have an additional translation, which is the one we used to translate TO^P formulae to SO in Subsection 4.1. Nevertheless, the translation of Subsection 3 yields a more clear and intuitive SO formula, and the maximum arity of the quantified SO relation variables in general seems to be much smaller. For the case of hypercube graphs the maximum arity obtained by the schema translation is 4, while the SO formulae obtained by the TO^P translation has maximum arity 8 ($X_\mathcal{C}$ has arity 4, since the degree is 1 and the type is $(1,2)$, and hence $X_\mathcal{O}$ has arity 8). And for the case of the Formula-Value query the maximum arity obtained by the schema translation is also 4, while the SO formulae obtained by the TO^P translation has maximum arity 22 ($X_\mathcal{C}$ has arity 11, since the degree is 1 and the type is $(1,2,1,1,1,1,1,1,1,1)$, and hence $X_\mathcal{O}$ has arity 22). Note that the maximum arity of a relation symbol in an SO formula *is relevant* for the complexity of its evaluation (see among others [13]). Hence, and not surprisingly it makes sense to study specific schemas of TO formulae that have equivalent SO formulae, aiming to find more efficient translations than the general strategy used for TO^P formulae (which had the purpose of proving equivalence, rather than looking for efficiency in the translation).

In [16] we showed that for any $i \geq 3$ the deterministic inflationary fixed-point quantifier (IFP) in HO^i (i.e., where the variable which is bound by the IFP quantifier is an $(i+1)$-th order variable) is expressible in $\exists HO^{i+1}$. Let $IFP|_P$ denote the restriction of IFP where there is a positive integer d such that in every structure $\mathbf{A}$, the number of stages of the fixed-point is bounded by $|dom(\mathbf{A})|^d$. And let $(SO + IFP)$ denote SO extended with the deterministic inflationary fixed-point quantifier, where the variable which is bound by the IFP quantifier is a third-order variable. Note that the addition of such IFP quantifier to SO means that we can express iterations of length exponential in $|dom(\mathbf{A})|$, so that it is strongly conjectured that $(SO + IFP)$ strictly includes SO as to expressive power. However, as a consequence of Theorem 2 above this is not the case with $IFP|_P$. Then, the following corollary is immediate:

Corollary 1. $(SO + IFP|_P)$ *collapses to SO. That is, for every formula in* $(SO + IFP|_P)$ *there is an equivalent SO formula.*

Finally, let us define ΣTO_n^P as the restriction of TO^P to prenex formulae of the form $Q_1 V_1 \ldots Q_k V_k(\varphi)$ such that:

- $Q_1, \ldots, Q_k \in \{\forall^{P,d}, \exists^{P,d}, \forall, \exists\}$.
- Each V_i for $1 \leq i \leq k$ is either a second or third-order variable (depending on Q_i).

- φ is a first order formula.
- The prefix $Q_1 V_1 \ldots Q_k V_k$ starts with an existential block of quantifiers and has at most n alternating (between universal and existential) blocks.

By the well known Fagin-Stockmeyer characterization [17] of the polynomial time hierarchy, we know that for every $n \geq 1$ the prenex fragment Σ_n of SO captures the level n of the polynomial time hierarchy (denoted Σ_n^{poly}). Using the strategy described in Subsection 4.1, it is not difficult to see that every formula in ΣTO_n^P can be translated into an equivalent SO formula in Σ_n. Thus we get the following result:

Theorem 3. ΣTO_n^P captures Σ_n^{poly}.

5 The Fragments HO^{i,P} of Higher-Order Logics

Let $d \geq 1$, and let $\tau = (r_1, \ldots, r_w)$ be a third-order relation type. A *third-order relation* $\mathcal{R}$ of type τ in a structure $\mathbf{A}$ is *downward polynomially bounded* by d if $|\mathcal{R}| \leq |dom(\mathbf{A})|^d$. Let $i \geq 4$, and let $\tau = (\rho_1, \ldots, \rho_w)$ be an i-th order relation type. An i-th order relation $\mathcal{R}$ of type τ in a structure $\mathbf{A}$ is *downward polynomially bounded* by d if $|\mathcal{R}| \leq |dom(\mathbf{A})|^d$, and for all $3 \leq j \leq i-1$, all the j-th order relations that form the tuples of $j+1$-th order relations, are in turn downward polynomially bounded by d.

We define inductively a restriction of HO^i for every $i \geq 4$. We denote it as HO^{i,P}, standing for *polynomial i-th order logic*. For $i = 4$, HO^{4,P} is the extension of TO^P, where the 4-th order quantifiers restrict the cardinality (i.e., the number of tuples of third-order relations of the valuating fourth order relations) to be bounded by a polynomial that depends on the quantifier. Likewise, for $i \geq 5$ we define HO^{i,P} as the extension of HO$^{i-1,P}$, where the i-th order quantifiers restrict the cardinality (i.e., the number of tuples of $(i-1)$-th order relations of the valuating i-th order relations) to be bounded by a polynomial that depends on the quantifier.

In the alphabet of HO^{i,P}, for every pair of positive integers d, and j, with $i \geq j \geq 4$, we have a j-th order quantifier $\exists^{j,P,d}$ and for every j-th order type τ, we have countably many j-th order variable symbols $\mathcal{X}^{j,d,\tau}$. Here, we will usually avoid the superindices d and τ for clarity. For simplicity we assume that the types of all relations of all orders $3 \leq j \leq i$ in every i-th order relation assigned by a valuation to an i-th order relation variable, have width s, for some $s \geq 1$, and that every such relation is *downward polynomially bounded* by $d \geq 1$.

A *valuation* in a structure $\mathbf{A}$ in this setting assigns to each i-th order relation variable $\mathcal{X}^{j,d,\tau}$ an i-th order relation $\mathcal{R}$ in A which is downward polynomially bounded by $|dom(\mathbf{A})|^d$. We additionally assume that the number of different relations of order $i-1$ that form the tuples of a valuating relation $\mathcal{R}$ is also bounded by $|dom(\mathbf{A})|^d$. This assumption is not strictly necessary, but it simplifies our encoding in marginal cases where the arity of $\mathcal{R}$ is strictly greater than $|dom(\mathbf{A})|^d$. As usual in Finite Model Theory, given that we study logics as a means to express queries to relational structures (which unless they are Boolean, they define a SO relation in each structure of the corresponding signature) we do not allow free SO or i-th order relation variables, for any $i \geq 3$, in $\mathrm{HO}^{i,P}$.

For any $3 \leq j \leq i$, the $\mathrm{HO}^{i,P}$ quantifier $\exists^{j,P,d}$ has the following semantics: let $\mathbf{A}$ be a structure, and let $\mathcal{X}^{j,d,\tau}$ be a j-th order relation variable; then $\mathbf{A} \models \exists^{j,P,d} \mathcal{X}^{j,d,\tau} \varphi(\mathcal{X})$ iff there is a j-th order relation $\mathcal{R}$ of type τ, such that $\mathbf{A} \models \varphi(\mathcal{X})[\mathcal{R}]$ and $\mathcal{R}$ is downward polynomially bounded by d in $\mathbf{A}$.

Collapse of The Fragment $\mathrm{HO}^{i,P}$ to SO

We discuss next how to build for every $\mathrm{HO}^{4,P}$ formula α an SO formula that is equivalent to α. To that end, we will have for every fourth order relation variable a set of SO relation variables that represent it. We do so by representing the fourth order relation variable by what in the field of Database Theory is known as a *normalized relational database*.

We use a rather cumbersome notation, mainly for the sub indices in the SO formulae. The aim is to make it very clear that both the names of the different variables needed and the structure of the formulae can be iterated in a straightforward way for *any order* $i \geq 5$, getting thus the corresponding translations for $\mathrm{HO}^{i,P}$ formulae to SO.

Suppose the $\mathrm{HO}^{4,P}$ formula α is of the form $\mathcal{X}^{4,d,\tau}(\mathcal{Y}_1^3, \ldots, \mathcal{Y}_s^3)$, with $|\tau| = s$, and where all the fourth order relations which valuate $\mathcal{X}$ are assumed to be polynomially bounded with degree $d \geq 1$. Next, we show that we can build an SO formula that is equivalent to α.

In Figure 3 we depict the (SO) relation variables that are used to represent the fourth order relation variable $\mathcal{X}^{4,d,\tau}$, and each of the third-order relation variables $\mathcal{Y}_j$ in the tuple $(\mathcal{Y}_1^3, \ldots, \mathcal{Y}_s^3)$. For $\mathcal{X}^{4,d,\tau}$, we use the following relation variables: 3-REL$_\mathcal{X}$, 2-REL$_\mathcal{X}$, $\mathrm{X}_{\mathcal{X}^4, \omega_{3,\mathcal{X}}}$ for each of the different patterns of empty third-order relations $\omega_{3,\mathcal{X}}$, and TUPLES-2-REL$_{\mathcal{X}, \omega_{2,\mathcal{X}}}$ for each of the different patterns of empty second-order relations $\omega_{2,\mathcal{X}}$. In turn, for each relation variable $\mathcal{Y}_j$ in the tuple $(\mathcal{Y}_1^3, \ldots, \mathcal{Y}_s^3)$, we use 2-REL$_{\mathcal{Y}_j}$, and $\mathrm{X}_{\mathcal{Y}^3_{j_{1_{3,\mathcal{X}}}}, \omega_{2, \mathcal{Y}_{j_{1_{3,\mathcal{X}}}}}}$ for each of the different patterns of empty second-order relations $\omega_{2, \mathcal{Y}_{j_{1_{3,\mathcal{X}}}}}$.

Note that for the individual (first order) variables and tuples we use the following convention. $\bar{x}^{3t}$, $\bar{x}^3$, $\bar{x}^{2t}$, $\bar{x}^2$ and $\bar{x}^{1t}$ are associated to the encoding of $\mathcal{X}^4$ and denote variables that range, respectively, over identifiers of tuples of TO relations, identifiers of TO relations, identifiers of tuples of SO relations, identifiers of SO relations, and identifiers of tuples of individual elements from the interpreting structure. Likewise, $\bar{y}^{2t}$, $\bar{y}^2$ and $\bar{y}^{1t}$ are associated to the encoding of $\mathcal{Y}^3$ and denote variables that range, respectively, over identifiers of tuples of SO relations, identifiers of SO relations, and identifiers of tuples of individual elements from the interpreting structure. The tuples of variables $\bar{x}^{3t}$, $\bar{x}^{2t}$, $\bar{y}^{2t}$ are of width d, while the tuples of variables $\bar{x}^{1t}$ and $\bar{y}^{1t}$ are of width s. Regarding the patterns of empty relations in tuples of relations, we use the same notation as in TO^P.

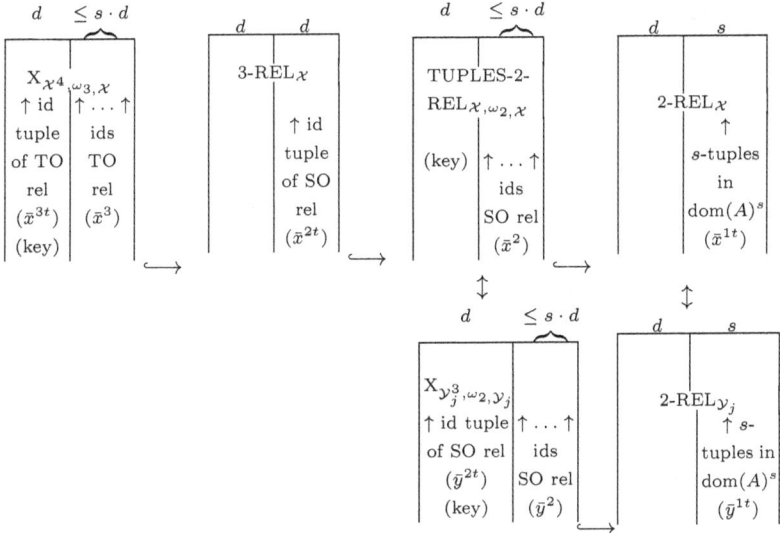

Fig. 3. SO variables encoding fourth-order variable $\mathcal{X}^{4,d,\tau}$.

As in the case of TO^P, the idea is to represent the fourth order relation variable $\mathcal{X}^{4,d,\tau}$ using 2^s (SO) relation variables, one for each pattern of empty TO relations in the tuples of TO relations in the given valuating fourth order relation. In each such relation variable we have the tuples of (non empty) TO relations whose pattern is the one which is used as sub index in the name of the relation variable. In 3-REL$_\mathcal{X}$ we have the identifiers of the tuples of SO relations that form each TO relation in each of the tuples in $X_{\mathcal{X}^4,\omega_3,\mathcal{X}}$. To represent the tuples of SO relations, we use 2^s relation variables, one for each pattern of empty SO relations in the tuples of SO relations in each TO relation in each of the tuples in the different $X_{\mathcal{X}^4,\omega_3,\mathcal{X}}$ relation variables. In each such relation variable we have the tuples of (non empty) SO relations whose pattern is the one that is used as sub index in the name of the relation. In the different relation variables TUPLES-2-REL$_{\mathcal{X},\omega_2,\mathcal{X}}$, we have the identifiers of the SO relations in each tuple of SO relations as above. Finally, in 2-REL$_\mathcal{X}$ we have all the tuples of individual elements from the interpreting structure that form each SO relation in the different relation variables TUPLES-2-REL$_{\mathcal{X},\omega_2,\mathcal{X}}$.

A detailed explanation of the proposed SO formulae for the nontrivial atomic and existential cases (as well as the actual SO formulae) can be found in Appendix A and B, respectively. It is not difficult to see that the SO formulae used in the translation above for $\text{HO}^{4,\text{P}}$ can be iterated, and thus we can build corresponding formulae for any order $i \geq 3$. So that we get the following result (as for TO^P, the cases for the translation on logical connectives are trivial).

Theorem 4. *For all $i \geq 3$, $\text{HO}^{i,\text{P}}$ collapses to SO. That is, for every formula in $\text{HO}^{i,\text{P}}$ there is an equivalent SO formula which can be built following the translation given above.* □

6 Conclusion

We think this is an interesting result, since beyond the practical applications mentioned in the paper, it means that in the framework of computable queries, where queries define (SO) relations on the input structures, *nesting* in *any arbitrary depth* is *irrelevant* as to expressive power. That is, the only reason why (unrestricted) higher-order quantification increases the expressive power of a logic is the fact that an $(i+1)$-th order relation can contain an exponential number of tuples of (i)-th order relations. As shown by the results on the expressive power of HO^i in [13], rising the order i essentially means rising the data complexity from non deterministic hyper exponential time of level $(i-2)$ to non deterministic hyper exponential time of level $(i-1)$.

Note that this fact also appears, among other subjects, in the study of the strict hierarchy induced in the class of primitive recursive functions, by bounding the minimal depth of nesting of the LOOP constructions needed by a LOOP program which can compute a given function in the context of Computability Theory (see [4]). The results suggest that, given that the contents of the variable which controls the LOOP iteration is fixed at the beginning of it, LOOP nesting seems to be the only way by which we can increase the running time of a program, on a given input.

References

1. Faisal N. Abu-Khzam and Michael A. Langston. Graph coloring and the immersion order. In *Computing and Combinatorics, 9th Annual International Conference, COCOON 2003, Big Sky, MT, USA, July 25-28, 2003, Proceedings*, pages 394–403, 2003.
2. Martin Beaudry and Pierre McKenzie. Cicuits, matrices, and nonassociative computation. In *Proceedings of the Seventh Annual Structure in Complexity Theory Conference, Boston, Massachusetts, USA, June 22-25, 1992*, pages 94–106, 1992.
3. Béla Bollobás. *Modern Graph Theory*. Graduate Texts in Mathematics (Book 184). Springer; Corrected edition, 2002.

4. Martin D. Davis and Elaine J. Weyuker. *Computability, complexity, and languages - fundamentals of theoretical computer science.* Computer science and applied mathematics. Academic Press, 1983.
5. Rodney G. Downey and Michael R. Fellows. *Parameterized Complexity.* Monographs in Computer Science. Springer, 1999.
6. Heinz-Dieter Ebbinghaus and Jörg Flum. *Finite Model Theory.* Perspectives in Mathematical Logic. Springer, Berlin Heidelberg New York, 2nd edition, 1999.
7. Ronald Fagin. Generalized first-order spectra and polynomial-time recognizable sets. In R. Karp, editor, *Complexity of Computations,* volume 7 of *SIAM-AMS Proc.*, pages 27–41. American Mathematical Society, 1974.
8. Flavio Ferrarotti. *Expressibility of Higher-Order Logics on Relational Databases: Proper Hierarchies.* PhD thesis, Department of Information Systems, Massey University, Wellington, New Zealand, 2008.
9. Flavio Ferrarotti, Senen Gonzalez, and Jose Maria Turull-Torres. On fragments of higher order logics that on finite structures collapse to second order. In J. Kennedy and R.J.G.B. de Queiroz, editors, *Logic, Language, Information, and Computation (to appear),* volume 10388 of *Lecture Notes in Computer Science.* Springer, 2017.
10. Flavio Ferrarotti, Wei Ren, and Jose Maria Turull-Torres. Expressing properties in second- and third-order logic: hypercube graphs and SATQBF. *Logic Journal of the IGPL,* 22(2):355–386, 2014.
11. Jörg Flum and Martin Grohe. *Parameterized Complexity Theory (Texts in Theoretical Computer Science. An EATCS Series).* Springer-Verlag New York, Inc., Secaucus, NJ, USA, 2006.
12. Martin Grohe, Kenichi Kawarabayashi, Dániel Marx, and Paul Wollan. Finding topological subgraphs is fixed-parameter tractable. In *Proceedings of the Forty-third Annual ACM Symposium on Theory of Computing,* STOC '11, pages 479–488, New York, NY, USA, 2011. ACM.
13. Lauri Hella and José María Turull-Torres. Computing queries with higher-order logics. *Theor. Comput. Sci.*, 355(2):197–214, April 2006.
14. Shachar Itzhaky, Sumit Gulwani, Neil Immerman, and Mooly Sagiv. A simple inductive synthesis methodology and its applications. *SIGPLAN Not.*, 45(10):36–46, October 2010.
15. Leonid Libkin. *Elements Of Finite Model Theory.* Texts in Theoretical Computer Science, EATCS. Springer, Berlin Heidelberg New York, 2004.
16. Klaus-Dieter Schewe and José María Turull-Torres. Fixed-point quantifiers in higher order logics. In *Proceedings of the 2006 Conference on Information Modelling and Knowledge Bases XVII,* pages 237–244, Amsterdam, The Netherlands, The Netherlands, 2006. IOS Press.
17. Larry J. Stockmeyer. The polynomial-time hierarchy. *Theoretical Computer Science,* 3(1):1 – 22, 1976.

A The case of Atomic Formulae in $HO^{4,P}$

In the SO formula below, in the feet descriptions of the big disjunctions and big conjunctions we use simplified expressions with the following meanings: In the big disjunction of the first line of the formula, with $\omega_{3,\mathcal{X}}$ we mean the following:

$$\omega_{3,\mathcal{X}} = (i_1,\ldots,i_{|\omega_{3,\mathcal{X}}|}), 1 \leq i_1 < i_2 < \ldots < i_{|\omega_{3,\mathcal{X}}|} \leq s, 0 \leq |\omega_{3,\mathcal{X}}| \leq s,$$
$$\bar{\omega}_{3,\mathcal{X}} = (j_1,\ldots,j_{|\bar{\omega}_{3,\mathcal{X}}|}), 1 \leq j_1 < j_2 < \ldots < j_{|\bar{\omega}_{3,\mathcal{X}}|} \leq s,$$
$$\{\bar{\omega}_{3,\mathcal{X}}\} \cup \{\omega_{3,\mathcal{X}}\} = \{1,\ldots,s\}, \{\bar{\omega}_{3,\mathcal{X}}\} \cap \{\omega_{3,\mathcal{X}}\} = \emptyset.$$

In the first big conjunction of the rest of the formula, with $\omega_{2,\mathcal{Y}_{j_{1_3,\mathcal{X}}}}$ we mean the following:

$$\omega_{2,\mathcal{Y}_{j_{1_3,\mathcal{X}}}} = (i_1,\ldots,i_{|\omega_{2,\mathcal{Y}_{j_{1_3,\mathcal{X}}}}|}), 1 \leq i_1 < i_2 < \ldots < i_{|\omega_{2,\mathcal{Y}_{j_{1_3,\mathcal{X}}}}|} \leq s,$$
$$0 \leq |\omega_{2,\mathcal{Y}_{j_{1_3,\mathcal{X}}}}| \leq s, \bar{\omega}_{2,\mathcal{Y}_{j_{1_3,\mathcal{X}}}} = (j_1,\ldots,j_{|\bar{\omega}_{2,\mathcal{Y}_{j_{1_3,\mathcal{X}}}}|}),$$
$$1 \leq j_1 < j_2 < \ldots < j_{|\bar{\omega}_{2,\mathcal{Y}_{j_{1_3,\mathcal{X}}}}|} \leq s, \{\bar{\omega}_{2,\mathcal{Y}_{j_{1_3,\mathcal{X}}}}\} \cup \{\omega_{2,\mathcal{Y}_{j_{1_3,\mathcal{X}}}}\} = \{1,\ldots,s\},$$
$$\{\bar{\omega}_{2,\mathcal{Y}_{j_{1_3,\mathcal{X}}}}\} \cap \{\omega_{2,\mathcal{Y}_{j_{1_3,\mathcal{X}}}}\} = \emptyset.$$

In the second big conjunction of the rest of the formula, with $\omega_{2,\mathcal{Y}_{j_{|\bar{\omega}_{3,\mathcal{X}}|}}}$ we mean the following:

$$\omega_{2,\mathcal{Y}_{j_{|\bar{\omega}_{3,\mathcal{X}}|}}} = (i_1,\ldots,i_{|\omega_{2,\mathcal{Y}_{j_{|\bar{\omega}_{3,\mathcal{X}}|}}}|}), 1 \leq i_1 < i_2 < \ldots < i_{|\omega_{2,\mathcal{Y}_{j_{|\bar{\omega}_{3,\mathcal{X}}|}}}|} \leq s,$$
$$0 \leq |\omega_{2,\mathcal{Y}_{j_{|\bar{\omega}_{3,\mathcal{X}}|}}}| \leq s, \bar{\omega}_{2,\mathcal{Y}_{j_{|\bar{\omega}_{3,\mathcal{X}}|}}} = (j_1,\ldots,j_{|\bar{\omega}_{2,\mathcal{Y}_{j_{|\bar{\omega}_{3,\mathcal{X}}|}}}|}),$$
$$1 \leq j_1 < j_2 < \ldots < j_{|\bar{\omega}_{2,\mathcal{Y}_{j_{|\bar{\omega}_{3,\mathcal{X}}|}}}|} \leq s, \{\bar{\omega}_{2,\mathcal{Y}_{j_{|\bar{\omega}_{3,\mathcal{X}}|}}}\} \cup \{\omega_{2,\mathcal{Y}_{j_{|\bar{\omega}_{3,\mathcal{X}}|}}}\} = \{1,\ldots,s\},$$
$$\{\bar{\omega}_{2,\mathcal{Y}_{j_{|\bar{\omega}_{3,\mathcal{X}}|}}}\} \cap \{\omega_{2,\mathcal{Y}_{j_{|\bar{\omega}_{3,\mathcal{X}}|}}}\} = \emptyset.$$

We use the same simplification in the feet descriptions of the big disjunctions and big conjunctions in the SO formula for the case of the $HO^{4,P}$ existential formula.

First we express the formula in a more intuitive way, using natural language, and then we present the actual SO formula with labels in the left side that correspond to the different sub formulae in the natural language expression.

Natural language expression of the SO formula for Atomic $HO^{4,P}$ Formulae:

For *some* pattern of non empty relations for tuples of TO relations $\omega_{3,\mathcal{X}}$, the tuple of TO relations $(\mathcal{Y}_1^3,\ldots,\mathcal{Y}_s^3)$ follows that pattern, and there is a tuple of TO relations in the relation $X_{\mathcal{X}^4,\omega_{3,\mathcal{X}}}$, and corresponding $|\bar{\omega}_{3,\mathcal{X}}|$ tuples of SO relations in the relation 3-$REL_\mathcal{X}$ such that

I) for *all* the patterns of non empty relations for tuples of SO relations $\omega_{2,\mathcal{Y}_{j_{1_3,\mathcal{X}}}}$ for the **first** non empty TO relation $\mathcal{Y}_{j_1}$ in the tuple $(\mathcal{Y}_1^3,\ldots,\mathcal{Y}_s^3)$ according to the specific pattern $\omega_{3,\mathcal{X}}$, it so happens that

1. **I.1)** whenever there is a tuple of SO relations in $X_{y^3_{j_{1_3,x}},\omega_2,y_{j_{1_3,x}}}$ (i.e., the SO relation that encodes the relation y_{j_1} mentioned above for the specific pattern $\omega_{2,y_{j_{1_3,x}}}$), *then* there is also a tuple of SO relations in the relation 3-REL$_x$ for the **first** non empty TO relation in the tuple of TO relations in the relation $X_{x^4,\omega_{3,x}}$ according to the specific pattern $\omega_{3,x}$, with a corresponding tuple in the relation TUPLES-2-REL$_x$, such that
   (a) i. whenever there is a tuple in the relation 2-REL$_{y_{j_{1_3,x}}}$ for the **first** non empty SO relation in the current tuple of SO relations in $X_{y^3_{j_{1_3,x}},\omega_2,y_{j_{1_3,x}}}$ according to the specific pattern $\omega_{2,y_{j_{1_3,x}}}$, then there is also a corresponding tuple in the relation 2-REL$_x$ for the **first** non empty SO relation in the current tuple of SO relations in TUPLES-2-REL$_x$ according to the specific pattern $\omega_{2,y_{j_{1_3,x}}}$, such that the tuples of elements in the relations 2-REL$_{y_{j_{1_3,x}}}$ and 2-REL$_x$ are the same,

   $\vdots$

   ii. whenever there is a tuple in the relation 2-REL$_{y_{j_{1_3,x}}}$ for the **last** non empty SO relation in the current tuple of SO relations in $X_{y^3_{j_{1_3,x}},\omega_2,y_{j_{1_3,x}}}$ according to the specific pattern $\omega_{2,y_{j_{1_3,x}}}$, then there is also a corresponding tuple in the relation 2-REL$_x$ for the **last** non empty SO relation in the current tuple of SO relations in TUPLES-2-REL$_x$ according to the specific pattern $\omega_{2,y_{j_{1_3,x}}}$, such that the tuples of elements in the relations 2-REL$_{y_{j_{1_3,x}}}$ and 2-REL$_x$ are the same,

   (b) and ⟨*viceversa*, i.e.,⟩

   i. whenever there is a tuple in the relation 2-REL$_x$ for the **first** non empty SO relation in the current tuple of SO relations in TUPLES-2-REL$_x$ according to the specific pattern $\omega_{2,y_{j_{1_3,x}}}$, then there is a corresponding tuple in the relation 2-REL$_{y_{j_{1_3,x}}}$, such that the tuples of elements in the relations 2-REL$_{y_{j_{1_3,x}}}$ and 2-REL$_x$ are the same,

   $\vdots$

   ii. whenever there is a tuple in the relation 2-REL$_x$ for the **last** non empty SO relation in the current tuple of SO relations in TUPLES-2-REL$_x$ according to the specific pattern $\omega_{2,y_{j_{1_3,x}}}$, then there is a corresponding tuple in the relation 2-REL$_{y_{j_{1_3,x}}}$, such that the tuples of elements in the relations 2-REL$y_{j_{1_3},x}$ and 2-RELx are the same,

and ⟨viceversa w.r.t. **(I.1)**, i.e.,⟩

2. **I.2)** whenever there is a tuple of SO relations in the relation 3-RELx for the **first** non empty TO relation in the tuple of TO relations in the relation X$_{x^4,\omega_3,x}$ according to the specific pattern $\omega_{3,x}$, with a corresponding tuple in the relation TUPLES-2-RELx then there is also a tuple of SO relations in the relation X$_{y^3_{j_{1_3},x},\omega_2,y_{j_{1_3},x}}$, such that

⟨the following sub formula in (a)i, (a)ii, (b)i and (b)ii is the exact repetition of the sub formula in (a)i, (a)ii, (b)i and (b)ii in **(I.1)** above, i.e.,⟩

(a) i. whenever there is a tuple in the relation 2-REL$y_{j_{1_3},x}$ for the **first** non empty SO relation in the current tuple of SO relations in X$_{y^3_{j_{1_3},x},\omega_2,y_{j_{1_3},x}}$ according to the specific pattern $\omega_{2,y_{j_{1_3},x}}$, then there is also a corresponding tuple in the relation 2-RELx for the **first** non empty SO relation in the current tuple of SO relations in TUPLES-2-RELx according to the specific pattern $\omega_{2,y_{j_{1_3},x}}$, such that the tuples of elements in the relations 2-REL$y_{j_{1_3},x}$ and 2-RELx are the same,

$\vdots$

ii. whenever there is a tuple in the relation 2-REL$y_{j_{1_3},x}$ for the **last** non empty SO relation in the current tuple of SO relations in X$_{y^3_{j_{1_3},x},\omega_2,y_{j_{1_3},x}}$ according to the specific pattern $\omega_{2,y_{j_{1_3},x}}$, then there is also a corresponding tuple in the relation 2-RELx for the **last** non empty SO relation in the current tuple of SO relations in TUPLES-2-RELx according to the specific pattern $\omega_{2,y_{j_{1_3},x}}$, such that the tuples of elements in the relations 2-REL$y_{j_{1_3},x}$ and 2-RELx are the same,

(b) and ⟨viceversa, i.e.,⟩

i. whenever there is a tuple in the relation 2-RELx for the **first** non empty SO relation in the current tuple of SO relations in TUPLES-2-RELx according to the specific pattern $\omega_{2,y_{j_{1_3},x}}$, then there is a corresponding tuple in the relation 2-REL$y_{j_{1_3},x}$, such that the tuples of elements in the relations 2-REL$y_{j_{1_3},x}$ and 2-RELx are the same,

⋮

ii. whenever there is a tuple in the relation 2-REL$_\mathcal{X}$ for the ***last*** non empty SO relation in the current tuple of SO relations in TUPLES-2-REL$_\mathcal{X}$ according to the specific pattern $\omega_{2,\mathcal{Y}_{j_{1_3,\mathcal{X}}}}$, then there is a corresponding tuple in the relation 2-REL$_{\mathcal{Y}_{j_{1_3,\mathcal{X}}}}$, such that the tuples of elements in the relations 2-REL$_{\mathcal{Y}_{j_{1_3,\mathcal{X}}}}$ and 2-REL$_\mathcal{X}$ are the same,

⋮

II) ⟨*the following sub formula is an **exact repetition** of the sub formula **(I)**, except that the subindex $j_{1_3,\mathcal{X}}$ must be replaced by $j_{|\bar\omega_{3,\mathcal{X}}|}$ in all its occurrences, i.e.,*⟩

for *all* the patterns of non empty relations for tuples of SO relations $\omega_{2,\mathcal{Y}_{j_{|\bar\omega_{3,\mathcal{X}}|}}}$ for the ***last*** non empty relation $\mathcal{Y}_{j_{|\bar\omega_{3,\mathcal{X}}|}}$ in the tuple $(\mathcal{Y}_1^3,\ldots,\mathcal{Y}_s^3)$ according to the specific pattern $\omega_{3,\mathcal{X}}$, it so happens that...

⟨*then, correspondingly, **the following labels would follow**: 1.II.1, 1(a)i, 1(a)ii, 1(b)i, 1(b)ii, 2.II.2, 2(a)i, 2(a)ii, 2(b)i, 2(b)ii*⟩.

SO formula for Atomic HO^{4,P} Formulae:

$$\bigvee_{\omega_{3,\mathcal{X}}} \exists \bar{x}^{3t} \bar{x}^3_{j_{1_3,\mathcal{X}}} \ldots \bar{x}^3_{j_{|\bar\omega_{3,\mathcal{X}}|}} \bar{x}^{2t}_{j_{1_3,\mathcal{X}}} \ldots \bar{x}^{2t}_{j_{|\bar\omega_{3,\mathcal{X}}|}} \left((\text{``}\mathcal{Y}^3_{i_{1_3,\mathcal{X}}} = \emptyset\text{''} \wedge \ldots \wedge \text{``}\mathcal{Y}^3_{i_{|\omega_{3,\mathcal{X}}|}} = \emptyset\text{''}) \wedge \right.$$

$$(\text{``}\mathcal{Y}^3_{j_{1_3,\mathcal{X}}} \neq \emptyset\text{''} \wedge \ldots \wedge \text{``}\mathcal{Y}^3_{j_{|\bar\omega_{3,\mathcal{X}}|}} \neq \emptyset\text{''}) \wedge X_{\mathcal{X}^4,\omega_{3,\mathcal{X}}}(\bar{x}^{3t},\bar{x}^3_{j_{1_3,\mathcal{X}}},\ldots,\bar{x}^3_{j_{|\bar\omega_{3,\mathcal{X}}|}}) \wedge$$

$$\text{3-REL}_\mathcal{X}(\bar{x}^3_{j_{1_3,\mathcal{X}}},\bar{x}^{2t}_{j_{1_3,\mathcal{X}}}) \wedge \ldots \wedge \text{3-REL}_\mathcal{X}(\bar{x}^3_{j_{|\bar\omega_{3,\mathcal{X}}|}},\bar{x}^{2t}_{j_{|\bar\omega_{3,\mathcal{X}}|}}) \wedge$$

⟨*I*⟩, ⟨*1.I.1*⟩

$$\bigwedge_{\omega_{2,\mathcal{Y}_{j_{1_3,\mathcal{X}}}}} \left[\forall \bar{y}^{2t} \bar{y}^2_{j_{1_2,\mathcal{Y}_{j_{1_3,\mathcal{X}}}}} \ldots \bar{y}^2_{j_{|\bar\omega_2,\mathcal{Y}_{j_{1_3,\mathcal{X}}}|}} \left[\left(X_{\mathcal{Y}^3_{j_{1_3,\mathcal{X}}},\omega_{2,\mathcal{Y}_{j_{1_3,\mathcal{X}}}}}(\bar{y}^{2t},\bar{y}^2_{j_{1_2,\mathcal{Y}_{j_{1_3,\mathcal{X}}}}},\ldots, \right. \right. \right.$$

$$\left. \bar{y}^2_{j_{|\bar\omega_2,\mathcal{Y}_{j_{1_3,\mathcal{X}}}|}} \right) \Rightarrow \exists \bar{x}'^{2t} \bar{x}^2_{j_{1_2,\mathcal{Y}_{j_{1_3,\mathcal{X}}}}} \ldots \bar{x}^2_{j_{|\bar\omega_2,\mathcal{Y}_{j_{1_3,\mathcal{X}}}|}} \left(\text{3-REL}_\mathcal{X}(\bar{x}^3_{j_{1_3,\mathcal{X}}},\bar{x}'^{2t}) \wedge \right.$$

$$\text{TUPLES-2-REL}_{\mathcal{X},\omega_{2,\mathcal{Y}_{j_{1_3,\mathcal{X}}}}}(\bar{x}'^{2t},\bar{x}^2_{j_{1_2,\mathcal{Y}_{j_{1_3,\mathcal{X}}}}},\ldots,\bar{x}^2_{j_{|\bar\omega_2,\mathcal{Y}_{j_{1_3,\mathcal{X}}}|}}) \wedge$$

⟨1(a)i⟩

$$\forall \bar{y}^{1t}_{j_{1_2,\mathcal{Y}_{j_{1_3,\mathcal{X}}}}} \ldots \bar{y}^{1t}_{j_{|\bar{\omega}_2,\mathcal{Y}_{j_{1_3,\mathcal{X}}}|}} \left[(2\text{-REL}_{\mathcal{Y}_{j_{1_3,\mathcal{X}}}}(\bar{y}^2_{j_{1_2,\mathcal{Y}_{j_{1_3,\mathcal{X}}}}}, \bar{y}^{1t}_{j_{1_2,\mathcal{Y}_{j_{1_3,\mathcal{X}}}}}) \Rightarrow \exists \bar{x}^{1t}_{j_{1_2,\mathcal{Y}_{j_{1_3,\mathcal{X}}}}} \right. ($$

$$2\text{-REL}_{\mathcal{X}}(\bar{x}^2_{j_{1_2,\mathcal{Y}_{j_{1_3,\mathcal{X}}}}}, \bar{x}^{1t}_{j_{1_2,\mathcal{Y}_{j_{1_3,\mathcal{X}}}}}) \wedge \text{``}\bar{y}^{1t}_{j_{1_2,\mathcal{Y}_{j_{1_3,\mathcal{X}}}}} = \bar{x}^{1t}_{j_{1_2,\mathcal{Y}_{j_{1_3,\mathcal{X}}}}}\text{''})) \wedge \ldots \wedge$$

⟨1(a)ii⟩

$$(2\text{-REL}_{\mathcal{Y}_{j_{1_3,\mathcal{X}}}}(\bar{y}^2_{j_{|\bar{\omega}_2,\mathcal{Y}_{j_{1_3,\mathcal{X}}}|}}, \bar{y}^{1t}_{j_{|\bar{\omega}_2,\mathcal{Y}_{j_{1_3,\mathcal{X}}}|}}) \Rightarrow \exists \bar{x}^{1t}_{j_{|\bar{\omega}_2,\mathcal{Y}_{j_{1_3,\mathcal{X}}}|}} ($$

$$2\text{-REL}_{\mathcal{X}}(\bar{x}^2_{j_{|\bar{\omega}_2,\mathcal{Y}_{j_{1_3,\mathcal{X}}}|}}, \bar{x}^{1t}_{j_{|\bar{\omega}_2,\mathcal{Y}_{j_{1_3,\mathcal{X}}}|}}) \wedge \text{``}\bar{y}^{1t}_{j_{|\bar{\omega}_2,\mathcal{Y}_{j_{1_3,\mathcal{X}}}|}} = \bar{x}^{1t}_{j_{|\bar{\omega}_2,\mathcal{Y}_{j_{1_3,\mathcal{X}}}|}}\text{''}))] \wedge$$

⟨1(b)i⟩

$$\forall \bar{x}'^{1t}_{j_{1_2,\mathcal{Y}_{j_{1_3,\mathcal{X}}}}} \ldots \bar{x}'^{1t}_{j_{|\bar{\omega}_2,\mathcal{Y}_{j_{1_3,\mathcal{X}}}|}} \left[(2\text{-REL}_{\mathcal{X}}(\bar{x}^2_{j_{1_2,\mathcal{Y}_{j_{1_3,\mathcal{X}}}}}, \bar{x}'^{1t}_{j_{1_2,\mathcal{Y}_{j_{1_3,\mathcal{X}}}}}) \Rightarrow \exists \bar{y}'^{1t}_{j_{1_2,\mathcal{Y}_{j_{1_3,\mathcal{X}}}}} \right. ($$

$$2\text{-REL}_{\mathcal{Y}_{j_{1_3,\mathcal{X}}}}(\bar{y}^2_{j_{1_2,\mathcal{Y}_{j_{1_3,\mathcal{X}}}}}, \bar{y}'^{1t}_{j_{1_2,\mathcal{Y}_{j_{1_3,\mathcal{X}}}}}) \wedge \text{``}\bar{y}'^{1t}_{j_{1_2,\mathcal{Y}_{j_{1_3,\mathcal{X}}}}} = \bar{x}'^{1t}_{j_{1_2,\mathcal{Y}_{j_{1_3,\mathcal{X}}}}}\text{''})) \wedge \ldots \wedge$$

⟨1(b)ii⟩

$$(2\text{-REL}_{\mathcal{X}}(\bar{x}^2_{j_{|\bar{\omega}_2,\mathcal{Y}_{j_{1_3,\mathcal{X}}}|}}, \bar{x}'^{1t}_{j_{|\bar{\omega}_2,\mathcal{Y}_{j_{1_3,\mathcal{X}}}|}}) \Rightarrow \exists \bar{y}'^{1t}_{j_{|\bar{\omega}_2,\mathcal{Y}_{j_{1_3,\mathcal{X}}}|}} ($$

$$2\text{-REL}_{\mathcal{Y}_{j_{1_3,\mathcal{X}}}}(\bar{y}^2_{j_{|\bar{\omega}_2,\mathcal{Y}_{j_{1_3,\mathcal{X}}}|}}, \bar{y}'^{1t}_{j_{|\bar{\omega}_2,\mathcal{Y}_{j_{1_3,\mathcal{X}}}|}}) \wedge \text{``}\bar{y}'^{1t}_{j_{|\bar{\omega}_2,\mathcal{Y}_{j_{1_3,\mathcal{X}}}|}} = \bar{x}'^{1t}_{j_{|\bar{\omega}_2,\mathcal{Y}_{j_{1_3,\mathcal{X}}}|}}\text{''}))] \right] \wedge$$

⟨2.I.2⟩

$$\forall \bar{x}'^{2t} \bar{x}^2_{j_{1_2,\mathcal{Y}_{j_{1_3,\mathcal{X}}}}} \ldots \bar{x}^2_{j_{|\bar{\omega}_2,\mathcal{Y}_{j_{1_3,\mathcal{X}}}|}} \left[\left(3\text{-REL}_{\mathcal{X}}(\bar{x}^3_{j_{1_3,\mathcal{X}}}, \bar{x}'^{2t}) \wedge \right. \right.$$

$$\text{TUPLES-2-REL}_{\mathcal{X}, \omega_2, \mathcal{Y}_{j_{1_3,\mathcal{X}}}}(\bar{x}'^{2t}, \bar{x}^2_{j_{1_2,\mathcal{Y}_{j_{1_3,\mathcal{X}}}}}, \ldots, \bar{x}^2_{j_{|\bar{\omega}_2,\mathcal{Y}_{j_{1_3,\mathcal{X}}}|}}) \right) \Rightarrow$$

$$\exists \bar{y}^{2t} \bar{y}^2_{j_{1_2,\mathcal{Y}_{j_{1_3,\mathcal{X}}}}} \ldots \bar{y}^2_{j_{|\bar{\omega}_2,\mathcal{Y}_{j_{1_3,\mathcal{X}}}|}} \left(\mathrm{X}_{\mathcal{Y}^3_{j_{1_3,\mathcal{X}}}, \omega_2, \mathcal{Y}_{j_{1_3,\mathcal{X}}}}(\bar{y}^{2t}, \bar{y}^2_{j_{1_2,\mathcal{Y}_{j_{1_3,\mathcal{X}}}}}, \ldots, \right.$$

$$\left. \bar{y}^2_{j_{|\bar{\omega}_2,\mathcal{Y}_{j_{1_3,\mathcal{X}}}|}}) \wedge \right.$$

⟨2(a)i⟩

$$\forall \bar{y}^{1t}_{j_{1_2,\mathcal{Y}_{j_{1_3,\mathcal{X}}}}} \ldots \bar{y}^{1t}_{j_{|\bar{\omega}_2,\mathcal{Y}_{j_{1_3,\mathcal{X}}}|}} \left[(2\text{-REL}_{\mathcal{Y}_{j_{1_3,\mathcal{X}}}}(\bar{y}^2_{j_{1_2,\mathcal{Y}_{j_{1_3,\mathcal{X}}}}}, \bar{y}^{1t}_{j_{1_2,\mathcal{Y}_{j_{1_3,\mathcal{X}}}}}) \Rightarrow \exists \bar{x}^{1t}_{j_{1_2,\mathcal{Y}_{j_{1_3,\mathcal{X}}}}} \right. ($$

$$2\text{-REL}_{\mathcal{X}}(\bar{x}^2_{j_{1_2,\mathcal{Y}_{j_{1_3,\mathcal{X}}}}}, \bar{x}^{1t}_{j_{1_2,\mathcal{Y}_{j_{1_3,\mathcal{X}}}}}) \wedge \text{``}\bar{y}^{1t}_{j_{1_2,\mathcal{Y}_{j_{1_3,\mathcal{X}}}}} = \bar{x}^{1t}_{j_{1_2,\mathcal{Y}_{j_{1_3,\mathcal{X}}}}}\text{''})) \wedge \ldots \wedge$$

⟨2(a)ii⟩
$$(\text{2-REL}_{\mathcal{Y}_{j_{1_{3,\mathcal{X}}}}}(\bar{y}^2_{j_{|\bar{\omega}_2,\mathcal{Y}_{j_{1_{3,\mathcal{X}}}}|}},\bar{y}^{1t}_{j_{|\bar{\omega}_2,\mathcal{Y}_{j_{1_{3,\mathcal{X}}}}|}}) \Rightarrow \exists \bar{x}^{1t}_{j_{|\bar{\omega}_2,\mathcal{Y}_{j_{1_{3,\mathcal{X}}}}|}}($$

$$\text{2-REL}_{\mathcal{X}}(\bar{x}^2_{j_{|\bar{\omega}_2,\mathcal{Y}_{j_{1_{3,\mathcal{X}}}}|}},\bar{x}^{1t}_{j_{|\bar{\omega}_2,\mathcal{Y}_{j_{1_{3,\mathcal{X}}}}|}}) \land \text{``}\bar{y}^{1t}_{j_{|\bar{\omega}_2,\mathcal{Y}_{j_{1_{3,\mathcal{X}}}}|}} = \bar{x}^{1t}_{j_{|\bar{\omega}_2,\mathcal{Y}_{j_{1_{3,\mathcal{X}}}}|}}\text{''}))] \land$$

⟨2(b)i⟩
$$\forall \bar{x'}^{1t}_{j_{1_2,\mathcal{Y}_{j_{1_{3,\mathcal{X}}}}}} \ldots \bar{x'}^{1t}_{j_{|\bar{\omega}_2,\mathcal{Y}_{j_{1_{3,\mathcal{X}}}}|}} \Big[(\text{2-REL}_{\mathcal{X}}(\bar{x}^2_{j_{1_2,\mathcal{Y}_{j_{1_{3,\mathcal{X}}}}}},\bar{x'}^{1t}_{j_{1_2,\mathcal{Y}_{j_{1_{3,\mathcal{X}}}}}}) \Rightarrow \exists \bar{y'}^{1t}_{j_{1_2,\mathcal{Y}_{j_{1_{3,\mathcal{X}}}}}} \Big($$

$$\text{2-REL}_{\mathcal{Y}_{j_{1_{3,\mathcal{X}}}}}(\bar{y}^2_{j_{1_2,\mathcal{Y}_{j_{1_{3,\mathcal{X}}}}}},\bar{y'}^{1t}_{j_{1_2,\mathcal{Y}_{j_{1_{3,\mathcal{X}}}}}}) \land \text{``}\bar{y'}^{1t}_{j_{1_2,\mathcal{Y}_{j_{1_{3,\mathcal{X}}}}}} = \bar{x'}^{1t}_{j_{1_2,\mathcal{Y}_{j_{1_{3,\mathcal{X}}}}}}\text{''})) \land \ldots \land$$

⟨2(b)ii⟩
$$(\text{2-REL}_{\mathcal{X}}(\bar{x}^2_{j_{|\bar{\omega}_2,\mathcal{Y}_{j_{1_{3,\mathcal{X}}}}|}},\bar{x'}^{1t}_{j_{|\bar{\omega}_2,\mathcal{Y}_{j_{1_{3,\mathcal{X}}}}|}}) \Rightarrow \exists \bar{y'}^{1t}_{j_{|\bar{\omega}_2,\mathcal{Y}_{j_{1_{3,\mathcal{X}}}}|}}($$

$$\text{2-REL}_{\mathcal{Y}_{j_{1_{3,\mathcal{X}}}}}(\bar{y}^2_{j_{|\bar{\omega}_2,\mathcal{Y}_{j_{1_{3,\mathcal{X}}}}|}},\bar{y'}^{1t}_{j_{|\bar{\omega}_2,\mathcal{Y}_{j_{1_{3,\mathcal{X}}}}|}}) \land \text{``}\bar{y'}^{1t}_{j_{|\bar{\omega}_2,\mathcal{Y}_{j_{1_{3,\mathcal{X}}}}|}} = \bar{x'}^{1t}_{j_{|\bar{\omega}_2,\mathcal{Y}_{j_{1_{3,\mathcal{X}}}}|}}\text{''}))\Big)\Big]\Big] \land$$

⟨II⟩, ⟨1.II.1⟩
$$\ldots \land \bigwedge_{\omega_2,\mathcal{Y}_{j_{|\bar{\omega}_3,\mathcal{X}}|}} \Big[\forall \bar{y}^{2t} \bar{y}^2_{j_{1_2,\mathcal{Y}_{j_{|\bar{\omega}_3,\mathcal{X}|}}}} \ldots \bar{y}^2_{j_{|\bar{\omega}_2,\mathcal{Y}_{j_{|\bar{\omega}_3,\mathcal{X}|}}|}} \Big[\Big(\mathrm{X}_{\mathcal{Y}^3_{j_{|\bar{\omega}_3,\mathcal{X}|}},\omega_2,\mathcal{Y}_{j_{|\bar{\omega}_3,\mathcal{X}|}}}(\bar{y}^{2t},\bar{y}^2_{j_{1_2,\mathcal{Y}_{j_{|\bar{\omega}_3,\mathcal{X}|}}}}$$

$$\ldots,\bar{y}^2_{j_{|\bar{\omega}_2,\mathcal{Y}_{j_{|\bar{\omega}_3,\mathcal{X}|}}|}})) \Rightarrow \exists \bar{x'}^{2t} \bar{x}^2_{j_{1_2,\mathcal{Y}_{j_{|\bar{\omega}_3,\mathcal{X}|}}}} \ldots \bar{x}^2_{j_{|\bar{\omega}_2,\mathcal{Y}_{j_{|\bar{\omega}_3,\mathcal{X}|}}|}} \Big(\text{3-REL}_{\mathcal{X}}(\bar{x}^3_{j_{|\bar{\omega}_3,\mathcal{X}|}},\bar{x'}^{2t}) \land$$

$$\text{TUPLES-2-REL}_{\mathcal{X},\omega_2,\mathcal{Y}_{j_{|\bar{\omega}_3,\mathcal{X}|}}}(\bar{x'}^{2t},\bar{x}^2_{j_{1_2,\mathcal{Y}_{j_{|\bar{\omega}_3,\mathcal{X}|}}}},\ldots,\bar{x}^2_{j_{|\bar{\omega}_2,\mathcal{Y}_{j_{|\bar{\omega}_3,\mathcal{X}|}}|}}) \land$$

⟨1(a)i⟩
$$\forall \bar{y}^{1t}_{j_{1_2,\mathcal{Y}_{j_{|\bar{\omega}_3,\mathcal{X}|}}}} \ldots \bar{y}^{1t}_{j_{|\bar{\omega}_2,\mathcal{Y}_{j_{|\bar{\omega}_3,\mathcal{X}|}}|}} \Big[(\text{2-REL}_{\mathcal{Y}_{j_{|\bar{\omega}_3,\mathcal{X}|}}}(\bar{y}^2_{j_{1_2,\mathcal{Y}_{j_{|\bar{\omega}_3,\mathcal{X}|}}}},\bar{y}^{1t}_{j_{1_2,\mathcal{Y}_{j_{|\bar{\omega}_3,\mathcal{X}|}}}}) \Rightarrow \exists \bar{x}^{1t}_{j_{1_2,\mathcal{Y}_{j_{|\bar{\omega}_3,}}}}$$

$$\text{2-REL}_{\mathcal{X}}(\bar{x}^2_{j_{1_2,\mathcal{Y}_{j_{|\bar{\omega}_3,\mathcal{X}|}}}},\bar{x}^{1t}_{j_{1_2,\mathcal{Y}_{j_{|\bar{\omega}_3,\mathcal{X}|}}}}) \land \text{``}\bar{y}^{1t}_{j_{1_2,\mathcal{Y}_{j_{|\bar{\omega}_3,\mathcal{X}|}}}} = \bar{x}^{1t}_{j_{1_2,\mathcal{Y}_{j_{|\bar{\omega}_3,\mathcal{X}|}}}}\text{''})) \land \ldots \land$$

⟨1(a)ii⟩
$$(\text{2-REL}_{\mathcal{Y}_{j_{|\bar{\omega}_3,\mathcal{X}|}}}(\bar{y}^2_{j_{|\bar{\omega}_2,\mathcal{Y}_{j_{|\bar{\omega}_3,\mathcal{X}|}}|}},\bar{y}^{1t}_{j_{|\bar{\omega}_2,\mathcal{Y}_{j_{|\bar{\omega}_3,\mathcal{X}|}}|}}) \Rightarrow \exists \bar{x}^{1t}_{j_{|\bar{\omega}_2,\mathcal{Y}_{j_{|\bar{\omega}_3,\mathcal{X}|}}|}}($$

$$\text{2-REL}_{\mathcal{X}}(\bar{x}^2_{j_{|\bar{\omega}_2,\mathcal{Y}_{j_{|\bar{\omega}_3,\mathcal{X}|}}|}},\bar{x}^{1t}_{j_{|\bar{\omega}_2,\mathcal{Y}_{j_{|\bar{\omega}_3,\mathcal{X}|}}|}}) \land \text{``}\bar{y}^{1t}_{j_{|\bar{\omega}_2,\mathcal{Y}_{j_{|\bar{\omega}_3,\mathcal{X}|}}|}} = \bar{x}^{1t}_{j_{|\bar{\omega}_2,\mathcal{Y}_{j_{|\bar{\omega}_3,\mathcal{X}|}}|}}\text{''}))] \land$$

⟨1(b)i⟩

$$\forall \bar{x}'^{1t}_{j_{1_2, y_{j_{|\bar{\omega}_3, \mathcal{X}|}}}} \ldots \bar{x}'^{1t}_{j_{|\bar{\omega}_2, y_{j_{|\bar{\omega}_3, \mathcal{X}|}}|}} \Big[(2\text{-REL}_{\mathcal{X}}(\bar{x}^2_{j_{1_2, y_{j_{|\bar{\omega}_3, \mathcal{X}|}}}}, \bar{x}'^{1t}_{j_{1_2, y_{j_{|\bar{\omega}_3, \mathcal{X}|}}}}) \Rightarrow \exists \bar{y}'^{1t}_{j_{1_2, y_{j_{|\bar{\omega}_3, \mathcal{X}|}}}}$$

$$2\text{-REL}_{y_{j_{|\bar{\omega}_3, \mathcal{X}|}}}(\bar{y}^2_{j_{1_2, y_{j_{|\bar{\omega}_3, \mathcal{X}|}}}}, \bar{y}'^{1t}_{j_{1_2, y_{j_{|\bar{\omega}_3, \mathcal{X}|}}}}) \wedge \text{``}\bar{y}'^{1t}_{j_{1_2, y_{j_{|\bar{\omega}_3, \mathcal{X}|}}}} = \bar{x}'^{1t}_{j_{1_2, y_{j_{|\bar{\omega}_3, \mathcal{X}|}}}}\text{''})) \wedge \ldots \wedge$$

⟨1(b)ii⟩

$$(2\text{-REL}_{\mathcal{X}}(\bar{x}^2_{j_{|\bar{\omega}_2, y_{j_{|\bar{\omega}_3, \mathcal{X}|}}|}}, \bar{x}'^{1t}_{j_{|\bar{\omega}_2, y_{j_{|\bar{\omega}_3, \mathcal{X}|}}|}}) \Rightarrow \exists \bar{y}'^{1t}_{j_{|\bar{\omega}_2, y_{j_{|\bar{\omega}_3, \mathcal{X}|}}|}} \Big($$

$$2\text{-REL}_{y_{j_{|\bar{\omega}_3, \mathcal{X}|}}}(\bar{y}^2_{j_{|\bar{\omega}_2, y_{j_{|\bar{\omega}_3, \mathcal{X}|}}|}}, \bar{y}'^{1t}_{j_{|\bar{\omega}_2, y_{j_{|\bar{\omega}_3, \mathcal{X}|}}|}}) \wedge \text{``}\bar{y}'^{1t}_{j_{|\bar{\omega}_2, y_{j_{|\bar{\omega}_3, \mathcal{X}|}}|}} = \bar{x}'^{1t}_{j_{|\bar{\omega}_2, y_{j_{|\bar{\omega}_3, \mathcal{X}|}}|}}\text{''}))\Big)\Big] \wedge$$

⟨2.II.2⟩

$$\forall \bar{x}'^{2t} \bar{x}^2_{j_{1_2, y_{j_{|\bar{\omega}_3, \mathcal{X}|}}}} \ldots \bar{x}^2_{j_{|\bar{\omega}_2, y_{j_{|\bar{\omega}_3, \mathcal{X}|}}|}} \Bigg[\Big(3\text{-REL}_{\mathcal{X}}(\bar{x}^3_{j_{|\bar{\omega}_3, \mathcal{X}|}}, \bar{x}'^{2t}) \wedge$$

$$\text{TUPLES-2-REL}_{\mathcal{X}, \omega_2, y_{j_{|\bar{\omega}_3, \mathcal{X}|}}}(\bar{x}'^{2t}, \bar{x}^2_{j_{1_2, y_{j_{|\bar{\omega}_3, \mathcal{X}|}}}}, \ldots, \bar{x}^2_{j_{|\bar{\omega}_2, y_{j_{|\bar{\omega}_3, \mathcal{X}|}}|}}) \Big) \Rightarrow$$

$$\exists \bar{y}^{2t} \bar{y}^2_{j_{1_2, y_{j_{|\bar{\omega}_3, \mathcal{X}|}}}} \ldots \bar{y}^2_{j_{|\bar{\omega}_2, y_{j_{|\bar{\omega}_3, \mathcal{X}|}}|}} \bigg(X_{\mathcal{Y}^3_{|\bar{\omega}_3, \mathcal{X}|}, \omega_2, y_{j_{|\bar{\omega}_3, \mathcal{X}|}}}(\bar{y}^{2t}, \bar{y}^2_{j_{1_2, y_{j_{|\bar{\omega}_3, \mathcal{X}|}}}}, \ldots,$$

$$\bar{y}^2_{j_{|\bar{\omega}_2, y_{j_{|\bar{\omega}_3, \mathcal{X}|}}|}}) \wedge$$

⟨2(a)i⟩

$$\forall \bar{y}^{1t}_{j_{1_2, y_{j_{|\bar{\omega}_3, \mathcal{X}|}}}} \ldots \bar{y}^{1t}_{j_{|\bar{\omega}_2, y_{j_{|\bar{\omega}_3, \mathcal{X}|}}|}} \Big[(2\text{-REL}_{y_{j_{|\bar{\omega}_3, \mathcal{X}|}}}(\bar{y}^2_{j_{1_2, y_{j_{|\bar{\omega}_3, \mathcal{X}|}}}}, \bar{y}^{1t}_{j_{1_2, y_{j_{|\bar{\omega}_3, \mathcal{X}|}}}}) \Rightarrow \exists \bar{x}^{1t}_{j_{1_2, y_{j_{|\bar{\omega}_3,}}}}$$

$$2\text{-REL}_{\mathcal{X}}(\bar{x}^2_{j_{1_2, y_{j_{|\bar{\omega}_3, \mathcal{X}|}}}}, \bar{x}^{1t}_{j_{1_2, y_{j_{|\bar{\omega}_3, \mathcal{X}|}}}}) \wedge \text{``}\bar{y}^{1t}_{j_{1_2, y_{j_{|\bar{\omega}_3, \mathcal{X}|}}}} = \bar{x}^{1t}_{j_{1_2, y_{j_{|\bar{\omega}_3, \mathcal{X}|}}}}\text{''})) \wedge \ldots \wedge$$

⟨2(a)ii⟩

$$(2\text{-REL}_{y_{j_{|\bar{\omega}_3, \mathcal{X}|}}}(\bar{y}^2_{j_{|\bar{\omega}_2, y_{j_{|\bar{\omega}_3, \mathcal{X}|}}|}}, \bar{y}^{1t}_{j_{|\bar{\omega}_2, y_{j_{|\bar{\omega}_3, \mathcal{X}|}}|}}) \Rightarrow \exists \bar{x}^{1t}_{j_{|\bar{\omega}_2, y_{j_{|\bar{\omega}_3, \mathcal{X}|}}|}} \Big($$

$$2\text{-REL}_{\mathcal{X}}(\bar{x}^2_{j_{|\bar{\omega}_2, y_{j_{|\bar{\omega}_3, \mathcal{X}|}}|}}, \bar{x}^{1t}_{j_{|\bar{\omega}_2, y_{j_{|\bar{\omega}_3, \mathcal{X}|}}|}}) \wedge \text{``}\bar{y}^{1t}_{j_{|\bar{\omega}_2, y_{j_{|\bar{\omega}_3, \mathcal{X}|}}|}} = \bar{x}^{1t}_{j_{|\bar{\omega}_2, y_{j_{|\bar{\omega}_3, \mathcal{X}|}}|}}\text{''})) \Big] \wedge$$

⟨2(b)i⟩

$$\forall \bar{x}'^{1t}_{j_{1_2, y_{j_{|\bar{\omega}_3, \mathcal{X}|}}}} \ldots \bar{x}'^{1t}_{j_{|\bar{\omega}_2, y_{j_{|\bar{\omega}_3, \mathcal{X}|}}|}} \Big[(2\text{-REL}_{\mathcal{X}}(\bar{x}^2_{j_{1_2, y_{j_{|\bar{\omega}_3, \mathcal{X}|}}}}, \bar{x}'^{1t}_{j_{1_2, y_{j_{|\bar{\omega}_3, \mathcal{X}|}}}}) \Rightarrow \exists \bar{y}'^{1t}_{j_{1_2, y_{j_{|\bar{\omega}_3, \mathcal{X}|}}}}$$

2-REL$_{\mathcal{Y}_{j|\bar{\omega}_3,\mathcal{X}|}}(\bar{y}^2_{j_{1_2},\mathcal{Y}_{j|\bar{\omega}_3,\mathcal{X}|}}, \bar{y}'^{1t}_{j_{1_2},\mathcal{Y}_{j|\bar{\omega}_3,\mathcal{X}|}}) \wedge \text{``}\bar{y}'^{1t}_{j_{1_2},\mathcal{Y}_{j|\bar{\omega}_3,\mathcal{X}|}} = \bar{x}'^{1t}_{j_{1_2},\mathcal{Y}_{j|\bar{\omega}_3,\mathcal{X}|}}\text{''})) \wedge \ldots \wedge$

⟨*2(b)ii*⟩

$(2\text{-REL}_{\mathcal{X}}(\bar{x}^2_{j|\bar{\omega}_2,\mathcal{Y}_{j|\bar{\omega}_3,\mathcal{X}|}|}, \bar{x}'^{1t}_{j|\bar{\omega}_2,\mathcal{Y}_{j|\bar{\omega}_3,\mathcal{X}|}|}) \Rightarrow \exists \bar{y}'^{1t}_{j|\bar{\omega}_2,\mathcal{Y}_{j|\bar{\omega}_3,\mathcal{X}|}|}($

2-REL$_{\mathcal{Y}_{j|\bar{\omega}_3,\mathcal{X}|}}(\bar{y}^2_{j|\bar{\omega}_2,\mathcal{Y}_{j|\bar{\omega}_3,\mathcal{X}|}|}, \bar{y}'^{1t}_{j|\bar{\omega}_2,\mathcal{Y}_{j|\bar{\omega}_3,\mathcal{X}|}|}) \wedge \text{``}\bar{y}'^{1t}_{j|\bar{\omega}_2,\mathcal{Y}_{j|\bar{\omega}_3,\mathcal{X}|}|} = \bar{x}'^{1t}_{j|\bar{\omega}_2,\mathcal{Y}_{j|\bar{\omega}_3,\mathcal{X}|}|}\text{''}))\Big)\Big]\Big]\Big])$

B The case of Existential Formulae in HO^{4,P}

The existential case $\exists \mathcal{X}^{4,d,\tau}(\varphi)$ with $|\tau| = s$, downward polynomially bounded, with degree $d \geq 1$ is as follows:

$$\exists \{X^{d+|\bar{f}_{\bar{\omega}_3,\mathcal{X}}|}_{\mathcal{X}^4,\omega_3,\mathcal{X}}\}_{\omega_3,\mathcal{X}} \; \exists 3\text{-REL}^{2d}_{\mathcal{X}} \; \exists \{\text{TUPLES-2-REL}^{d+|\bar{f}_{\bar{\omega}_2,\mathcal{X}}|}_{\mathcal{X},\omega_2,\mathcal{X}}\}_{\omega_2,\mathcal{X}}$$

$$\exists 2\text{-REL}^{d+s}_{\mathcal{X}} \Bigg[(\text{``database } \mathcal{X}^4 \text{ has referential integrity''}) \wedge$$

$$\Bigg[\forall \bar{x}^{3t} \Bigg[\bigwedge_{\omega_3,\mathcal{X}} \forall \bar{f}_{\bar{\omega}_3,\mathcal{X}} \Bigg[X_{\mathcal{X}^4,\omega_3,\mathcal{X}}(\bar{x}^{3t}, \bar{f}_{\bar{\omega}_3,\mathcal{X}}) \Rightarrow$$

$$\Bigg(\bigwedge_{\omega'_3,\mathcal{X} \neq \omega_3,\mathcal{X}} \forall \bar{f}'_{\bar{\omega}'_3,\mathcal{X}} \Big(\neg X_{\mathcal{X}^4,\omega'_3,\mathcal{X}}(\bar{x}^{3t}, \bar{f}'_{\bar{\omega}'_3,\mathcal{X}}) \Big) \Bigg) \Bigg]\Bigg]\Bigg] \wedge$$

$$\Bigg[\forall \bar{x}^{2t} \Bigg[\bigwedge_{\omega_2,\mathcal{X}} \forall \bar{f}_{\bar{\omega}_2,\mathcal{X}} \Bigg[\text{TUPLES-2-REL}_{\mathcal{X},\omega_2,\mathcal{X}}(\bar{x}^{2t}, \bar{f}_{\bar{\omega}_2,\mathcal{X}}) \Rightarrow$$

$$\Bigg(\bigwedge_{\omega'_2,\mathcal{X} \neq \omega_2,\mathcal{X}} \forall \bar{f}'_{\bar{\omega}'_2,\mathcal{X}} \Big(\neg \text{TUPLES-2-REL}_{\mathcal{X},\omega'_2,\mathcal{X}}(\bar{x}^{2t}, \bar{f}'_{\bar{\omega}'_2,\mathcal{X}}) \Big) \Bigg) \Bigg]\Bigg]\Bigg] \wedge \hat{\varphi}\Bigg],$$

where $\hat{\varphi}$ is the HO^{4,P} formula φ, obtained by inductively applying the translations described above.

"database $\mathcal{X}^4$ has referential integrity":
"in $X_{\mathcal{X}^4,\omega_3,\mathcal{X}}$ there are no two tuples of TO relations with the same id of TO relations tuple":

$$\Bigg[\bigwedge_{\omega_3,\mathcal{X}} \forall \bar{x}^{3t} \neg \exists \bar{x}^3_{j_{1_3},\mathcal{X}} \bar{x}^3_{j|\bar{\omega}_3,\mathcal{X}|} \ldots \bar{x}^3_{j|\bar{\omega}_3,\mathcal{X}|} \bar{x}'^3_{j|\bar{\omega}_3,\mathcal{X}|} \Big((\bar{x}^3_{j_{1_3},\mathcal{X}} \neq \bar{x}'^3_{j_{1_3},\mathcal{X}} \vee \ldots$$

$$\vee \bar{x}^3_{j|\bar{\omega}_3,\mathcal{X}|} \neq \bar{x}'^3_{j|\bar{\omega}_3,\mathcal{X}|}) \wedge X_{\mathcal{X}^4,\omega_3,\mathcal{X}}(\bar{x}^{3t}, \bar{x}^3_{j_{1_3},\mathcal{X}}, \ldots, \bar{x}^3_{j|\bar{\omega}_3,\mathcal{X}|}) \wedge$$

$$X_{\mathcal{X}^4,\omega_{3,\mathcal{X}}}(\bar{x}^{3t},\bar{x'}^3_{j_{1_{3,\mathcal{X}}}},\ldots,\bar{x'}^3_{j_{|\bar{\omega}_{3,\mathcal{X}}|}})\Big)\wedge$$

"all TO relations in the tuples in $X_{\mathcal{X}^4,\omega_{3,\mathcal{X}}}$ are in 3-REL$_\mathcal{X}$":

$$\forall \bar{x}^{3t}\bar{x}^3_{j_{1_{3,\mathcal{X}}}}\ldots\bar{x}^3_{j_{|\bar{\omega}_{3,\mathcal{X}}|}}\bigg(X_{\mathcal{X}^4,\omega_{3,\mathcal{X}}}(\bar{x}^{3t},\bar{x}^3_{j_{1_{3,\mathcal{X}}}},\ldots,\bar{x}^3_{j_{|\bar{\omega}_{3,\mathcal{X}}|}})\Rightarrow$$

$$\exists \bar{x}^{2t}_{j_{1_{3,\mathcal{X}}}}\ldots\bar{x}^{2t}_{j_{|\bar{\omega}_{3,\mathcal{X}}|}}\bigg(\text{3-REL}_\mathcal{X}(\bar{x}^3_{j_{1_{3,\mathcal{X}}}},\bar{x}^{2t}_{j_{1_{3,\mathcal{X}}}})\wedge\ldots\wedge\text{3-REL}_\mathcal{X}(\bar{x}^3_{j_{|\bar{\omega}_{3,\mathcal{X}}|}},\bar{x}^{2t}_{j_{|\bar{\omega}_{3,\mathcal{X}}|}})\bigg)\bigg)\bigg]\wedge$$

"every TO relation in 3-REL$_\mathcal{X}$ is in some tuple in $X_{\mathcal{X}^4,\omega_{3,\mathcal{X}}}$":

$$\bigg[\forall \bar{x}^3\bar{x}^{2t}\bigg(\text{3-REL}_\mathcal{X}(\bar{x}^3,\bar{x}^{2t})\Rightarrow\exists\bar{x}^{3t}\bigg(\bigvee_{\omega_{3,\mathcal{X}}}\exists\bar{x}^3_{j_{1_{3,\mathcal{X}}}}\ldots\bar{x}^3_{j_{|\bar{\omega}_{3,\mathcal{X}}|}}\big((\text{``}\bar{x}^3=\bar{x}^3_{j_{1_{3,\mathcal{X}}}}\text{''}\vee\ldots\vee$$

$$\text{``}\bar{x}^3=\bar{x}^3_{j_{|\bar{\omega}_{3,\mathcal{X}}|}}\text{''}\big)\wedge X_{\mathcal{X}^4,\omega_{3,\mathcal{X}}}(\bar{x}^{3t},\bar{x}^3_{j_{1_{3,\mathcal{X}}}},\ldots,\bar{x}^3_{j_{|\bar{\omega}_{3,\mathcal{X}}|}})\big)\bigg)\bigg)\bigg]\wedge$$

"all tuples of SO relations in 3-REL$_\mathcal{X}$ are in some TUPLES-2-REL$_{\mathcal{X},\omega_{2,\mathcal{X}}}$":

$$\forall \bar{x}^3\bar{x}^{2t}\bigg[\text{3-REL}_\mathcal{X}(\bar{x}^3,\bar{x}^{2t})\Rightarrow\bigvee_{\omega_{2,\mathcal{X}}}\exists\bar{x}^2_{j_{1_{2,\mathcal{X}}}}\ldots\bar{x}^2_{j_{|\bar{\omega}_{2,\mathcal{X}}|}}\bigg(\text{TUPLES-2-REL}_{\mathcal{X},\omega_{2,\mathcal{X}}}(\bar{x}^{2t},$$

$$\bar{x}^2_{j_{1_{2,\mathcal{X}}}},\ldots,\bar{x}^2_{j_{|\bar{\omega}_{2,\mathcal{X}}|}})\bigg)\bigg]\wedge$$

"all SO relations in the tuples in TUPLES-2-REL$_{\mathcal{X},\omega_{2,\mathcal{X}}}$ are in 2-REL$_\mathcal{X}$":

$$\bigg[\bigwedge_{\omega_{2,\mathcal{X}}}\forall\bar{x}^{2t}\bar{x}^2_{j_{1_{2,\mathcal{X}}}}\ldots\bar{x}^2_{j_{|\bar{\omega}_{2,\mathcal{X}}|}}\bigg[\text{TUPLES-2-REL}_{\mathcal{X},\omega_{2,\mathcal{X}}}(\bar{x}^{2t},\bar{x}^2_{j_{1_{2,\mathcal{X}}}},\ldots,\bar{x}^2_{j_{|\bar{\omega}_{2,\mathcal{X}}|}})\Rightarrow$$

$$\exists \bar{x}^{1t}_{j_{1_{2,\mathcal{X}}}}\ldots\bar{x}^{1t}_{j_{|\bar{\omega}_{2,\mathcal{X}}|}}\bigg(\text{2-REL}_\mathcal{X}(\bar{x}^2_{j_{1_{2,\mathcal{X}}}},\bar{x}^{1t}_{j_{1_{2,\mathcal{X}}}})\wedge\ldots\wedge\text{2-REL}_\mathcal{X}(\bar{x}^2_{j_{|\bar{\omega}_{2,\mathcal{X}}|}},\bar{x}^{1t}_{j_{|\bar{\omega}_{2,\mathcal{X}}|}})\bigg)\bigg]\bigg]\wedge$$

"every SO relation in 2-REL$_\mathcal{X}$ is in some tuple in TUPLES-2-REL$_{\mathcal{X},\omega_{2,\mathcal{X}}}$":

$$\bigg[\forall\bar{x}^2\forall\bar{x}^{1t}\bigg[\text{2-REL}_\mathcal{X}(\bar{x}^2,\bar{x}^{1t})\Rightarrow\exists\bar{x}^{2t}\bigg(\bigvee_{\omega_{2,\mathcal{X}}}\exists\bar{x}^2_{j_{1_{2,\mathcal{X}}}}\ldots\bar{x}^2_{j_{|\bar{\omega}_{2,\mathcal{X}}|}}\big((\text{``}\bar{x}^2=\bar{x}^2_{j_{1_{2,\mathcal{X}}}}\text{''}\vee\ldots\vee$$

$$\text{``}\bar{x}^2=\bar{x}^2_{j_{|\bar{\omega}_{2,\mathcal{X}}|}}\text{''}\big)\wedge\text{TUPLES-2-REL}_{\mathcal{X},\omega_{2,\mathcal{X}}}(\bar{x}^{2t},\bar{x}^2_{j_{1_{2,\mathcal{X}}}},\ldots,\bar{x}^2_{j_{|\bar{\omega}_{2,\mathcal{X}}|}})\big)\bigg)\bigg]\bigg]\wedge$$

"in TUPLES-2-REL$_{\mathcal{X},\omega_{2,\mathcal{X}}}$ there are no two tuples of SO relations with the same id of SO relation tuple":

$$\bigg[\bigwedge_{\omega_{2,\mathcal{X}}}\forall\bar{x}^{2t}\neg\exists\bar{x}^2_{j_{1_{2,\mathcal{X}}}}\bar{x'}^2_{j_{1_{2,\mathcal{X}}}}\ldots\bar{x}^2_{j_{|\bar{\omega}_{2,\mathcal{X}}|}}\bar{x'}^2_{j_{|\bar{\omega}_{2,\mathcal{X}}|}}\big((\text{``}\bar{x}^2_{j_{1_{2,\mathcal{X}}}}\neq\bar{x'}^2_{j_{1_{2,\mathcal{X}}}}\text{''}\vee\ldots\vee$$

$$\text{``}\bar{x}^2_{j_{|\bar{\omega}_{2,\mathcal{X}}|}} \neq \bar{x}'^2_{j_{|\bar{\omega}_{2,\mathcal{X}}|}}\text{''}) \wedge \text{TUPLES-2-REL}_{\mathcal{X},\omega_{2,\mathcal{X}}}(\bar{x}^{2t}, \bar{x}^2_{j_{1_{2,\mathcal{X}}}}, \ldots, \bar{x}^2_{j_{|\bar{\omega}_{2,\mathcal{X}}|}}) \wedge$$

$$\text{TUPLES-2-REL}_{\mathcal{X},\omega_{2,\mathcal{X}}}(\bar{x}^{2t}, \bar{x}'^2_{j_{1_{2,\mathcal{X}}}}, \ldots, \bar{x}'^2_{j_{|\bar{\omega}_{2,\mathcal{X}}|}}))\Big] \wedge$$

"every tuple of SO relations in TUPLES-2-REL$_{\mathcal{X},\omega_{2,\mathcal{X}}}$ is in some TO relation in 3-REL$_\mathcal{X}$":

$$\left[\forall \bar{x}^{2t} \bigwedge_{\omega_{2,\mathcal{X}}} \forall \bar{x}^2_{j_{1_{2,\mathcal{X}}}} \ldots \bar{x}^2_{j_{|\bar{\omega}_{2,\mathcal{X}}|}} \Big[\text{TUPLES-2-REL}_{\mathcal{X},\omega_{2,\mathcal{X}}}(\bar{x}^{2t}, \bar{x}^2_{j_{1_{2,\mathcal{X}}}}, \ldots, \bar{x}^2_{j_{|\bar{\omega}_{2,\mathcal{X}}|}}) \Rightarrow\right.$$

$$\left.\exists \bar{x}^3 (\text{3-REL}_\mathcal{X}(\bar{x}^3, \bar{x}^{2t}))\Big]\right].$$

Tolerant Constraint-Preserving Snapshot Isolation: Extended Concurrency for Interactive Transactions

Stephen J. Hegner

Umeå University
Department of Computing Science
SE-901 87 Umeå Sweden
hegner@cs.umu.se
http://hegner.people.cs.umu.se

Abstract

In a database setting involving writers, there is a delicate balance between allowing sufficient concurrency for adequate performance and adequate isolation to prevent transactions from interfering with each other. This problem is particularly acute for interactive transactions; that is, ones which involve human input, since state-of-the-art approaches, such as serializable snapshot isolation (SSI), which rely upon an abort-and-restart strategy to resolve conflicts, do not provide a suitable solution. In this work, an extension of constraint-preserving snapshot isolation (CPSI) is provided. As does CPSI, this extension provides snapshot-isolation (SI) plus constraint preservation. By employing a model in which the values of data objects, and not just their identities, are used, it provides a significantly higher level of concurrency without sacrificing isolation than do approaches without such value modelling. In addition to the theory, an operational model of transaction execution is provided.

1 Introduction

Support for concurrent database transactions has long been recognized as a difficult problem. In the *ACID* characterization [6, pp. 166-167], [7, Sec. 1.1], transactions must run in *isolation*; that is, they must not interfere with one another. In practice, enforcing the theoretical ideal *view serializability* [13, Sec. 2.4] of isolation has proven difficult to enforce efficiently. Consequently, lower levels of isolation are commonly found in practice, including *snapshot isolation* (*SI*) at an intermediate level and *read committed* (*RC*) among the lowest levels, with the level of acceptability dependent upon the application.

Strategies for managing concurrency may be classified along a pessimistic-optimistic dimension. In a *pessimistic* policy, a transaction must wait until it may be given access to the resource(s) which it needs. In an *optimistic* strategy, a transaction is given access to a (copy of a) resource immediately, with conflicts resolved before the writes of the transaction become part of the persistent database. Roughly speaking, pessimistic policies use waiting to avoid problems, while optimistic policies require transactions to abort and restart. All real policies involve both pessimism and optimism, but the classification is nevertheless a useful one.

For the purposes of this paper, an *interactive transaction* is one which requires interactive human input at certain stages in order to continue. Business processes, in which humans authorize tasks and provide input, such as the allocation of funds for a business trip [8] [11], are prime examples. The problem of providing suitable isolation while supporting adequate concurrency is particularly acute for such transactions. Human input for transaction T may not be available at the time at which it is needed, or additional time may be needed in order to reach a decision before input can be provided. If a pessimistic strategy for concurrency is employed, then another transaction T' which needs resources which cannot be allocated until T commits or at least continues, must wait, perhaps for days, until T finally completes the interaction and proceeds with its execution. If T' is also interactive, such a long delay is likely unacceptable. On the other hand, if an optimistic strategy for concurrency is employed, then in the case of such a conflict, one of T and T' must be aborted and restarted. In many cases, that may not be suitable, since further input may be required, and even if not, the state of the database will likely have changed, and the inputs may have depended upon the state. In short, neither pessimistic nor optimistic policies are appropriate for interactive transactions. On the other hand, since human interaction is many orders of magnitude slower than computer operations, there is ample time to take more complex measures in order to minimize conflicts in the first place.

Virtually all mainstream work on transaction concurrency is based upon an *object-level* model, in which conflict between two transactions is characterized entirely with respect to access to data objects, without any regard to their current values or to how the transactions might alter those values. The prospects for supporting interactive transactions within such a framework are limited. Using a finer granularity for the data objects (for example, fields instead of tuples in a relational context) may help occasionally, but often conflicts between two transactions are inherent to atomic, indecomposable attributes, such as balances in accounts. The resulting conflicts must still be addressed, by waiting or else by aborting at least one of the transactions involved, neither of which is desirable.

To maintain an adequate level of isolation while minimizing waits and aborts, another approach is to employ a finer-grained notion of conflict. In a *value-level* model, not only the identity of a data item but also its current value, as well as how the transactions involved intend to change that value, may be taken into account. In this case, the transaction manager will be more complex, but the payoffs in terms of increased concurrency may be substantial. The idea is not new; it was proposed more than thirty years ago for long-running transactions in computer-aided design [1], with a more comprehensive theory along the same lines presented in [12]. Those works, however, are focused upon sets of nested transactions, assembled to realize a single goal via cooperation, and have not seen widespread use outside of that context.

The focus of this paper is the development of a simple value-level model for concurrency of transactions, called *tolerant constraint-preserving snapshot isolation*, or *TCPSI*. It builds upon *constraint-preserving snapshot isolation* (*CPSI*) [10], a framework based upon object-level modelling, which provides the isolation and efficiency of SI, together with a guarantee of constraint preservation[1]. Although not providing true serializability, the constraint preservation of CPSI avoids well-known anomalies such as write skew [2, A5B], [5, Ex. 2.2]. This is significant because, while lack of true serializability may be tolerable, violation of integrity constraints almost never is. Additionally, substantially fewer conflict situations arise with CPSI than with *serializable SI* (*SSI*) [4], [5]. TCPSI is a true extension of CPSI, in that all concurrency allowed under CPSI is also allowed under TCPSI.

To illustrate the main idea of this paper, consider the classical write-skew example [2, A5B], which illustrates how SI can fail to preserve constraints. Let $\mathbf{E}_0$ be a schema with two integer-valued data objects x_1 and x_2, related by the constraint $x_1 + x_2 > 0$. Let T_{01} be the transaction which reduces the value of x_1 by 1 if the result satisfies the constraint and otherwise does nothing; *i.e.*, if $(x_1 + x_2 > 1)$ then $x_1 \leftarrow x_1 - 1$; here the values of x_1 and x_2 seen by T_{01} are those found in the snapshot taken when T_{01} starts; after that point, it never sees updates performed by other transactions. Similarly, let T_{02} be the analogous transaction for x_2, *i.e.*, if $(x_1 + x_2 > 1)$ then $x_2 \leftarrow x_2 - 1$. Under SSI, and even under CPSI, these transactions are prevented from running concurrently, since an illegal state may arise under certain conditions. Indeed, if $x_1 = x_2 = 1$ initially, then each transaction may run successfully in isolation, but if they run concurrently under SI, the resulting state, with $x_1 = x_2 = 0$, is not legal. Each model identifies T_{01} as a writer of x_1 and a reader of x_2, as well as T_{02} a writer of x_2 and a reader of x_1. Thus, each transaction reads an object written by the other, indicating a potential conflict. However, that does not mean that there will be a conflict, only that there might be. Whether or not a violation of integrity occurs depends upon the initial state. Under TCPSI, for T_{01}, rather than prohibiting a concurrent transaction from writing x_2, a tolerance on the range of writes is stipulated. If the initial snapshot of T_{01} has $x_1 = x_2 = 2$, for

---

[1] Most modern relational DBMSs enforce all built-in constraints, such as primary-, secondary-, and foreign-key dependencies, internally, in real time, regardless of the level of isolation. However, this is not the case for *extended* constraints, defined using triggers or via application programs. CPSI guarantees the preservation of all constraints, including extended ones. See [10, Summary 2.4] for further explanation.

example, then T_{01} will tolerate updates to x_2, as long as the final value of x_2 is at least 0. Likewise, T_{02} will tolerate updates to x_1, as long as the final value of x_1 is at least 0. In that situation, T_{01} and T_{02} would be allowed to execute concurrently under TCPSI, although such concurrency would be blocked both under CPSI and under SSI.

For more than two concurrent transactions, additional issues arise. Let $\mathbf{E}_1$ be the extension of $\mathbf{E}_0$ to three data objects. More precisely, $\mathbf{E}_1$ has three integer-valued data objects x_1, x_2, and x_3, constrained by $x_1+x_2+x_3 > 0$. For $i \in \{1,2,3\}$, define the transaction T_{1i} by if $(x_1+x_2+x_3 > 1)$ then $x_i \leftarrow x_i - 1$. For an initial state with $x_1+x_2+x_3 = 3$ (e.g., M_{10} with $x_1 = x_2 = x_3 = 1$), it is easy to see that at most two of the three transactions in $\{T_{11}, T_{12}, T_{13}\}$ may execute concurrently without a constraint violation. For T_{11} (for example), the admissibility of its update for the initial state M_{10} requires that $x_2 + x_3 > 0$, a combined condition of the objects to be updated by T_{12} and T_{13}. To retain pairwise testing, the solution forwarded in this work is to require T_{11} to place separate conditions on x_2 and x_3 which imply that $x_2 + x_3 > 0$ remains true during its lifetime. For example, it may require that $x_2 \geq 0$ and $x_3 > 0$ (in which case T_{13} is blocked from concurrent execution while T_{12} may proceed), or it may choose that $x_2 > 0$ while $x_3 \geq 0$, (in which case T_{12} is blocked but T_{13} may proceed). In return for creating some false positives and thus reducing potential concurrency, a far simpler test for admissible concurrency results. For $k > 3$ a positive integer and $0 < m < k$, this form of example extends to to k transactions, of which at most k may execute concurrently; thus, no test which does not involve all transactions simultaneously is sufficient to guarantee concurrency without false positives. The reader who is curious about these details now is invited to look at 4.4 for a detailed example of the concurrency problems, at 4.10 for details of the proposed solution sketched above, and at 5.7 for a detailed operational example of how the approach works. Although some aspects will require reading other parts of the paper, it should be possible to grasp the main ideas with only the above example as background.

The paper is organized as follows. Section 2 provides the database framework used, while CPSI is reviewed in Sec. 3. Both sections are summaries of ideas developed in detail in [10]. In Sec. 4, the theory of TCPSI is developed in detail, while Sec. 5 provides an operational model of how the entire process proceeds. Finally, Sec. 6 contains conclusions and further directions.

2 The Database Framework

In this section, the database framework which is used throughout this paper is sketched. With few exceptions, the concepts are taken from [10], to which the reader is referred for details and examples. Although [10] is based upon the earlier paper [9], the frameworks differ substantially; the journal article [10] should in all cases be taken as the primary reference.

2.1 Notation $f(x)\downarrow$ indicates that the partial function f is defined on x. $f(x)\downarrow \in Y$ indicates that both $f(x)\downarrow$ and $f(x) \in Y$. $\mathbb{Z}$ denotes the set of integers. For $i, j \in \mathbb{Z}$, $[i, j] = \{n \in \mathbb{Z} \mid i \leq n \leq j\}$.

2.2 Data objects, constraints, and schemata A *data object* x is a mutable object; that is, an object whose value may be altered. A *simple data object* x is indecomposable; it is characterized by the set $\mathsf{States}\langle x \rangle$ of its *states*. A *compound data object* (or just *data object*) is a set $\mathbf{x}$ of simple data objects. A *database* over $\mathbf{x}$ is a function $M : \mathbf{x} \to \bigcup_{x \in \mathbf{x}} \mathsf{States}\langle x \rangle$ with the property that $M(x) \in \mathsf{States}\langle x \rangle$ for each $x \in \mathbf{x}$, with the set of all databases over $\mathbf{x}$ denoted $\mathsf{DB}(\mathbf{x})$. Put another way, M defines an $\mathbf{x}$ tuple $\mathbf{s} \in \prod_{x \in \mathbf{x}} \mathsf{States}\langle x \rangle$ of values, with $\pi_x(\mathbf{s}) = M(x)$.

An *unconstrained database schema* $\mathbf{d}$ is just a data object $\mathsf{DObj}\langle \mathbf{d} \rangle$. A *database* of $\mathbf{d}$ is a database over $\mathsf{DObj}\langle \mathbf{d} \rangle$. The set of all databases of $\mathbf{d}$ is denoted $\mathsf{DB}(\mathbf{d})$. Thus, $\mathsf{DB}(\mathbf{d})$ is shorthand for $\mathsf{DB}(\mathsf{DObj}\langle \mathbf{d} \rangle)$.

A *constrained database schema* is a triple $\mathbf{D} = \langle \mathsf{DObj}\langle \mathbf{D} \rangle, \mathsf{LDB}(\mathbf{D}), \mathsf{ELDB}(\mathbf{D}) \rangle$ in which $\mathsf{DObj}\langle \mathbf{D} \rangle$ is a set of data objects, $\mathsf{LDB}(\mathbf{D})$ is a subset of $\mathsf{DB}(\mathsf{DObj}\langle \mathbf{D} \rangle)$, the set of *legal databases* of $\mathbf{D}$, and $\mathsf{ELDB}(\mathbf{D})$ is a subset of $\mathsf{LDB}(\mathbf{D})$, the set of *extended legal databases*, or *x-legal databases*, of $\mathbf{D}$. Think of $\mathsf{DB}(\mathbf{D})$ as the set of all databases, regardless of constraints. The reason for the distinction between $\mathsf{LDB}(\mathbf{D})$ and $\mathsf{ELDB}(\mathbf{D})$ is that, as noted in the footnote of Sec. 1, most modern relational DBMSs enforce all built-in constraints. $\mathsf{LDB}(\mathbf{D})$ represents the databases which satisfy all such internal constraints, while $\mathsf{ELDB}(\mathbf{D})$ represents those databases which satisfy all constraints, including those defined by means such as triggers; for example, $x_1 + x_2 > 0$ of $\mathbf{E}_0$, described in Sec. 1. The work in this paper, as well as the earlier work on CPSI, [10], is concerned with isolation which preserves not only membership in $\mathsf{LDB}(\mathbf{D})$ but also in $\mathsf{ELDB}(\mathbf{D})$.

Define $\mathsf{SubObj}\langle \mathbf{D} \rangle = \{\mathbf{y} \mid \mathbf{y} \subseteq \mathsf{DObj}\langle \mathbf{D} \rangle\}$ to be the set of *subobjects* of $\mathbf{D}$. Let $\mathbf{x}, \mathbf{y} \in \mathsf{SubObj}\langle \mathbf{D} \rangle$. For $M \in \mathsf{DB}(\mathbf{x})$, the *restriction* of M to $\mathbf{y}$ is the database on $\mathbf{x} \cap \mathbf{y}$ defined by $M_{|\mathbf{x} \cap \mathbf{y}}$, the function M restricted to $\mathbf{x} \cap \mathbf{y}$. As a slight abuse of notation, this restriction will also be written as simply $M_{|\mathbf{y}}$, with the understanding that subobjects in $\mathbf{y}$ which do not apply to M (*i.e.*, which are not in $\mathbf{x}$ also) are ignored. For $\mathbf{M} \subseteq \mathsf{DB}(\mathbf{x})$, $\mathbf{M}_{|\mathbf{y}} = \{M_{|\mathbf{y}} \mid M \in \mathbf{M}\}$. The database schema $[\![\mathbf{D}|\mathbf{y}]\!] = \langle \mathbf{y}, \mathsf{LDB}\langle \mathbf{D}|\mathbf{y}\rangle, \mathsf{ELDB}\langle \mathbf{D}|\mathbf{y}\rangle\rangle$, in which $\mathsf{LDB}\langle \mathbf{D}|\mathbf{y}\rangle = \{M_{|\mathbf{y}} \mid M \in \mathsf{LDB}(\mathbf{D})\}$ and $\mathsf{ELDB}\langle \mathbf{D}|\mathbf{y}\rangle = \{M_{|\mathbf{y}} \mid M \in \mathsf{ELDB}(\mathbf{D})\}$.

A compound data object may be the empty set $\emptyset$. In that case $\mathsf{DB}(\emptyset)$ is a function on domain $\emptyset$. There is only one such function, so the empty database object has just one possible database, which will be denoted by Φ_{DB}. It is always the case that $\Phi_{\mathsf{DB}} \in \mathsf{ELDB}\langle \mathbf{D}|\emptyset\rangle$.

2.3 Notational convention Throughout the rest of this paper, unless stated specifically to the contrary, take $\mathbf{D} = \langle \mathsf{DObj}\langle \mathbf{D} \rangle, \mathsf{LDB}(\mathbf{D}), \mathsf{ELDB}(\mathbf{D}) \rangle$

to be a (constrained) database schema.

2.4 Updates and updateable objects A *(syntactic) update* on $\mathbf{D}$ is a pair $u = \langle u^{(1)}, u^{(2)} \rangle \in \mathsf{LDB}(\mathbf{D}) \times \mathsf{DB}(\mathbf{D})$. $u^{(1)}$ is the current or old state before the update; $u^{(2)}$ the new state afterwards. $\mathsf{SynUpdates}(\mathbf{D})$ denotes the set of all syntactic updates on $\mathbf{D}$. $u \in \mathsf{SynUpdates}(\mathbf{D})$ is *legal* if $u^{(2)} \in \mathsf{LDB}(\mathbf{D})$, with the set of all legal updates on $\mathbf{D}$ denoted $\mathsf{LUpdates}(\mathbf{D})$. It is *extended legal* (or *x-legal*) if $u^{(1)}, u^{(2)} \in \mathsf{ELDB}(\mathbf{D})$. The set of all extended legal updates on $\mathbf{D}$ is denoted $\mathsf{ELUpdates}(\mathbf{D})$. Note that $\mathsf{ELUpdates}(\mathbf{D}) \subseteq \mathsf{LUpdates}(\mathbf{D}) \subseteq \mathsf{SynUpdates}(\mathbf{D})$. $\mathbf{u} \subseteq \mathsf{SynUpdates}(\mathbf{D})$ is *complete* if for every $M \in \mathsf{LDB}(\mathbf{D})$, there is a $u \in \mathbf{u}$ with $u^{(1)} = M$. Define $u_1 \circ u_2 = \{(M_1, M_3) \mid (\exists M_2 \in \mathsf{LDB}(\mathbf{D}))((M_1, M_2) \in u_1 \wedge (M_2, M_3) \in u_2)\}$. The *identity update* on $\mathbf{x} \in \mathsf{DObj}\langle\mathbf{D}\rangle$ is $1_\mathbf{x} = \{(N, N) \mid N \in \mathsf{DB}(\mathbf{x})\}$.

For $u \in \mathsf{SynUpdates}(\mathbf{D})$, $\mathbf{y} \subseteq \mathsf{DObj}\langle\mathbf{D}\rangle$, define $u_{|\mathbf{y}} = \langle u^{(1)}_{|\mathbf{y}}, u^{(2)}_{|\mathbf{y}} \rangle$. For $\mathbf{u} \subseteq \mathsf{SynUpdates}(\mathbf{D})$, define $\mathbf{u}_{|\mathbf{y}} = \{u_{|\mathbf{y}} \mid u \in \mathbf{u}\}$. The *trimming* of $\mathbf{u}$ to $M \in \mathsf{LDB}(\mathbf{D})$ is $\mathsf{Trim}_M\langle\mathbf{u}\rangle = \{u \in \mathbf{u} \mid u^{(1)} = M\}$. If $\mathbf{x}, \mathbf{y} \subseteq \mathsf{DObj}\langle\mathbf{D}\rangle$, $M \in \mathsf{LDB}\langle\mathbf{D}|\mathbf{x}\rangle$, and $\mathbf{u} \subseteq \mathsf{SynUpdates}(\llbracket\mathbf{D}|\mathbf{y}\rrbracket)$, then $\mathsf{Trim}_M\langle\mathbf{u}\rangle$ is shorthand for $\mathsf{Trim}_{M_{|\mathbf{y}}}\langle\mathbf{u}\rangle = \mathsf{Trim}_{M_{|\mathbf{x} \cap \mathbf{y}}}\langle\mathbf{u}\rangle$. In general, for any $M \in \mathsf{LDB}(\mathbf{D})$, $\mathbf{u}(M)$ denotes $\{u^{(2)} \mid (u \in \mathbf{u}) \wedge (u^{(1)} = M)\} = \{u^{(2)} \mid u \in \mathsf{Trim}_M\langle\mathbf{u}\rangle\}$. Call $\mathbf{u} \subseteq \mathsf{SynUpdates}(\mathbf{D})$ *functional* if for every $M \in \mathsf{LDB}(\mathbf{D})$, $\mathbf{u}(M)$ contains at most one element, and for $\mathbf{M} \subseteq \mathsf{LDB}(\mathbf{D})$, define $\mathbf{u}(\mathbf{M}) = \{\mathbf{u}(M) \mid M \in \mathbf{M}\}$.

An *updateable object* over $\mathbf{D}$ is a pair $\langle \mathbf{c}, \mathbf{u} \rangle$ in which $\mathbf{c} \subseteq \mathsf{DObj}\langle\mathbf{D}\rangle$, and $\mathbf{u} \subseteq \mathsf{SynUpdates}(\llbracket\mathbf{D}|\mathbf{c}\rrbracket)$, with $\langle \mathbf{c}, \mathbf{u} \rangle$ *functional*, (resp. *complete*, resp. *legal*, resp. *x-legal*) precisely in the case that $\mathbf{u}$ has that property; it is called *singleton* if $\mathbf{u}$ consists of just one update. The updateable object $\langle \mathsf{DObj}\langle\mathbf{D}\rangle, \mathbf{u} \rangle$ is abbreviated to $\langle \mathbf{D}, \mathbf{u} \rangle$. For $\langle \mathbf{c}, \mathbf{u} \rangle$, $\mathbf{x} \subseteq \mathsf{DObj}\langle\mathbf{D}\rangle$ with $\mathbf{c} \subseteq \mathbf{x}$, and $M \in \mathsf{LDB}\langle\mathbf{D}|\mathbf{x}\rangle$, define $\mathsf{Trim}_M\langle\langle \mathbf{c}, \mathbf{u} \rangle\rangle = \langle \mathbf{c}, \mathsf{Trim}_M\langle\mathbf{u}\rangle \rangle$; and for $\mathbf{y} \subseteq \mathbf{c}$, define $\langle \mathbf{c}, \mathbf{u} \rangle_{|\mathbf{y}} = \langle \mathbf{y}, \mathbf{u}_{|\mathbf{y}} \rangle$.

2.5 Transactions A *black-box transaction* T over $\mathbf{D}$ is represented by an updateable object $\langle \mathbf{D}, \mathcal{U}_T \rangle$ which is functional, complete, and x-legal. The set of all black-box transactions over $\mathbf{D}$ is denoted $\mathsf{BBTrans}_\mathbf{D}$. The notation $\langle \mathbf{D}, \mathcal{U}_T \rangle$ will be used throughout the rest of this paper to denote the updateable object which underlies the transaction T.

2.6 The contexts of a transaction The contexts of a transaction are central to this work. Here, only brief illustration via example is provided. For a full discussion, see [10, Disc. 3.18].

Let $\mathbf{E}_2$ be the schema with four integer-valued data objects: x_1, x_2, y_1, and y_2, constrained by $x_1 + x_2 > 0$. Let T_{21} be the transaction defined by the rule if $(x_1 + x_2) - y_1 > 0$ then $x_1 \leftarrow x_1 - y_1$; in other words, the value of x_1 is reduced by the value of y_1, provided the result will satisfy the constraints. Otherwise, no update is performed. For M_{20}, defined by $(x_1 = 3, x_2 = 3, y_1 = 1, y_2 = 0)$, the *grounded write* of this update is $\{3 \stackrel{x_1}{\leadsto} 2\}$, representing the update on $\mathsf{ELDB}(\mathbf{E}_2)$ which changes x_1 from 3 to 2, leaving the other three data objects fixed. The *write context* is the set of all data objects which are written; in this case $\{x_1\}$. The grounded write specifies a change to the write context, without referring to other data objects. For M_{21}, defined by $(x_1 = 3, x_2 = -1, y_1 = 2, y_2 = 2)$, the ground write is $\emptyset$; *i.e.*, there is no change to the state, with the write context is $\emptyset$ as well. For full details on ground updates, see [10, Def. 3.9].

Continuing with the example, the *read context* is $\{x_2, y_1\}$, and the members of this set are further subclassified into the the *integrity context*, consisting of $\{x_2\}$, and the *grounding context*, consisting of $\{y_1\}$. The integrity context consists of those reads which are necessary to verify that the integrity constraints will be satisfied after the update; it will be formalized via the notion of guard in 3.4. The grounding context consists of those reads which are necessary to determine the grounded write. The grounding and integrity contexts need not be disjoint. Formally, the read and write contexts are always taken to be disjoint. If a data object is written by a transaction, its old value (before the update) is irrelevant to constraint satisfaction after the update, while the new value is automatically considered in checking integrity constraints.

2.7 Write sets and trimming The *write set* $\mathsf{WSet}\langle\langle\mathbf{c},\mathbf{u}\rangle\rangle$ of the updateable object $\langle\mathbf{c},\mathbf{u}\rangle$ is the set of all $y \in \mathbf{c}$ for which there is a legal $u \in \mathbf{u}$ which alters the state of y. Formally, $\mathsf{WSet}\langle\langle\mathbf{c},\mathbf{u}\rangle\rangle = \{y \in \mathbf{c} \mid (\forall x \in \mathbf{x})(\exists u \in \mathbf{u} \cap \mathsf{LUpdates}(\llbracket\mathbf{D}|\mathbf{c}\rrbracket))(u^{(1)}{}_{|\{x\}} \neq u^{(2)}{}_{|\{x\}})\}$. $\mathsf{WUpd}\langle\langle\mathbf{c},\mathbf{u}\rangle\rangle = \mathbf{u}_{|\mathsf{WSet}\langle\langle\mathbf{c},\mathbf{u}\rangle\rangle}$ is the set of *write updates* of $\langle\mathbf{c},\mathbf{u}\rangle$, while $\mathsf{WObj}\langle\langle\mathbf{c},\mathbf{u}\rangle\rangle = \langle\mathsf{WSet}\langle\langle\mathbf{c},\mathbf{u}\rangle\rangle, \mathsf{WUpd}\langle\langle\mathbf{c},\mathbf{u}\rangle\rangle\rangle$ is its set of *write objects*. $\langle\mathbf{c},\mathbf{u}\rangle$ is a *full write object* if $\langle\mathbf{c},\mathbf{u}\rangle = \mathsf{WObj}\langle\langle\mathbf{c},\mathbf{u}\rangle\rangle$.

For $M \in \mathsf{LDB}(\mathbf{D})$, the *write trim* of $\langle\mathbf{c},\mathbf{u}\rangle$ to M is $\mathsf{WTrim}_M\langle\langle\mathbf{c},\mathbf{u}\rangle\rangle = \langle\mathbf{c},\mathbf{u}\rangle_{|\mathsf{WSet}\langle\mathsf{Trim}_M\langle\langle\mathbf{c},\mathbf{u}\rangle\rangle\rangle}$. $\mathsf{WTrim}_M\langle\langle\mathbf{c},\mathbf{u}\rangle\rangle$ is a full write object, and if $\langle\mathbf{c},\mathbf{u}\rangle$ is functional and complete, then $\mathsf{WUpd}\langle\langle\mathbf{c},\mathbf{u}\rangle\rangle$ consists of exactly one update. If $\mathbf{c} = \emptyset$, that update must be $\langle\Phi_{\mathsf{DB}}, \Phi_{\mathsf{DB}}\rangle$.

For a transaction T whose initial snapshot is $M \in \mathsf{LDB}(\mathbf{D})$, the update which it performs is represented by the (single) update of the updateable object $\mathsf{WTrim}_M\langle\langle\mathbf{D}, \mathcal{U}_T\rangle\rangle$. Data objects not in $\mathsf{WTrim}_M\langle\langle\mathbf{D}, \mathcal{U}_T\rangle\rangle$ are left unchanged.

2.8 Lifting of updates Let $\langle\mathbf{c},\mathbf{u}\rangle$ be an updateable object on $\mathbf{D}$, and let $\mathbf{x} \subseteq \mathsf{DObj}\langle\mathbf{D}\rangle$ with $\mathbf{c} \subseteq \mathbf{x}$. $\langle\mathbf{c},\mathbf{u}\rangle$ may be *lifted* to an update $\mathbf{x}$ by requiring that it be the identity on all data objects in $\mathbf{c} \setminus \mathbf{x}$. Formally, $\mathsf{Lift}_{\llbracket\mathbf{D}|\mathbf{x}\rrbracket}\langle\langle\mathbf{c},\mathbf{u}\rangle\rangle = \{v \in \mathsf{SynUpdates}(\llbracket\mathbf{D}|\mathbf{x}\rrbracket) \mid (v_{|\mathbf{c}} \in \mathbf{u})$ and $(v^{(1)}{}_{|\mathbf{x}\setminus\mathbf{c}} = v^{(2)}{}_{|\mathbf{x}\setminus\mathbf{c}})\}$ is the *lifting* of $\langle\mathbf{c},\mathbf{u}\rangle$ (or just of $\mathbf{u}$) from $\llbracket\mathbf{D}|\mathbf{c}\rrbracket$ to $\llbracket\mathbf{D}|\mathbf{x}\rrbracket$. If $\langle\mathbf{c},\mathbf{u}\rangle$ is functional, then so too is $\mathsf{Lift}_{\llbracket\mathbf{D}|\mathbf{x}\rrbracket}\langle\langle\mathbf{c},\mathbf{u}\rangle\rangle$. In that case, $\mathsf{FLift}_{\llbracket\mathbf{D}|\mathbf{x}\rrbracket}\langle\langle\mathbf{c},\mathbf{u}\rangle\rangle : \mathsf{LDB}\langle\mathbf{D}|\mathbf{x}\rangle \to \mathsf{DB}(\mathbf{x})$ is the partial function defined by $M \mapsto M'$ if $\langle M, M'\rangle \in \mathsf{Lift}_{\llbracket\mathbf{D}|\mathbf{c}\rrbracket}\langle\langle\mathbf{c},\mathbf{u}\rangle\rangle$ and is undefined otherwise. If $\mathbf{u}$ consists of a single update, then $\langle\mathbf{c},\mathbf{u}\rangle$ is functional, so $\mathsf{Lift}_{\llbracket\mathbf{D}|\mathbf{x}\rrbracket}\langle\langle\mathbf{c},\mathbf{u}\rangle\rangle$ is functional as well. When $x = \mathsf{DObj}\langle\mathbf{D}\rangle$, a simpler notation is used: $\mathsf{Lift}_{[\mathbf{D}|\mathsf{DObj}\langle\mathbf{D}\rangle]}\langle\langle\mathbf{c},\mathbf{u}\rangle\rangle$ is abbreviated to $\mathsf{Lift}_\mathbf{D}\langle\langle\mathbf{c},\mathbf{u}\rangle\rangle$, and $\mathsf{FLift}_{[\mathbf{D}|\mathsf{DObj}\langle\mathbf{D}\rangle]}\langle\langle\mathbf{c},\mathbf{u}\rangle\rangle$ is abbreviated to $\mathsf{FLift}_\mathbf{D}\langle\langle\mathbf{c},\mathbf{u}\rangle\rangle$.

3 SI and CPSI

Tolerant CPSI is an extension of ordinary CPSI; thus, it is first necessary to present the core ideas of the latter in a manner which is precise enough to allow their application and extension to the former. Due to space limitations, the presentation is abbreviated; for full details, the reader is referred to [10].

3.1 Transaction Schedules and Snapshot Isolation

Both ordinary and tolerant CPSI are built upon *snapshot isolation* (*SI*) [14, 12.5], [15, Sec. 10.6.2], a commonly used level of isolation in modern DBMSs which employ *multiversion concurrency control* (*MVCC*) [3, Ch. 5], [14, Ch. 12], [15, Ch. 5]. Under SI, each transaction T receives, when it starts, a private copy of the database, called its *snapshot*. All operations by T, both reads and writes, during its duration, are on this private copy. T is allowed to commit its writes to the *global database* (the version of the database consisting of committed values, which is visible to other transactions) only if none of them writes a data object which another, concurrent transaction has already written.[2]

Each transaction T has a start time and an end time. Actual times are not important; rather, it is only the order of events which is significant. Let $\mathbf{T}$ be a finite set of transactions, and let $\mathsf{SCSet}\langle\mathbf{T}\rangle = \{T^s \mid T \in \mathbf{T}\} \cup \{T^c \mid T \in \mathbf{T}\}$, with the members of $\mathsf{SCSet}\langle\mathbf{T}\rangle$ just symbols. An *SI-schedule* on $\mathbf{T}$ is a total order $\leq_\mathbf{T}$ on $\mathsf{SCSet}\langle T\rangle$ with the property that for each $T \in \mathbf{T}$, $T^s <_\mathbf{T} T^c$. ($x <_\mathbf{T} y$ denotes that $x \leq_\mathbf{T} y$ but $x \neq y$.) The schedule indicates the temporal order of events, with T^s (resp. T^c) representing the relative start time (resp. commit time) of T. For $T_1, T_2 \in \mathbf{T}$, if neither $T_1^c <_\mathbf{T} T_2^s$ nor $T_2^c <_\mathbf{T} T_1^s$ holds, then T_1 and T_2 *execute concurrently* and $\{T_1, T_2\}$ forms a *concurrent pair*.

Fix an SI-schedule $\leq_\mathbf{T}$ for $\mathbf{T} \subseteq \mathsf{BBTrans}_\mathbf{D}$ and let $M \in \mathsf{ELDB}(\mathbf{D})$. As the transactions in $\mathbf{T}$ execute according to $\leq_\mathbf{T}$, the global database is updated by the transactions, as they commit. Given $M \in \mathsf{ELDB}(\mathbf{D})$ as the initial global database, (*i.e.*, before any transaction in $\mathbf{T}$ commits), there are three classes of values of interest for the global database which occur as the transactions run. $\mathsf{InitSnap}_{\langle\leq_\mathbf{T};M\rangle}\langle T\rangle$ is the initial snapshot of T, the database which T reads at the beginning of its execution. Observe that $\mathsf{InitSnap}_{\langle\leq_\mathbf{T};M\rangle}\langle T\rangle = M$ iff no transaction of $\mathbf{T}$ commits before T starts. $\mathsf{BeforeCmt}_{\langle\leq_\mathbf{T};M\rangle}\langle T\rangle$ is the state of the global database immediately before T commits. It may differ from $\mathsf{InitSnap}_{\langle\leq_\mathbf{T};M\rangle}\langle T\rangle$

---

[2] In practice, some details may differ; in particular, internal constraints are always enforced immediately, and checks for concurrent writes to the same data object may be made earlier. However, those details do not affect the theory developed here materially. See [10, Summaries 2.2-2.4] for details.

because other transactions, running concurrently with T, may have committed before T and made changes to the global database. $\mathsf{AfterCmt}_{\langle \leq_\mathbf{T} : M \rangle}\langle T \rangle$ is the state of the global database immediately after T commits. It differs from $\mathsf{BeforeCmt}_{\langle \leq_\mathbf{T} : M \rangle}\langle T \rangle$ to the extent that T made changes to the global database upon its commit. It suffices to model only transactions which commit, since those which do not commit have no effect upon the global database, and so can simply be removed. Call $\leq_\mathbf{T}$ *constraint preserving for (initial state)* $M \in \mathsf{ELDB}(\mathbf{D})$ if for every $T \in \mathbf{T}$, $\mathsf{AfterCmt}_{\langle \leq_\mathbf{T} : M \rangle}\langle T \rangle \in \mathsf{ELDB}(\mathbf{D})$. In other words, *constraint preservation*, as used in the remainder of this paper, entails preservation of both internal and external constraints, as distinguished in 2.2.

For a more comprehensive presentation of SI, see [10, Sec. 2 and 4].

3.2   Notational convention In 3.3 and 3.4, take $\mathbf{T}$ to be a finite subset of $\mathsf{BBTrans}_\mathbf{D}$ and $\leq_\mathbf{T}$ an SI-schedule for $\mathbf{T}$.

3.3   State assignment Let $\mathbf{T}' \subseteq \mathbf{T}$. A *state assignment* for $\mathbf{T}'$ is a function $\iota : \mathbf{T}' \to \mathsf{LDB}(\mathbf{D})$. The most important use is the *state assignment for* $\leq_\mathbf{T}$ *with initial state* $M \in \mathsf{LDB}(\mathbf{D})$, the function $\mathsf{StAssign}_{\langle \leq_\mathbf{T} : M \rangle} : \mathbf{T} \to \mathsf{LDB}(\mathbf{D})$ given on elements by $T \mapsto \mathsf{InitSnap}_{\langle \leq_\mathbf{T} : M \rangle}\langle T \rangle$, identifying the initial snapshot of each $T \in \mathbf{T}$, when the entire schedule begins with database state M. For $\mathbf{T}' \subseteq \mathbf{T}$, $\mathsf{StAssign}_{\langle \leq_\mathbf{T} : M \rangle}^{|\mathbf{T}'} : \mathbf{T}' \to \mathsf{LDB}(\mathbf{D})$ is the function $\mathsf{StAssign}_{\langle \leq_\mathbf{T} : M \rangle}$ restricted to $\mathbf{T}'$. In particular, for $T_i, T_j \in \mathbf{T}$, $\mathsf{StAssign}_{\langle \leq_\mathbf{T} : M \rangle}^{|\{T_i, T_j\}}$ is the function $\mathsf{StAssign}_{\langle \leq_\mathbf{T} : M \rangle}$ restricted to $\{T_i, T_j\}$. When working with a general state assignment ι, think of it as a (possible) restriction of $\mathsf{StAssign}_{\langle \leq_\mathbf{T} : M \rangle}$ for an SI-schedule $\leq_\mathbf{T}$ with initial state M.

The state assignment $\iota : \mathbf{T}' \to \mathsf{LDB}(\mathbf{D})$ is nonoverlapping if none of its transactions writes a common data object (as required for concurrent transactions under SI). Formally, ι is *nonoverlapping* if for every $T_1, T_2 \in \mathbf{T}'$ with $T_1 \neq T_2$, $\mathsf{WSet}\langle \mathsf{Trim}_{\iota(T_1)}\langle \langle \mathbf{D}, \mathcal{U}_{T_1} \rangle \rangle \rangle \cap \mathsf{WSet}\langle \mathsf{Trim}_{\iota(T_2)}\langle \langle \mathbf{D}, \mathcal{U}_{T_2} \rangle \rangle \rangle = \emptyset$.

Given $M \in \mathsf{ELDB}(\mathbf{D})$, ι is *extendedly legal* (or *x-legal*) for M if, for every $T \in \mathbf{T}'$, $\mathsf{FLift}_\mathbf{D}\langle \mathsf{WTrim}_{\iota(T)}\langle \langle \mathbf{D}, \mathcal{U}_T \rangle \rangle \rangle(M)\!\downarrow \in \mathsf{ELDB}(\mathbf{D})$.

3.4   Guards and guard functions Database constraints are frequently quite localized in nature. If a data object $\mathbf{c}$ is to be written, only a small subset of the remaining data objects $\mathbf{y}$ must be checked to determine whether or not that write will preserve the integrity constraints. The notion of a guard object formalizes this. Let $\langle \mathbf{c}, \{u\} \rangle$ be a singleton full write object over $\mathbf{D}$; that is, a full write object with just one update. A *guard object* for $\langle \mathbf{c}, \{u\} \rangle$ is a $\mathbf{y} \in \mathsf{DObj}\langle \mathbf{D} \rangle$ which satisfies the following two properties.

(go-i) $\mathbf{y} \cap \mathbf{c} = \emptyset$.

(go-ii) For every $M \in \mathsf{ELDB}(\mathbf{D})$ with $M_{|\mathbf{c}} = u^{(1)}$,
$\mathsf{FLift}_{\mathbf{D}} \langle\langle \mathbf{c}, \{u\}\rangle\rangle (M) \downarrow \in \mathsf{ELDB}(\mathbf{D}) \Leftrightarrow$
$\mathsf{FLift}_{[\![\mathbf{D}|\mathbf{y}\cup\mathbf{c}]\!]} \langle\langle \mathbf{c}, \{u\}\rangle\rangle (M_{|\mathbf{y}\cup\mathbf{c}}) \downarrow \in \mathsf{ELDB}([\![\mathbf{D}|\mathbf{y}\cup\mathbf{c}]\!])$.

Condition (go-ii) states that the update $\langle \mathbf{c}, \{u\}\rangle$ is x-legal when lifted to all of $\mathbf{D}$ iff it is x-legal when lifted to just $[\![\mathbf{D}|\mathbf{y}\cup\mathbf{c}]\!]$, Thus, a global test for constraint satisfaction may be replaced by a much more localized one.

A guard function for a transaction provides a guard object for the write defined by each snapshot. Formally, a *guard function* for a transaction T is a $g : \mathsf{LDB}(\mathbf{D}) \to \mathsf{SubObj}\langle \mathbf{D}\rangle$ which assigns to each $N \in \mathsf{ELDB}(\mathbf{D})$ a guard object $g(N)$ for $\mathsf{WTrim}_N \langle\langle \mathbf{D}, \mathcal{U}_T\rangle\rangle$, subject to the additional condition that the guard depends only upon the update, and not the initial snapshot which induced it:
$(\forall N_1, N_2 \in \mathsf{ELDB}(\mathbf{D}))((\mathsf{WTrim}_{N_1} \langle\langle \mathbf{D}, \mathcal{U}_T\rangle\rangle = \mathsf{WTrim}_{N_2} \langle\langle \mathbf{D}, \mathcal{U}_T\rangle\rangle)$
$$\Rightarrow (g(N_1) = g(N_2))).$$
Given $N \in \mathsf{ELDB}(\mathbf{D})$, $g(N)$ is called the *guard object* for N. The set of all guard functions for T is denoted $\mathsf{Guards}_{\mathbf{D}}\langle T\rangle$. This set is always nonempty; i.e., a guard function always exists. See [10, 5.10] for details.

A *guarded black-box transaction* T is represented by a pair $\langle\langle \mathbf{D}, \mathcal{U}_T\rangle, \mathcal{G}_T\rangle$ in which $\langle \mathbf{D}, \mathcal{U}_T\rangle \in \mathsf{FUpdObj}(\mathbf{D})$ and $\mathcal{G}_T \in \mathsf{Guards}_{\mathbf{D}}\langle T\rangle$, with T represented by $\langle \mathbf{D}, \mathcal{U}_T\rangle$ in the latter. The set of all guarded black-box transactions over $\mathbf{D}$ is denoted $\mathsf{GBBTrans}_{\mathbf{D}}$.

The notion of a guard is closely related to, but not the same as, the idea of integrity context, as presented in 2.6. Consider again the schema $\mathbf{E}_2$ and the transaction T_{21} of 2.6. For snapshot M_{20}, the guard object is exactly the integrity context $\{x_2\}$. However, for snapshot M_{21}, although the integrity context is also $\{x_2\}$, the guard is $\emptyset$. The difference is that while the integrity context is used to determine whether or not a candidate ground write will preserve the integrity constraints (*i.e.*, whether that update would satisfy the *consistency* property of ACID [6, p. 166]), the purpose of the guard is to identify a read set which must be protected from change in order to ensure *isolation* of ACID, once a consistent ground write has been selected. For an extensive set of examples of guards, see [10, 5.11-5.14].

3.5 Notational convention From now on, unless stated explicitly to the contrary, augment 3.2 so that $\mathbf{T}$ is taken to be a finite subset of $\mathsf{GBBTrans}_{\mathbf{D}}$, and not just of $\mathsf{BBTrans}_{\mathbf{D}}$. In other words, assume that every transaction in $\mathbf{T}$ has a guard function associated with it. Furthermore, as explained in 3.4 above, the guard function of $T \in \mathbf{T}$ will be denoted $\mathcal{G}_T$, while $<_{\mathbf{T}}$ continues to be an SI-schedule for $\mathbf{T}$.

3.6 Independent pairs of guarded transactions A pair of nonoverlapping concurrent transactions is guard independent if at least one does not write the guard of the other. Formally, given $\{T_1, T_2\} \subseteq \mathsf{GBBTrans}_{\mathbf{D}}$, a state assignment $\iota : \{T_1, T_2\} \to$ ELDB(**D**) is *guard independent* if it is nonoverlapping and at least one of WSet$\langle$Trim$_{\iota(T_1)}\langle\langle\mathbf{D}, \mathcal{U}_{T_1}\rangle\rangle\rangle \cap \mathcal{G}_{T_2}(\iota(T_2)) = \emptyset$ or WSet$\langle$Trim$_{\iota(T_2)}\langle\langle\mathbf{D}, \mathcal{U}_{T_2}\rangle\rangle\rangle \cap \mathcal{G}_{T_1}(\iota(T_1)) = \emptyset$ holds, See [10, Thm. 5.17] for a proof of the following result, which is central to the results of Sec. 4.

3.7 Theorem — guard independence ⇒ constraint preservation
Let $M \in$ ELDB(**D**). *If* StAssign$_{\langle\leq_{\mathbf{T}} : M\rangle}^{|\{T, T'\}|}$ *is guard independent for every concurrent pair* $\{T, T'\}$ *of* **T***, then* $\leq_{\mathbf{T}}$ *is constraint preserving for initial state* M. □

3.8 Guard-write dependencies and CPSI
Constraint-preserving snapshot isolation, or *CPSI*, is the application of guard independence to establish that a schedule of transactions, run under SI, will not result in any constraint violations, internal or external. In other words, under CPSI, for the schedule $\leq_{\mathbf{T}}$ to be constraint preserving, every pair $\{T_1, T_2\}$ of distinct concurrent transactions must satisfy one of the disjointness conditions identified in 3.6.

This may be expressed in another way; let $T_1, T_2 \in \mathbf{T}$. There is a *gw-dependency* from T_1 to T_2 for $\leq_{\mathbf{T}}$ with initial state $M \in$ ELDB(**D**), written $T_1 \xrightarrow{\mathsf{gw}} T_2$, if T_2 writes the guard of T_1; that is, if
$$\mathsf{WSet}\langle\mathsf{Trim}_{\mathsf{InitSnap}_{\langle\leq_{\mathbf{T}} : M\rangle}\langle T_2\rangle}\langle\langle\mathbf{D}, \mathcal{U}_{T_2}\rangle\rangle\rangle \cap \mathcal{G}_{T_1}(\mathsf{InitSnap}_{\langle\leq_{\mathbf{T}} : M\rangle}\langle T_1\rangle) \neq \emptyset.$$
$\{T_1, T_2\}$ is a *guard-write pair*, or *gw-pair*, in GDSG$\langle\leq_{\mathbf{T}} : M\rangle$ if it forms a concurrent pair for which both $T_1 \xrightarrow{\mathsf{gw}} T_2$ and $T_2 \xrightarrow{\mathsf{gw}} T_1$ hold. The result of 3.7 may be restated to say that if $\leq_{\mathbf{T}}$ does not contain any guard-write pairs, then it is constraint preserving. For extensive examples surrounding CPSI, including in particular ones for which SSI flags conflict but CPSI does not, see [10, Examples 5.18 and 5.22]. For implementation issues, see [10, Disc. 5.25].

3.9 Examples of guard pairs and gw-dependency
Consider again the transactions T_{01} and T_{02} on the schema $\mathbf{E}_0$, introduced in Sec. 1. Let $n_1, n_2 \in \mathbb{Z}$ with $n_1 + n_2 > 0$, and let $M_{0\langle n_1, n_2\rangle} \in$ ELDB($\mathbf{E}_0$) be the database with $x_1 = n_1$ and $x_2 = n_2$. It is easy to see that, for $i \in \{1, 2\}$, $\mathcal{G}_{T_i}(M_{0\langle n_1, n_2\rangle}) = \{x_{3-i}\}$ if the update $x_i \leftarrow x_i - 1$ is allowed; i.e., if $n_1 + n_2 > 1$. If $n_1 + n_2 \leq 1$, the update would violate the integrity constraint $x_1 + x_2 > 0$, so the transaction executes the identify update instead, and $\mathcal{G}_{T_i}(M) = \emptyset$. Thus, under ordinary CPSI, T_{01} and T_{02} may not execute concurrently on the same initial snapshot $M_{0\langle n_1, n_2\rangle}$ with $n_1 + n_2 > 1$, since the gw-dependencies $T_{01} \xrightarrow{\mathsf{gw}} T_{02}$ and $T_{02} \xrightarrow{\mathsf{gw}} T_{01}$ both hold, identifying a gw-dependency (see 3.8). Nevertheless, it is clear that if $n_1 + n_2 > 2$, the two may execute concurrently, on the same initial snapshot, with no integrity violation. In order to permit such concurrent execution, the guards need to be made *tolerant*, as developed in the next section.

4 Tolerant CPSI

In this section, the main ideas of tolerant CPSI are developed, as an extension of the ideas of CPSI outlined in Sec. 3.

4.1 Tolerant guard pairs, functions, and transactions Under CPSI, given two concurrent transactions, at least one is not allowed to write the guard of the other. Under tolerant CPSI, on the other hand, for each transaction, a set of allowable writes to its guard is specified. A concurrent transaction is allowed to write the guard, provided those writes lie within the specification. Formally, a *tolerant guard function* h for $T \in \mathsf{BBTrans_D}$ assigns to each $N \in \mathsf{ELDB}(\mathbf{D})$ a pair $\langle \mathsf{GObj}_h(N), \mathsf{GTol}_h(N)\rangle$, with the following properties.

(tgp-i) GObj_h is a guard function for T; *i.e.*, $\mathsf{GObj}_h(N)$ is a guard object for $\mathsf{WTrim}_N\langle\langle \mathbf{D}, \mathcal{U}_T\rangle\rangle$.

(tgp-ii) $\mathsf{GTol}_h(N) \subseteq \mathsf{DB}(\mathsf{GObj}_h(N))$ with $N_{|\mathsf{GObj}_h(N)} \in \mathsf{GTol}_h(N)$.

(tgp-iii) $((N_{|\mathsf{WSet}\langle \mathsf{WTrim}_N\langle\langle \mathbf{D}, \mathcal{U}_T\rangle\rangle\rangle} = \mathsf{WTrim}_N\langle\langle \mathbf{D}, \mathcal{U}_T\rangle\rangle^{(1)})$
$\wedge\, ((\mathsf{WTrim}_N\langle\langle \mathbf{D}, \mathcal{U}_T\rangle\rangle^{(2)})_{|\mathsf{GObj}_h(N)} \in \mathsf{GTol}_h(N))$
$\Rightarrow (\mathsf{FLift}_{\mathsf{GObj}_h(N) \cup \mathsf{WSet}\langle\langle \mathbf{D},\mathcal{U}_T\rangle\rangle}\langle\langle \mathbf{D}, \mathcal{U}_T\rangle\rangle(N)\!\downarrow$
$\in \mathsf{ELDB}([\![\mathbf{D}|(\mathsf{GObj}_h(N) \cup \mathsf{WSet}\langle\langle \mathbf{D}, \mathcal{U}_T\rangle\rangle)]\!]))$.

(tgp-iv) $(\forall N_1, N_2 \in \mathsf{ELDB}(\mathbf{D}))((\mathsf{WTrim}_{N_1}\langle\langle \mathbf{D}, \mathcal{U}_T\rangle\rangle = \mathsf{WTrim}_{N_2}\langle\langle \mathbf{D}, \mathcal{U}_T\rangle\rangle)$
$\Rightarrow (h(N_1) = h(N_2)))$.

GObj_h is called the *guard-object assignment*, while GTol_h is called the *tolerance assignment*. For a fixed $N \in \mathsf{ELDB}(\mathbf{D})$, $\mathsf{GObj}_h(N)$ is called the *guard object* and $\mathsf{GTol}_h(N)$ is called the *tolerance* (of h) for N. As a slight abuse of notation, h may be written as $\langle \mathsf{GObj}_h, \mathsf{GTol}_h\rangle$.

Condition (tgp-i) identifies GObj_h as the associated guard function. Condition (tgp-ii) asserts that GTol_h assigns to each $N \in \mathsf{ELDB}(\mathbf{D})$ a set of databases of the guard object which includes in particular the projection of N onto the guard object. Condition (tgp-iii) liberalizes (go-ii) of 3.4. Instead of requiring that a concurrent transaction T' not write the guard object at all, the the tolerance identifies a range of values, within which a write of T' may lie. Since $N_{|\mathsf{GObj}_h(N)} \in \mathsf{GTol}_h(N)$, no change to the guard state is allowed, recapturing the requirement of an ordinary guard function. In view of (go-ii) and the fact that GObj_h is a guard function in the sense of 3.4, (tgp-iii) is equivalent to the following, simpler assertion.

(tgp-iii′) $((N_{|\mathsf{WSet}\langle \mathsf{WTrim}_N\langle\langle \mathbf{D},\mathcal{U}_T\rangle\rangle\rangle} = \mathsf{WTrim}_N\langle\langle \mathbf{D}, \mathcal{U}_T\rangle\rangle^{(1)})$
$\wedge\, ((\mathsf{WTrim}_N\langle\langle \mathbf{D}, \mathcal{U}_T\rangle\rangle^{(2)})_{|\mathsf{GObj}_h(N)} \in \mathsf{GTol}_h(N))$
$\Rightarrow (\mathsf{FLift}_\mathbf{D}\langle\langle \mathbf{D}, \mathcal{U}_T\rangle\rangle(N)\!\downarrow \in \mathsf{ELDB}(\mathbf{D}))$.

Finally, (tgp-iv) corresponds to the similar condition of 3.4; the tolerant guard depends only upon the ground update, not upon how it was obtained.

It might seem that $\mathsf{GTol}_h(N)$ should be limited to databases in $\mathsf{ELDB}(\mathsf{GObj}_h(N))$, but it is harmless to allow those which do not satisfy the constraints, and that flexibility will prove to be of use.

The tolerant guard function h is *zero tolerance* if $\mathsf{GTol}_h(N) = \{N_{|\mathsf{GObj}_h(N)}\}$ for every $N \in \mathsf{ELDB}(\mathbf{D})$. In that case, h is effectively just an ordinary guard function, as defined in 3.4, since GTol_h has no additional effect.

The set of all tolerant guard functions for T is denoted $\mathsf{TolGuards}_\mathbf{D}\langle T\rangle$. A *tolerantly guarded black-box transaction* T is a pair $\langle\langle \mathbf{D}, \mathcal{U}_T\rangle, \mathcal{H}_T\rangle$ in which $\langle \mathbf{D}, \mathcal{U}_T\rangle \in \mathsf{FUpdObj}(\mathbf{D})$ and $\mathcal{H}_T \in \mathsf{TolGuards}_\mathbf{D}\langle T\rangle$. As a notational convenience, let $\mathcal{H}_T = \langle \mathcal{H}_T^{\mathsf{GFn}}, \mathcal{H}_T^{\mathsf{Tol}}\rangle$. In other words, $\mathcal{H}_T^{\mathsf{GFn}}$ is the function which assigns guard objects, while $\mathcal{H}_T^{\mathsf{Tol}}$ assigns tolerances. The set of all tolerantly guarded black-box transactions over $\mathbf{D}$ is denoted $\mathsf{TolGBBTrans}_\mathbf{D}$.

4.2 Examples of tolerant guard pairs and functions Return to the context of $\mathbf{E}_2$, particularly as developed in 3.9. For $i \in \{1,2\}$, the ordinary guard function $\mathcal{G}_{T_i}$ is renamed $\mathcal{H}_{T_i}^{\mathsf{GFn}}$ in the tolerant setting. To obtain a tolerant guard which permits as much concurrency as possible, for any $n_1, n_2 \in \mathbb{Z}$, define $\mathcal{H}_{T_{2i}}^{\mathsf{Tol}}(M_{0\langle n_1,n_2\rangle}) = \{N \in \mathsf{DB}(x_{3-i}) \mid x_{3-i} > 1 - n_i\}$ if $n_1 + n_2 > 1$, with $\mathcal{H}_{T_{2i}}^{\mathsf{Tol}}(M_{0\langle n_1,n_2\rangle}) = \{\Phi_{\mathsf{DB}}\}$ otherwise. For example, $\mathcal{H}_{T_{21}}^{\mathsf{Tol}}(M_{0\langle 3,2\rangle}) = \{N \in \mathsf{DB}(x_2) \mid x_2 > -2\}$, and $\mathcal{H}_{T_{22}}^{\mathsf{Tol}}(M_{0\langle 3,2\rangle}) = \{N \in \mathsf{DB}(x_1) \mid x_1 > -1\}$. In that case, T_{21} and T_{22} are each tolerant of the other for the state assignment of $M_{0\langle n_1,n_2\rangle}$ to each, and so they may execute concurrently. Indeed, T_{21} sets x_1 to 2, which is within the tolerance range of T_{22}, and T_{22} sets x_2 to 1, which is within the tolerance range of T_{21}.

To recapture an ordinary guard function within the tolerant framework, a tolerant guard of zero tolerance is used. Specifically, for any $n_1, n_2 \in \mathbb{Z}$, if $n_1 + n_2 > 1$, for $i \in \{1,2\}$ define $\mathcal{H}_{T_{2i}}^{\mathsf{Tol}}(M_{0\langle n_1,n_2\rangle}) = \{(M_{0\langle n_1,n_2\rangle})_{|x_{3-i}}\}$. If $n_1 + n_2 \leq 1$, $\mathcal{H}_{T_{2i}}^{\mathsf{Tol}}(M_{0\langle n_1,n_2\rangle}) = \{\Phi_{\mathsf{DB}}\}$ The effect of this tolerant guard is exactly the same as that of the ordinary guard, since no change of value of x_{3-i} is tolerated by T_{2i}.

4.3 Tolerance among transactions Let $\mathbf{T}' \subseteq \mathsf{TolGBBTrans}_\mathbf{D}$ and let $\iota: \mathbf{T}' \to \mathsf{LDB}(\mathbf{D})$ be a nonoverlapping state assignment for $\mathbf{T}'$. Given an ordered pair $\langle T_1, T_2\rangle \in \mathbf{T}' \times \mathbf{T}'$, ι is *tolerant* for $\langle T_1, T_2\rangle$ if
$$((\mathsf{FLift}_\mathbf{D}\langle \mathsf{WTrim}_{\iota(T_1)}\langle\langle \mathbf{D}, \mathcal{U}_{T_1}\rangle\rangle\rangle)^{(2)}(\iota(T_1)))_{|\mathcal{H}_{T_2}^{\mathsf{GFn}}(\iota(T_2))} \in \mathcal{H}_{T_2}^{\mathsf{Tol}}(\iota(T_2)).$$
In words, the result of the update of T_1, when restricted to the guard object of T_2, lies within the guard tolerance of T_2. For an unordered pair $\{T_1, T_2\} \subseteq \mathbf{T}'$ of distinct transactions, ι is *tolerant* for $\{T_1, T_2\}$ if it is tolerant for both $\langle T_1, T_2\rangle$ and $\langle T_2, T_1\rangle$.

4.4 The inadequacy of pairwise tolerance While it would be desirable to be able to formulate a constraint-preservation result based upon the pairwise property of 4.3, in a manner similar to the way in which 3.6 is used to establish 3.7, this is unfortunately not possible. To illustrate by example, let $k > 1$ be a positive integer, and let $\mathbf{E}_{3_k}$ be the database schema with k integer-valued data objects $\{x_i \mid i \in [1,k]\}$, constrained by $(\sum_{j=1}^{k} x_j) > 0$. Extending the notation of 3.9, for $n_1, n_2, \ldots, n_k \in \mathbb{Z}$, define $M_{3_k\langle n_1, n_2, \ldots, n_k \rangle} \in \mathsf{DB}(\mathbf{E}_{3_k})$ to have $x_i = n_i$ for $i \in [1,k]$. For $i \in [1,k]$, define $M_{3_k\langle n_1, n_2, \ldots, n_k \rangle}^{\overline{x_i}} \in \mathsf{DB}(\{x_j \mid j \in [1,k] \setminus \{i\}\})$ to be $M_{3_k\langle n_1,n_2,\ldots,n_k\rangle}|_{\{x_j \mid j\in[1,k]\setminus\{i\}\}}$; i.e., $M_{3_k\langle n_1,n_2,\ldots,n_k\rangle}$ with the component for x_i removed. Since n_i in the subscript is irrelevant, and may be written -.

For $i \in [1,k]$, define the transaction $T_{3_k i}$ by the conditional if $(\sum_{j=1}^{k} x_j) > 1$ then $x_i \leftarrow x_i - 1$. If $(\sum_{j=1}^{k} n_j) > 1$, the guard object $\mathcal{H}_{T_{3_k i}}^{\mathsf{GFn}}(M_{3_k\langle n_1,n_2,\ldots,n_k\rangle}) = \{x_j \mid (j \in [1,k] \setminus \{i\})\}$, with the tolerance set $\mathcal{H}_{T_{3_k i}}^{\mathsf{Tol}}(M_{3_k\langle n_1,n_2,\ldots,n_k\rangle}) = \{M_{3_k\langle m_1,m_2,\ldots,m_k\rangle}^{\overline{x_i}} \mid (\sum_{j\in[1,k]\setminus\{i\}} m_j) > (2 - n_i)\}$. If $(\sum_{j=1}^{k} n_j) \leq 1$, the identity update is performed, so $\mathcal{H}_{T_{3_k i}}^{\mathsf{GFn}}(M_{3_k\langle n_1,n_2,\ldots,n_k\rangle}) = \emptyset$ with $\mathcal{H}_{T_{3_k i}}^{\mathsf{Tol}}(M_{3_k\langle n_1,n_2,\ldots,n_k\rangle}) = \{\Phi_{\mathsf{DB}}\}$.

For any integer p with $1 < p < k+1$, and integers $\langle n_1, n_2, \ldots, n_k \rangle$ satisfying $(\sum_{j=1}^{k} n_j) = p$, at most $p - 1$ of the transactions in $\{T_{3i} \mid 1 \leq i \leq k\}$ may be run concurrently, using the same initial snapshot $M_{3_k\langle n_1,n_2,\ldots,n_k\rangle}$, without a constraint violation. If p or more are run concurrently, with that same initial snapshot, a constraint violation will result. This is true despite the fact that, if $p > 2$, then for any distinct pair $\{T_{j_1}, T_{j_2}\} \subseteq \{T_{3i} \mid 1 \leq i \leq k\}$ the state assignment which assigns $M_{3_k\langle n_1,n_2,\ldots,n_k\rangle}$ to each, is tolerant for $\{T_{j_1}, T_{j_2}\}$. Thus, the pairwise definition of tolerance of 4.3 is not adequate, in the general case, to characterize constraint preservation.

It is, however, possible to obtain a pairwise characterization of tolerance, provided that special conditions are imposed upon the structure of the guard object and guard tolerance. The main idea is to partition the schema $\mathbf{D}$ into complex data objects, called Π-objects, to require the each guard object be a union of Π-objects, and, most importantly, to require that each tolerance set be defined by a product of database states, one factor for each Π-object within the guard object. Although this limits somewhat how liberal the tolerance may be, far simpler tests suffice to determine whether an update by a transaction lies within the tolerances specified by the other concurrent transactions. The details constitute the remainder of this section.

4.5 Schema partitions and partition-compatible subobjects

A *schema partition* for $\mathbf{D}$ is a partition Π on $\mathsf{DObj}\langle\mathbf{D}\rangle$. Each $\mathbf{x} \in \Pi$ is called a *block* of Π, with the set of all blocks of Π denoted $\mathsf{Blocks}\langle\Pi\rangle$. A subobject $\mathbf{x} \subseteq \mathsf{DObj}\langle\mathbf{D}\rangle$ is a Π-*object* if it is the union of some of the blocks of Π. In that case, $\mathsf{Blocks}_\Pi\langle\mathbf{x}\rangle$ denotes the set of blocks of which it is the union; i.e., $\mathbf{x} = \bigcup \mathsf{Blocks}_\Pi\langle\mathbf{x}\rangle$. The set of all Π-objects of $\mathbf{D}$ is denoted $\mathsf{DObj}_\Pi\langle\mathbf{D}\rangle$.

The Π-*closure* of a data object $\mathbf{z} \subseteq \mathsf{DObj}\langle\mathbf{D}\rangle$ is the smallest Π-object containing $\mathbf{z}$. Formally, $\mathsf{Closure}^\Pi\langle\mathbf{z}\rangle = \bigcup\{\mathbf{x} \in \mathsf{Blocks}_\Pi\langle\Pi\rangle \mid \mathbf{x} \cap \mathbf{z} \neq \emptyset\}$. Given an updateable object $\langle\mathbf{c},\mathbf{u}\rangle$, $\mathsf{WSet}^\Pi\langle\langle\mathbf{c},\mathbf{u}\rangle\rangle$ denotes $\mathsf{Closure}^\Pi\langle\mathsf{WSet}\langle\langle\mathbf{c},\mathbf{u}\rangle\rangle\rangle$. For $N \in \mathsf{LDB}(\mathbf{D})$, the Π-*closed write trim* of $\langle\mathbf{c},\mathbf{u}\rangle$ to N, denoted $\mathsf{WTrim}^\Pi_N\langle\langle\mathbf{c},\mathbf{u}\rangle\rangle$, is $\langle\mathbf{c},\mathbf{u}\rangle_{|\mathsf{WSet}^\Pi\langle\mathsf{Trim}_N\langle\langle\mathbf{c},\mathbf{u}\rangle\rangle\rangle}$. It is the smallest Π-object containing $\mathsf{WSet}^\Pi\langle\mathsf{Trim}_N\langle\langle\mathbf{c},\mathbf{u}\rangle\rangle\rangle$; that is, the smallest Π-object which contains the write set of $\langle\mathbf{c},\mathbf{u}\rangle$ when trimmed to N.

4.6 Product sets of states Let Π be a schema partition for $\mathbf{D}$, and let $\mathbf{x} \in \mathsf{DObj}_\Pi\langle\mathbf{D}\rangle$. A subset $\mathbf{M} \subseteq \mathsf{DB}(\mathbf{x})$ is a Π-*product set* for $\mathbf{x}$ if there is a $\mathsf{Blocks}\langle\Pi\rangle$-indexed family $\{\mathbf{M}_\mathbf{z} \subseteq \mathsf{DObj}\langle\mathbf{x}\rangle \mid \mathbf{z} \in \mathsf{Blocks}\langle\Pi\rangle\}$ such that for any $N \in \mathsf{DB}(\mathbf{x})$, $N \in \mathbf{M}$ iff $N_{|\mathbf{z}} \in \mathbf{M}_\mathbf{z}$ for every $\mathbf{z} \in \mathsf{Blocks}\langle\Pi\rangle$. In other words, a Π-product set for $\mathbf{x}$ is a product of sets, one for each $\mathbf{z} \in \mathsf{Blocks}_\Pi\langle\mathbf{x}\rangle$.

4.7 Π-compatibleuard objects and functions Using the concepts surrounding schema partitions, a definition of tolerant guard function which will admit a pairwise characterization of independence may be made. Formally, given a schema partition Π for $\mathbf{D}$, a tolerant guard function $h = \langle\mathsf{GObj}_h, \mathsf{GTol}_h\rangle$ for T is Π-*compatible* if the following three conditions are satisfied.

(Π-tg-i) For each $N \in \mathsf{ELDB}(\mathbf{D})$,
$$\mathsf{Blocks}_\Pi\langle\mathsf{WSet}^\Pi\langle\mathsf{Trim}_{\iota(T_1)}\langle\langle\mathbf{D},\mathcal{U}_{T_1}\rangle\rangle\rangle\rangle \cap \mathsf{Blocks}_\Pi\langle\mathsf{GObj}_h(N)\rangle = \emptyset.$$

(Π-tg-ii) For each $N \in \mathsf{ELDB}(\mathbf{D})$, $\mathsf{GObj}_h(N) \in \mathsf{DObj}_\Pi\langle\mathbf{D}\rangle$.

(Π-tg-iii) $\mathsf{GTol}_h(N)$ is a Π-product set for $\mathsf{GObj}_h(N)$.

Condition (Π-tg-i) extends (go-i) of 3.4 by requiring that the Π-closure of the write set not intersect the guard. (Π-tg-ii) and (Π-tg-iii) mandate that the guard object be Π-compatible and the tolerance be a Π-product set, respectively.

The tolerant guard function h is Π-*minimal* just in case the guard object cannot be reduced in size, and the tolerance cannot be increased in size, while still retaining the property of a guard. Formally, call h Π-*minimal* if for every $N \in \mathsf{ELDB}(\mathbf{D})$, no proper Π-compatible subset $H \subsetneq \mathsf{GObj}_h(N)$ is a guard object for T, and for no proper Π-product superset $\mathsf{GTol}_h(N) \subsetneq \mathbf{P}$ is condition (tgp-iii) of 4.1 satisfied when, in that formula, $\mathsf{GTol}_h(N)$ is replaced by $\mathbf{P}$.

A tolerantly guarded transaction $T \in \mathsf{TolGBBTrans}_\mathbf{D}$ is called Π-*compatible* if $\mathcal{H}_T$ has that property. The set of all Π-compatible tolerantly guarded transactions is denoted Π-$\mathsf{TolGBBTrans}_\mathbf{D}$.

Extending the definition of 3.3 to the Π-compatible setting, for $\mathbf{T}' \subseteq \Pi$-$\mathsf{TolGBBTrans}_\mathbf{D}$, call a state assignment $\iota : \mathbf{T}' \to \mathsf{ELDB}(\mathbf{D})$ Π-*nonoverlapping* for if for every $T_1, T_2 \in \mathbf{T}'$ with $T_1 \neq T_2$,
$$\mathsf{WSet}^\Pi\langle\mathsf{Trim}_{\iota(T_1)}\langle\langle\mathbf{D},\mathcal{U}_{T_1}\rangle\rangle\rangle \cap \mathsf{WSet}^\Pi\langle\mathsf{Trim}_{\iota(T_2)}\langle\langle\mathbf{D},\mathcal{U}_{T_2}\rangle\rangle\rangle = \emptyset.$$

Extending the definition of 4.3 to the Π-compatible setting, given an ordered pair $\langle T_1, T_2 \rangle \in \mathbf{T'} \times \mathbf{T'}$ and ι Π-nonoverlapping as above, ι is Π-*tolerant for* $\langle T_1, T_2 \rangle$ if, for each
$$\mathbf{x} \in \mathsf{Blocks}_\Pi \langle \mathsf{WSet}^\Pi \langle \mathsf{Trim}_{\iota(T_1)} \langle \langle \mathbf{D}, \mathcal{U}_{T_1} \rangle \rangle \rangle \rangle \cap \mathsf{Blocks}_\Pi \langle \mathcal{H}_{T_2}^{\mathsf{GFn}}(\iota(T_2)) \rangle,$$
it is the case that $((\mathsf{FLift}_{\mathbf{D}} \langle \mathsf{WTrim}_{\iota(T_1)}^\Pi \langle \langle \mathbf{D}, \mathcal{U}_{T_1} \rangle \rangle \rangle)^{(2)}(\iota(T_1))_{|\mathbf{x}} \in \mathcal{H}_{T_2}^{\mathsf{Tol}}(\iota(T_2))_{|\mathbf{x}}$.
Finally, ι is Π-tolerant for $\{T_1, T_2\}$ if is Π-tolerant for both (T_1, T_2) and (T_2, T_1).

4.8 Notational convention Extending 3.5, from now on, unless stated explicitly to the contrary, assume that a partition Π of $\mathsf{DObj}\langle \mathbf{D} \rangle$ is fixed, and that $\mathbf{T}$ is a finite subset of Π-$\mathsf{TolGBBTrans}_\mathbf{D}$. Assume further that the guard for transaction T is denoted $\mathcal{H}_T = \langle \mathcal{H}_T^{\mathsf{GFn}}, \mathcal{H}_T^{\mathsf{Tol}} \rangle$.

4.9 Theorem — constraint preservation under tolerance *Let $\leq_\mathbf{T}$ be an SI-schedule for $\mathbf{T}$ and let $M \in \mathsf{ELDB}(\mathbf{D})$. If $\mathsf{StAssign}_{\langle \leq_\mathbf{T} \,:\, M \rangle}$ is Π-nonoverlapping and $\{T_1, T_2\}$-tolerant for Π for every pair of distinct concurrent transactions $\{T_1, T_2\} \subseteq \mathbf{T}$, then $\leq_\mathbf{T}$ is constraint preserving for initial state M.*

PROOF: The proof is by induction on the number of transactions in $\mathbf{T}$. Let T_i represent the ith transaction which commits; for n transactions, the commit order is therefore $T_1, T_2, \ldots, T_{n-1}, T_n$. The basis step of the induction, for $n \in \{0, 1\}$, is trivial. For the inductive step, let $n > 1$ and assume that the result is true whenever the number of transactions in $\mathbf{T}$ is no more than n, and consider the case that $\mathbf{T}$ consists of $n + 1$ transactions. Upon removing T_{n+1}, the schedule consisting of the transactions in $\mathbf{T} \setminus \{T_{n+1}\}$ is constraint preserving for M by the inductive hypothesis. To verify constraint preservation for the entire set $\mathbf{T}$, it suffices to verify that committing the writes of T_{n+1} to $\mathsf{AfterCmt}_{\langle \leq_\mathbf{T} \,:\, M \rangle}\langle T_n \rangle$ does not violate any integrity constraints. This is guaranteed by the requirement that, for every T_i with the property that T_i and T_{n+1} run concurrently, $\{T_i, T_{n+1}\}$ form Π-tolerant pair for $\mathsf{StAssign}_{\langle \leq_\mathbf{T} \,:\, M \rangle}$. If the Π-closure of the update set of T_i overlaps any part of the guard tolerance T_{n+1}, then for each $\mathbf{x} \in \mathsf{Blocks}\langle \Pi \rangle$ which lies in both, the update by T_{n+1}, restricted to $\mathbf{x}$ is guaranteed to lie within the tolerance set $\mathcal{H}_{T_{n+1}}^{\mathsf{Tol}}(\mathsf{InitSnap}_{\langle \leq_\mathbf{T} \,:\, M \rangle}\langle T_{n+1} \rangle)$ of T_{n+1}, as stipulated in 4.3. The key point is that (in contrast to the general case, as illustrated in 4.4), the actions of other transactions, even concurrent ones, cannot change this. Thus, the pairwise check for tolerance suffices. □

4.10 Examples of constraint preservation under tolerance Continue with the setting and examples of 4.4, specifically the schema $\mathbf{E}_{3_k}$ for some $k > 3$. Let the partition $\Pi_{\mathbf{E}_{3_k}}$ have each data object of $\{x_j \mid j \in [1, k]\}$ in its own block. It is useful to extend the notation of 4.4 for states to $\Pi_{\mathbf{E}_{3_k}}$-product sets. To this end, define
$$\mathbf{M}_{3_k \langle n_1, n_2, \ldots, n_k \rangle} = \{M_{3_k \langle m_1, m_2, \ldots, m_k \rangle} \in \mathsf{DB}(\mathbf{E}_{3_k}) \mid (\forall j \in [1, k])(m_i \geq n_i)\},$$
the product set in which, for each database, the value of x_i is at least as large

as n_i. Similarly, define
$$\mathbf{M}^{\overline{x_i}}_{3_k\langle n_1,n_2,\ldots,n_k\rangle} = \{M^{\overline{x_i}}_{3_k\langle m_1,m_2,\ldots,m_k\rangle} \in \mathsf{DB}(\mathbf{E}_{3_k}) \mid (\forall j \in [1,k])(m_i \geq n_i)\}.$$
Take the initial state to be $M_{3_k\langle p_1,p_2,\ldots,p_k\rangle}$ with $(\sum_{j\in[1,k]} p_i) = p$, for some integer p with $0 < p < k+1$. As argued in 4.4, exactly $p-1$ of the transactions in $\{T_{3i} \mid 1 \leq i \leq k\}$ may be run concurrently with initial snapshot $M_{3\langle p_1,p_2,\ldots,p_k\rangle}$. However, in contrast to the general setting, in the setting of partition compatibility, which $p+1$ transactions may run depends upon how tolerance sets are chosen. Consider the transaction $T_{3_k i}$. In the $\Pi_{\mathbf{E}_{3_k}}$-compatible context, while $\mathcal{H}^{\mathsf{GFn}}_{T_{3_k i}}(M_{3_k\langle p_1,p_2,\ldots,p_k\rangle}) = \{x_j \mid j \in [1,k]\setminus\{i\}\}$, as in 4.4, it is no longer permissible to define its guard tolerance via $(\sum_{j=2}^k x_j) > 1$ alone, since a $\Pi_{\mathbf{E}_{3_k}}$-product set must be chosen for the guard. A set of the form $\mathbf{M}^{\overline{x_1}}_{3_k\langle -,n_2,\ldots,n_k\rangle}$ must have $(\sum_{i\in[2,k]} n_i) > 2-p_1$ (in order to allow the update $x_1 \leftarrow x_1-1$ to occur without violating the integrity constraint), as well as $n_i \leq p_i$ for $i \in [2,k]$ (since the current value of x_i must always be present in the range).

To keep things concrete, choose $k=3$, with initial snapshot $M_{3_3\langle 1,1,1\rangle}$. For transaction $T_{3_3 1}$, a minimal Π_{3_4}-compatible guard is of the form $\mathbf{M}^{\overline{x_1}}_{3_3\langle -,q_2,q_3\rangle}$ with $q_2 + q_3 > 1$ and both $q_2 \leq 1$ and $q_3 \leq 1$. This means that the only two possibilities are $\mathbf{M}^{\overline{x_1}}_{3_3\langle -,0,1\rangle}$ and $\mathbf{M}^{\overline{x_1}}_{3_3\langle -,1,0\rangle}$. If $\mathbf{M}^{\overline{x_1}}_{3_3\langle -,0,1\rangle}$ is chosen, then $T_{3_3 3}$ may not run concurrently, since $x_2 \geq 1$ is required by $T_{3_3 1}$. Similarly, if $\mathbf{M}^{\overline{x_1}}_{3_3\langle -,1,0\rangle}$ is chosen, then $T_{3_3 2}$ may not run concurrently. Thus, $T_{3_3 1}$ must in effect choose a "victim" which cannot run concurrently. Suppose that victim is $T_{3_2 3}$, and $T_{3_3 2}$ attempts to run concurrently with $T_{3_2 1}$. It will succeed only if it chooses its tolerance correctly. Specifically, it must choose $\mathbf{M}^{\overline{x_1}}_{3_3\langle 0,-,1\rangle}$. If it chooses $\mathbf{M}^{\overline{x_1}}_{3_3\langle 1,-,0\rangle}$, its update will not lie in the tolerance specified by $T_{3_2 1}$. This illustrates that transactions should be given the opportunity to know the guard functions of other concurrent transactions, and to choose their own guard functions to ensure success. This idea is explored more thoroughly in Sec. 5.

Although Π-compatibility may seem limiting, two points should be kept in mind. First, constraints which tie many data objects together are relatively uncommon. For the most part, constraints relate just two data objects, The examples shown here are specifically designed to show the theoretical limitations. Second, these problems only occur in border cases. For example, if the initial state is $M_{3_3\langle 2,2,2\rangle}$, then all three transactions may execute concurrently, without any problem, provided the guards are chosen reasonably.

4.11 Tolerant CPSI — TPCSI

By *tolerant CPSI*, or *TCPSI* for short, is meant the strategy described in this section, with Π-compatibility.

5 An Operational Description of TCPSI

As noted in 4.10, under TCPSI, there is an advantage in allowing transactions to be aware of the actions of each other, in order to choose guards dynamically. To this end, an operational version of TCPSI is described, in the spirit of FUW (first-updater wins) of SI [10, Sum. 2.3], which allows such dynamic choices.

5.1 Declaration-augmented SI schedules and transactions To begin, the simple model of SI, as summarized in 3.1, is extended to a third time point of each transaction, the *declaration time*. It is at this time that a transaction declares its (proposed) update as well as its guard pair to the system (and to the other transactions), and a check is made to determine whether these declarations are consistent with the current global database state, as well as the declarations of the other, concurrent transactions which have already declared. Formally, given $\mathbf{T} \subseteq \mathsf{BBTrans_D}$, define $\mathsf{PSCSet}\langle \mathbf{T}\rangle = \mathsf{SCSet}\langle \mathbf{T}\rangle \cup \{T^d \mid T \in \mathbf{T}\} = \{T^s \mid T \in \mathbf{T}\} \cup \{T^d \mid T \in \mathbf{T}\} \cup \{T^c \mid T \in \mathbf{T}\}$. A *declaration-augmented SI-schedule* on $\mathbf{T}$ is a total order $\leq_{\mathbf{T}}$ on $\mathsf{PSCSet}\langle T\rangle$ with the property that for each $T \in \mathbf{T}$, $T^s <_{\mathbf{T}} T^d <_{\mathbf{T}} T^c$. T^d represents the time at which T declares.

A transaction which declares has the same black-box representation, as given in 4.1, as one which does not declare. Thus, the conditions for constraint preservation, as presented in 4.9, apply equally well to transactions which declare. However, the interpretation is a bit different. In a transaction T which declares, the update $\langle \mathbf{D}, \mathcal{U}_T\rangle$ and the guard pair $\mathcal{H}_T = \langle \mathcal{H}_T^{\mathsf{GFn}}, \mathcal{H}_T^{\mathsf{Tol}}\rangle$ need not be specified when the transaction begins; rather, they may be determined at declaration time, allowing T to make real-time decisions about the choice of guard and even the choice of update. To highlight this difference, transactions with an explicit declaration point will be termed *grey box*. The set of all tolerantly guarded Π-compatible grey-box transactions over $\mathbf{D}$ is denoted Π-$\mathsf{TolGBBTrans_D}$.

5.2 Notational convention In addition to the conventions of 3.5 and 4.8, from now on, also assume that $\leq_{\mathbf{T}}$ is a declaration-augmented SI schedule, and that $\mathbf{T}$ is a finite subset of Π-$\mathsf{TolGBBTrans_D}$.

5.3 Global values for declarations The steps which are taken during the execution of a declaration-augmented SI schedule are based upon the values of certain objects which evolve during the execution. Each object listed below has a value at each point in time during the lifetime of the schedule $\leq_{\mathbf{T}}$.

$\mathsf{ActiveTrans}_{\langle \leq_{\mathbf{T}}:M\rangle}$ is the set of all transactions which are *active*; that is, which have started but have not yet committed.

$\mathsf{DeclTrans}_{\langle \leq_{\mathbf{T}}:M\rangle}$ is the set of all active transactions which have declared.

The remaining three object families have values which are defined at each point in time for each $\mathbf{z} \in \mathsf{Blocks}\langle\Pi\rangle$.

$\mathsf{CmtVal}^{\mathbf{z}}_{\langle\leq_{\mathbf{T}}:M\rangle}$ is the current committed value of data object $\mathbf{z}$. It is already present in every system as a record or set of records in the global database. However, it is convenient to have this explicit notation for it.

$\mathsf{PndVal}^{\mathbf{z}}_{\langle\leq_{\mathbf{T}}:M\rangle}$ is the value, if any, which a transaction $T \in \mathsf{DeclTrans}_{\langle\leq_{\mathbf{T}}:M\rangle}$ has proposed as an update via a declaration (see 5.5 below). It may be realized as a link to a record or records in the snapshot of T. If no running transaction has proposed an update to $\mathbf{x}$, the value is that of $\mathsf{CmtVal}^{\mathbf{z}}_{\langle\leq_{\mathbf{T}}:M\rangle}$.

$\mathsf{CurrTol}^{\mathbf{z}}_{\langle\leq_{\mathbf{T}}:M\rangle}$ is defined to be
$$\bigcap \{(\mathcal{H}^{\mathsf{Tol}}_{T'}(\mathsf{InitSnap}_{\langle\leq_{\mathbf{T}}:M\rangle}\langle T'\rangle))_{|\mathbf{z}} \mid (\mathbf{z} \in \mathcal{H}^{\mathsf{GFn}}_{T}(\mathsf{InitSnap}_{\langle\leq_{\mathbf{T}}:M\rangle}\langle T'\rangle)) \wedge (T' \in \mathsf{DeclTrans}_{\langle\leq_{\mathbf{T}}:M\rangle})\}.$$
It expresses the combined tolerance on $\mathbf{z}$ of all transactions which have declared updates. If there is no active transaction T with $\mathbf{z} \in \mathcal{H}^{\mathsf{GFn}}_{T}(\mathsf{InitSnap}_{\langle\leq_{\mathbf{T}}:M\rangle}\langle T\rangle)$, the value of $\mathsf{CurrTol}^{\mathbf{z}}_{\langle\leq_{\mathbf{T}}:M\rangle}$ is $\mathsf{DB}(\mathbf{z})$.

It should be noted that an explicit value need not be stored for each such $\mathbf{z}$, any more than an initial snapshot under SI contains an explicit record for each data object. Rather, in the above, explicit values for $\mathbf{z} \in \mathsf{Blocks}\langle\Pi\rangle$ are only required in the case that some active transaction which has declared uses that data object, either as a writer or else in a guard.

5.4 Transaction-specific instances of global values For each transaction, the values of the last three objects listed in 5.3, at the point of declaration, are central to the process. A transaction cannot change the value of $\mathsf{CmtVal}^{\mathbf{z}}_{\langle\leq_{\mathbf{T}}:M\rangle}$ until it commits, so a single value at the point of declaration suffices. Since the transaction may alter $\mathsf{PndVal}^{\mathbf{z}}_{\langle\leq_{\mathbf{T}}:M\rangle}$ and $\mathsf{CurrTol}^{\mathbf{z}}_{\langle\leq_{\mathbf{T}}:M\rangle}$ when it declares, separate before and after values are necessary. The formal definitions follow. Each applies for each $T \in \mathbf{T}$ and each $\mathbf{z} \in \mathsf{Blocks}\langle\Pi\rangle$.

$\mathsf{PrCmtVal}^{\mathbf{z}}_{\langle\leq_{\mathbf{T}}:M\rangle}\langle T\rangle$: the value of $\mathsf{CmtVal}^{\mathbf{z}}_{\langle\leq_{\mathbf{T}}:M\rangle}$ at T^d.

$\mathsf{BeforePrPndVal}^{\mathbf{z}}_{\langle\leq_{\mathbf{T}}:M\rangle}\langle T\rangle$: the value of $\mathsf{PndVal}^{\mathbf{z}}_{\langle\leq_{\mathbf{T}}:M\rangle}$ immediately before T^d.

$\mathsf{AfterPrPndVal}^{\mathbf{z}}_{\langle\leq_{\mathbf{T}}:M\rangle}\langle T\rangle$: the value of $\mathsf{PndVal}^{\mathbf{z}}_{\langle\leq_{\mathbf{T}}:M\rangle}$ immediately after T^d.

$\mathsf{BeforePrTolVal}^{\mathbf{z}}_{\langle\leq_{\mathbf{T}}:M\rangle}\langle T\rangle$: the value of $\mathsf{CurrTol}^{\mathbf{z}}_{\langle\leq_{\mathbf{T}}:M\rangle}$ immediately before T^d.

$\mathsf{AfterPrTolVal}^{\mathbf{z}}_{\langle\leq_{\mathbf{T}}:M\rangle}\langle T\rangle$: the value of $\mathsf{CurrTol}^{\mathbf{z}}_{\langle\leq_{\mathbf{T}}:M\rangle}$ immediately after T^d.

5.5 Pending-update maintenance under SI The steps (pum-i)-(pum-ix), identified below, when applied to the schedule $\leq_{\mathbf{T}}$ with initial database $M \in \mathsf{ELDB}(\mathbf{D})$, are called collectively *pending-update maintenance*.

Schedule initialization: At the beginning of the execution of $\leq_\mathbf{T}$ with initial database M, for each $\mathbf{z} \in \mathsf{Blocks}_\Pi\langle\Pi\rangle$, the following three assignments are executed, in order to give the objects the appropriate initial values.

(pum-i) $\mathsf{CmtVal}^\mathbf{z}_{\langle\leq_\mathbf{T}:M\rangle} \leftarrow M_{|\mathbf{z}}$.

(pum-ii) $\mathsf{PndVal}^\mathbf{z}_{\langle\leq_\mathbf{T}:M\rangle} \leftarrow M_{|\mathbf{z}}$.

(pum-iii) $\mathsf{CurrTol}^\mathbf{z}_{\langle\leq_\mathbf{T}:M\rangle} \leftarrow \mathsf{DB}(\mathbf{z})$.

Transaction declaration: When a transaction T declares (at T^d), the following assignments are executed, provided the conditions (decl-i) to (decl-iv) of 5.6 below are satisfied. If those conditions are not satisfied, the transaction may either wait until the conditions are satisfied, or else abort.

(pum-iv) $\mathsf{DeclTrans}_{\langle\leq_\mathbf{T}:M\rangle} \leftarrow \mathsf{DeclTrans}_{\langle\leq_\mathbf{T}:M\rangle} \cup \{T\}$.

(pum-v) For each $\mathbf{z} \in \mathsf{Blocks}\langle\Pi\rangle$
$\cap\; \mathsf{WSet}\langle\mathsf{WTrim}^\Pi_{\mathsf{InitSnap}_{\langle\leq_\mathbf{T}:M\rangle}\langle T\rangle}\langle\langle\mathbf{D},\mathcal{U}_T\rangle(\mathsf{InitSnap}_{\langle\leq_\mathbf{T}:M\rangle}\langle T\rangle)\rangle\rangle$,
execute $\mathsf{PndVal}^\mathbf{z}_{\langle\leq_\mathbf{T}:M\rangle} \leftarrow$
$(\mathsf{WTrim}^\Pi_{\mathsf{InitSnap}_{\langle\leq_\mathbf{T}:M\rangle}\langle T\rangle}\langle\langle\mathbf{D},\mathcal{U}_T\rangle(\mathsf{InitSnap}_{\langle\leq_\mathbf{T}:M\rangle}\langle T\rangle)\rangle^{(2)})_{|\mathbf{z}}$.

(pum-vi) For each $\mathbf{z} \in \mathsf{Blocks}_\Pi\langle\mathcal{H}^{\mathsf{GFn}}(\mathsf{InitSnap}_{\langle\leq_\mathbf{T}:M\rangle}\langle T\rangle)\rangle$,
execute $\mathsf{CurrTol}^\mathbf{z}_{\langle\leq_\mathbf{T}:M\rangle} \leftarrow \mathsf{CurrTol}^\mathbf{z}_{\langle\leq_\mathbf{T}:M\rangle} \cap \mathcal{H}^{\mathsf{Tol}}(\mathsf{InitSnap}_{\langle\leq_\mathbf{T}:M\rangle}\langle T\rangle)_{|\mathbf{z}}$.

Action (pum-v) sets the pending value $\mathsf{PndVal}^\mathbf{z}_{\langle\leq_\mathbf{T}:M\rangle}$ of each data object $\mathbf{z}$ which is updated by T to the new, updated value, while (pum-vi) adds the tolerance for that update to $\mathsf{CurrTol}^\mathbf{z}_{\langle\leq_\mathbf{T}:M\rangle}$.

Transaction commit: For any $T \in \mathbf{T}$, when T commits, the following three assignments are executed.

(pum-vii) $\mathsf{DeclTrans}_{\langle\leq_\mathbf{T}:M\rangle} \leftarrow \mathsf{DeclTrans}_{\langle\leq_\mathbf{T}:M\rangle} \setminus \{T\}$.

(pum-viii) For each $\mathbf{z} \in \mathsf{WSet}\langle\mathsf{WTrim}^\Pi_{\mathsf{InitSnap}_{\langle\leq_\mathbf{T}:M\rangle}\langle T\rangle}\langle\langle\mathbf{D},\mathcal{U}_T\rangle(M)\rangle\rangle$,
execute $\mathsf{CmtVal}^\mathbf{z}_{\langle\leq_\mathbf{T}:M\rangle} \leftarrow \mathsf{PndVal}^\mathbf{z}_{\langle\leq_\mathbf{T}:M\rangle}$.

(pum-ix) For each $\mathbf{z} \in \mathsf{Blocks}\langle\Pi\rangle \cap \mathcal{H}^{\mathsf{GFn}}_T(\mathsf{InitSnap}_{\langle\leq_\mathbf{T}:M\rangle}\langle T\rangle)$, update $\mathsf{CurrTol}^\mathbf{z}_{\langle\leq_\mathbf{T}:M\rangle}$ to reflect that T is no longer active nor declared.

Step (pum-viii) commits the pending update to the global database; (pum-ix) removes the tolerance required by transaction T, since it is now finished.

5.6 Tolerance-compliant and nonoverlapping schedules

The following four conditions must be verified for each $T \in \mathbf{T}$, at T^d.

(decl-i) $(\forall \mathbf{z} \in \mathsf{Blocks}\langle\Pi\rangle$
$\cap \mathsf{WSet}\langle\mathsf{WTrim}^{\Pi}_{\mathsf{InitSnap}_{\langle\leq_\mathbf{T}:M\rangle}\langle T\rangle}\langle\langle\mathbf{D},\mathcal{U}_T\rangle(\mathsf{InitSnap}_{\langle\leq_\mathbf{T}:M\rangle}\langle T\rangle)\rangle\rangle$
$(\mathsf{InitSnap}_{\langle\leq_\mathbf{T}:M\rangle}\langle T\rangle_{|\mathbf{z}} = \mathsf{BeforePrPndVal}^\mathbf{z}_{\langle\leq_\mathbf{T}:M\rangle}\langle T\rangle = \mathsf{PrCmtVal}^\mathbf{z}_{\langle\leq_\mathbf{T}:M\rangle}\langle T\rangle).$

(decl-ii) $(\forall \mathbf{z} \in \mathsf{Blocks}\langle\Pi\rangle \cap \mathcal{H}^{\mathsf{GFn}}(\mathsf{InitSnap}_{\langle\leq_\mathbf{T}:M\rangle}\langle T\rangle))$
$((\mathsf{CmtVal}^\mathbf{z}_{\langle\leq_\mathbf{T}:M\rangle} \neq (\mathsf{InitSnap}_{\langle\leq_\mathbf{T}:M\rangle}\langle T\rangle)_{|\mathbf{z}})$
$\Rightarrow (\mathsf{CmtVal}^\mathbf{z}_{\langle\leq_\mathbf{T}:M\rangle} \in \mathcal{H}^{\mathsf{Tol}}(\mathsf{InitSnap}_{\langle\leq_\mathbf{T}:M\rangle}\langle T\rangle)_{|\mathbf{z}})).$

(decl-iii) $(\forall \mathbf{z} \in \mathsf{Blocks}\langle\Pi\rangle \cap \mathcal{H}^{\mathsf{GFn}}(\mathsf{InitSnap}_{\langle\leq_\mathbf{T}:M\rangle}\langle T\rangle))$
$((\mathsf{PndVal}^\mathbf{z}_{\langle\leq_\mathbf{T}:M\rangle} \neq (\mathsf{InitSnap}_{\langle\leq_\mathbf{T}:M\rangle}\langle \mathbf{z}\rangle)_{|\mathbf{z}})$
$\Rightarrow (\mathsf{PndVal}^\mathbf{z}_{\langle\leq_\mathbf{T}:M\rangle} \in \mathcal{H}^{\mathsf{Tol}}(\mathsf{InitSnap}_{\langle\leq_\mathbf{T}:M\rangle}\langle T\rangle)_{|\mathbf{z}})),$

(decl-iv) $(\forall \mathbf{z} \in \mathsf{Blocks}\langle\Pi\rangle$
$\cap \mathsf{WSet}\langle\mathsf{WTrim}^{\Pi}_{\mathsf{InitSnap}_{\langle\leq_\mathbf{T}:M\rangle}\langle T\rangle}\langle\langle\mathbf{D},\mathcal{U}_T\rangle(\mathsf{InitSnap}_{\langle\leq_\mathbf{T}:M\rangle}\langle T\rangle)\rangle\rangle$
$((\mathsf{WTrim}^{\Pi}_{\mathsf{InitSnap}_{\langle\leq_\mathbf{T}:M\rangle}\langle T\rangle}\langle\langle\mathbf{D},\mathcal{U}_T\rangle(\mathsf{InitSnap}_{\langle\leq_\mathbf{T}:M\rangle}\langle T\rangle)\rangle\rangle^{(2)}_{|\mathbf{z}} \in \mathsf{CurrTol}^\mathbf{z}_{\langle\leq_\mathbf{T}:M\rangle}).$

Condition (decl-i) checks for write overlap; this test must be performed for any implementation of SI, with or without constraint preservation. Testing at declaration time is nothing more than FUW (first updater wins) [10, Sum. 2.3] — a transaction T is allowed to continue if its declared update does not conflict with those updates by other transactions which have been declared since T started.

The remaining three tests concern constraint preservation. When transaction T declares, it must be established that for each concurrent transaction T' which has already declared, the update of T' does not violate the tolerance of T. (decl-ii) verifies this in the case that T' has already committed, while (decl-iii) serves the same purpose in the case that T' has declared but not yet committed. (decl-iv) verifies that no read-write conflict exists in the opposite direction; it checks whether the proposed writes of T violate the tolerance limits of the other declared but not yet committed transactions.

5.7 Example An example of operational TCPSI, as described in this section, is summarized in Table 1. It uses the schema $\mathbf{E}_{3_3}$, as well as the associated transactions and notation, described in 4.4 and 4.10, On transactions, 3_3 is omitted in the subscript; thus, for $1 \in [1,3]$, $T_{3_3 i}$ becomes just T_i. The transaction S_2 is new; it executes the update $x_2 \leftarrow x_2 + 1$ unconditionally. Its guard object is always $\emptyset$, and its tolerance always $\{\Phi_{\mathsf{DB}}\}$; it can never induce a constraint violation when run in isolation, since increasing the value of x_2 can never decrease the value of a sum of which it is a summand. The initial state is $M_{3_3\langle 1,1,1\rangle}$. In the first two lines of the table, T_1 and T_2 begin, in that order. Note that the CmtVal and the PndVal for each x_i carries the values of the initial state. Next, T_1 declares its intent to execute the ground update

Tr	x_1 CmtVal	PndVal	CurrTol	x_2 CmtVal	PndVal	CurrTol	x_3 CmtVal	PndVal	CurrTol	$\geq$ TrTol
T_1^s	1	1	-	1	1	-	1	1	-	
T_2^s	1	1	-	1	1	-	1	1	-	
T_1^d	1	0	-	1	1	≥ 0	1	1	≥ 1	(-,0,1)
T_2^d	1	0	≥ 0	1	0	≥ 0	1	1	≥ 1	(0,-,1)
T_3^s	1	0	≥ 0	1	0	≥ 0	1	1	≥ 1	
T_3^d	1	0	≥ 0	1	0	≥ 0	1	1	≥ 1	blocked
T_2^c	1	0	-	0	0	≥ 0	1	1	≥ 1	
S_2^s	1	0	-	0	0	≥ 0	1	1	≥ 1	
S_2^d	1	0	-	0	1	≥ 0	1	1	≥ 1	(-,-,-)
S_2^c	1	0	-	1	1	≥ 0	1	1	≥ 1	
$[T_3^d]$	1	0	≥ 0	1	1	≥ 1	1	0	≥ 1	(0,1,-)
T_1^c	0	0	-	1	1	≥ 1	1	0	-	
T_3^c	0	0	-	1	1	-	0	0	-	

Table 1: Tabular summary of the example of 5.7

$1 \stackrel{x_1}{\rightsquigarrow} 0$ (see 2.6) using the tolerance $\mathbf{M}_{33\langle -,0,1\rangle}^{\overline{x_1}}$. Note that PndVal^{x_1} has been set to zero, but that CmtVal^{x_2} remains unchanged. In the next step, T_2 declares its intent to execute the ground update $1 \stackrel{x_2}{\rightsquigarrow} 0$. For this not to conflict with T_1, the tolerance $\mathbf{M}_{33\langle 0,-,1\rangle}^{\overline{x_2}}$ is chosen. Had $\mathbf{M}_{33\langle 1,-,0\rangle}^{\overline{x_2}}$ been chosen instead, T_1 and T_2 could not both commit without a constraint violation. This illustrates the desirability of T_2 choosing its guard tolerance interactively, with knowledge of the pending updates of other, concurrent transactions. In the next two lines, T_3 begins and declares. However, it is blocked by the combination of T_1 and T_2; there is no choice of tolerance on x_1 and x_2 which would allow it to run. At this point, it can either abort or wait; assume that it waits. Next, T_2 commits. This sets CmtVal^{x_2} to 0 and the $\mathsf{CurrTol}^{x_1}$ and $\mathsf{CurrTol}^{x_3}$ constraints which it applied are removed. There is now no constraint on $\mathsf{CurrTol}^{x_1}$, but the constraint on $\mathsf{CurrTol}^{x_3}$ remains since T_1 requires it also. Next, S_2 starts, declares, and commits, executing the update $0 \stackrel{x_2}{\rightsquigarrow} 1$. No changes are made to the value of any $\mathsf{CurrTol}^{x_i}$, since the guard object is $\emptyset$ for this transaction. The waiting T_3 may now continue, asserting the tolerance $\mathbf{M}_{33\langle 0,1,-\rangle}^{\overline{x_3}}$, since the value of x_2 has increased to 1. Its second declaration point is shown as $[T_3^d]$. This illustrates the advantage of a transaction which is blocked at declaration to wait; in an interactive setting this is possible. Finally, T_1 and then T_3 commit, to complete the execution of the schedule.

5.8 Extended and multiple declaration As illustrated in 5.7, it is advantageous to allow a transaction to determine its declarations later than at the very beginning of execution, and to wait if conditions are not suitable. It is furthermore advantageous to allow it to make different parts of its declarations at different times, and even to alter its declarations. Indeed, this approach has the potential to blend well with cooperative update [8], [11]. The extension is straightforward, but space limitations preclude a further elaboration here.

6 Conclusions and Further Directions

The ideas of CPSI have been extended to the case in which one transaction may write the guard of another while preserving satisfaction of integrity constraints. By using a value-oriented model, in which transactions announce a tolerance on updates to their read sets by other transactions, significantly greater concurrency is possible. In addition to an abstract model, a more operational model has been presented as well. There are at least two key areas for further work.

EXTENSION TO CONCURRENT WRITES: The extension of CPSI presented here supports updates to read sets, but does not allow concurrent writes. An approach which supports write concurrency under certain conditions is the next step in the theoretical development.

PROTOTYPE IMPLEMENTATION: A prototype implementation is essential, especially to evaluate data structures and algorithms for the management of the declaration phase. The preferable platform would be to build upon an existing open-source systems, such as PostgreSQL or MariaDB.

References

[1] Bancilhon, F., Kim, W., Korth, H.F.: A model of CAD transactions. In: A. Pirotte, Y. Vassiliou (eds.) VLDB'85, Proceedings of 11th International Conference on Very Large Data Bases, August 21-23, 1985, Stockholm, Sweden, pp. 25–33. Morgan Kaufmann (1985)

[2] Berenson, H., Bernstein, P.A., Gray, J., Melton, J., O'Neil, E.J., O'Neil, P.E.: A critique of ANSI SQL isolation levels. In: Proceedings of the 1995 ACM SIGMOD International Conference on Management of Data, San Jose, California, May 22-25, 1995, pp. 1–10 (1995)

[3] Bernstein, P.A., Hadzilacos, V., Goodman, N.: Concurrency Control and Recovery in Database Systems. Addison-Wesley (1987)

[4] Cahill, M.J., Röhm, U., Fekete, A.D.: Serializable isolation for snapshot databases. ACM Trans. Database Syst. **34**(4) (2009)

[5] Fekete, A., Liarokapis, D., O'Neil, E.J., O'Neil, P.E., Shasha, D.: Making snapshot isolation serializable. ACM Trans. Database Syst. **30**(2), 492–528 (2005)

[6] Gray, J., Reuter, A.: Transaction Processing: Concepts and Techniques. Morgan Kaufmann (1993)

[7] Härder, T., Reuter, A.: Principles of transaction-oriented database recovery. ACM Comput. Surv. **15**(4), 287–317 (1983)

[8] Hegner, S.J.: A simple model of negotiation for cooperative updates on database schema components. In: Y. Kiyoki, T. Tokuda, A. Heimbürger, H. Jaakkola, N. Yoshida. (eds.) Frontiers in Artificial Intelligence and Applications XX11, pp. 154–173. IOS Press (2011)

[9] Hegner, S.J.: Guard independence and constraint-preserving snapshot isolation. In: C. Bierle, C. Meghini (eds.) Foundations of Information and Knowledge Systems: Eighth International Symposium, FoIKS 2014, Bordeaux, France, March 3-7, 2014, Proceedings, *Lecture Notes in Computer Science*, vol. 8367, pp. 231–250. Springer-Verlag (2014)

[10] Hegner, S.J.: Constraint-preserving snapshot isolation. Ann. Math. Art. Intell. **76**(3), 281–326 (2016)

[11] Hegner, S.J., Schmidt, P.: Update support for database views via cooperation. In: Y. Ioannis, B. Novikov, B. Rachev (eds.) Advances in Databases and Information Systems, 11th East European Conference, ADBIS 2007, Varna, Bulgaria, September 29 - October 3, 2007, Proceedings, *Lecture Notes in Computer Science*, vol. 4690, pp. 98–113. Springer-Verlag (2007)

[12] Korth, H.F., Speegle, G.D.: Formal aspects of concurrency control in long-duration transaction systems using the NT/PV model. ACM Trans. Database Syst. **19**(3), 492–535 (1994)

[13] Papadimitriou, C.: The Theory of Database Concurrency Control. Computer Science Press (1986)

[14] Sippu, S., Soisalon-Soininen, E.: Transaction Processing: Management of the Logical Database and its Underlying Physical Structure. Springer (2014)

[15] Weikum, G., Vossen, G.: Transactional Information Systems. Morgan Kaufmann (2002)

Viewpoint-Oriented Data Management in Collaborative Research Projects

Yannic Ole Kropp and Bernhard Thalheim

Department of Computer Science, Christian-Albrechts University of Kiel, CRC 1266
(yk | thalheim)@is.informatik.uni-kiel.de

Abstract Data management in research projects is a challenging task. Different participants are likely to have individual requirements in various contexts, which result in a heterogeneity concerning data, (data-) structures, (data-)quality, (data-)utilisation, preferred representations, platforms and points of view. These discrepancies become even larger when different academic disciplines are involved. Thus, the data schemata tend to become very large, while the single classes are often used only by few people and are therefore rather small. Manifold ways of data collection, exploitation and usages exist as well as different levels of access rights.

At the same time, a simple collaboration and data exchange within a project has to be supported in a sophisticated way by data management activities. The proposal in this paper provides support for distribution of right data in right quality in right representation to right person.

We present a novel strategy for data management that overcomes those central challenges and achieves these objective. It combines the strength of a central fully integrated storage with the flexibility of local working environments. Users are offered personal data frontends for CRUD-operations while in the backend data is stored in a centralised form with views for each user or partner.

1 Introduction

Managing data is part of almost every scientific project. Several people in inter-disciplinary projects have to collaborate and typically base their collaboration on data. So, exchange of data and appropriate integration of partner data are an essential part of their communication, cooperation, and collaboration. Depending on the complexity and size of the research project, supporting activities can reach from creation of e-mail lists or arranging shared network folders to implementation of quite complex (distributed) systems. Usually sustainable data storage is also required by the funding agencies such as the German Research Council. All data that are collected in a research project have to be kept in such a form that later usage of data is supported independently when the data request occurs.

The classical approach is full integration of data in a federated database. All data are going to be integrated into a common data store. Querying and in general data processing is then executed in the central data store. This approach failed, however, in most projects. The main obstacle is the heterogeneity in data acquisition, data structuring, infrastructures used, viewpoints preferred, and also granularity and maturity of data. A rather pessimistic approach is the development of metadata for each database. The collaboration partners have to develop their own collaboration environment. Any new partnership requires new collaboration features. Therefore, this approach fails too. In the past, interoperability of databases has been intensively investigated. Technical interoperability is based on standard exchange procedures. Semantic interoperability has been aiming at exchange of data in an effective way within and across organisations and institutions. Organisational interoperability refers to supporting processes for collaboration. The success stories for interoperability are rather seldom. It seems that research communities cannot properly use this approach.

Research communities challenge these approaches by agile evolution, by continuous change and velocity of research questions and partnerships, by addressing changing data needs, by heterogeneity of platforms, by large variety of viewpoints which data must be considered, by short-term employment of partners with requirements to publish and to finish in time, and by heterogeneous viability of data. Some research projects face also the data volume challenge. So, the collaboration with databases becomes a real challenge.

Very high heterogeneity of the community databases and of collaboration is also an issue in our research project. At the same time, data-based collaboration only starts. Research groups have their own data, mainly execute their queries in their own data massive, and use relatively small data sets. Similar to other larger (collaborative research) projects, we are able to develop our own way of handling data. Collaboration can thus be explicitly considered right from the start of the project. This prevents confusion by a system of data handling which developed more or less by chance and allows to apply suitable procedures and infrastructure. Those explicit considerations and activities are commonly called *(explicit) data management*.

In this paper we will present an approach which has been developed for medium sized (interdisciplinary) research projects. The approach is designed to deal with typical problems of such projects, like heterogeneous data (of multiple domains) or evolution of data models. In contrast to other application fields of data management we expect research projects to be rather in a qualitative form than in a quantitative one concerning their used data. In other words we do not expect huge amounts of uniform data with fixed (data) structures, but many diverse groups of data which tend to have small sizes and dynamically change their structures multiple times during the life span of the project. The management of those dynamic and heterogeneous structures and their mapping with the personal requirements of the project participants is therefore main focus of this paper. Additionally we will show how collaboration can be supported in such an environment.

Our approach is based on some observations that are made in many collaboration research projects and thus applicable to these projects as well. Data computation is either performed at the local level or performed by explicitly agreed data exchange. Data storage can be separated from data computation. Data can be kept redundant on a basis of a data repository, data replication, and data exchange among partners. Redundancy is then controlled redundancy and conforms to agreed in advance standard procedures and data handling styles such as master-slave. At the same time, users do not want to adapt themselves to different data structures and formats. They want to see their data in the same way they are used to.

Therefore, we can use an approach that consists of six assumptions:

1. *Separation of data storage and data computation*: Data stores become more powerful. The computational facilities are already used technologies beyond database technology. So, we tackle the collaboration challenges by separating data storage from data computation.
2. *Viewpoints of users on their own data*: View technology has reached the maturity to consider local data structures as a tightly integrated view schema on the global schema. Therefore, local data can be considered to be input data to the global storage. At the same time, local data can be extracted from global data through view extraction.
3. *Universal data storage*: Data can be universally stored by generalising the tuple space approach as long as data can be properly identified and their provenance is maintained. Each data has thus its source and can be tracked back to the local databases.
4. *Collaboration view schemata for collaborative data exploitation*: Collaboration can also be supported by the universal data storage. A collaborating party specifies the view schema on top of the own view schema. These second-order view schemata for data exchange allow to retrieve the collaboration data from the universal storage.
5. *Maintenance of local data at the local level*: Proprietors of data are the best evaluators for data integrity and quality. Therefore, local maintenance is more efficient than integrated global maintenance. We keep then local maintenance as the best strategy for data modification as long as it is possible.
6. *Global versioning of data*: It often happens in research projects that old data must be kept as well after new data sets arrive. Local maintenance of data versions is an activity that must be installed for all databases. A better approach is global versioning on the basis of provenance tools and replication tools.

This paper will describe the first four of these main ideas in more detail. But especially the *separation of storage and computation* and the *universal storage* will be in focus.

In general, our approach focuses on creating an adaptive joint dataset in a collective (central) storage, while preserving the individual working environments of the researchers. We assume that - especially in interdisciplinary projects - the different project participants anyway create working environments that better suit their individual and discipline specific (local) requirements for successful research than any project wide (global) standard could. Therefore the computation, analysis and processing of data shall take place in these local environments, which can be individually designed to support workflows or to match technical requirements.

We thus separate data computation from data storage. In general, we currently assume that the global storage must not support query execution as long as it can be performed on the local level. The storage is rather the platform for exchange of data that are devoted to be used in a collaboration[1]. Additionally, publication of collaboration data can be supported by creation of specific collaboration view schemata which are defined on the universal data store. Therefore the universal model of the storage allows the deposition of data in any (relational) schema model and supports the mapping of data models from the local environments with the global structure. In contrast to other approaches the universal storage does not manage the data from the different local environments as different datasets but integrates each datatuple and hereby creates a joint dataset. This eases/enables the automated integration into other environments and allows ('normal'[2]) queries on the complete available data pool of the project.

The approach taken allows multiple use of collaboration view schemata. A team might have the same data demand as another one. So, second-order view schemata can be reused. The data are kept anyway in the universal storage and must not be locally exchanged among different partners. Any partner may thus use shared data. Moreover, we envision for the next realisation step that partners may modify their data and share these modifications with other partners on the basis of the protocol that has been agreed, e.g. eager enforced modification for all participating parties or only notification on modifications made by somebody.

Integrated database systems develop each data provision service only once. This efficiency is one of the main arguments for integration. While separating storage and computation we can be flexible to provide global support for all services that are required by many sites but are rather similar.

Our approach is based on advanced view technology for object-relational database systems. In the past, views have only be considered at the logical or

---

[1] For example temporary data, which is created during computation and is of no further interest, remains in the local environments.

[2] in contrast to metadata-based queries

physical layers of database system architectures. For instance, relational views are single-table relations. Conceptual views may, however, be defined as a database schema themselves. Views can support the modification, e.g. update, of the base tables. We can extend this modifiability through views also to view schemata. As a theoretical underpinning a theory of sophisticated views can be developed. We already applied elements of this approach to website support and to culture-adaptive interfacing for systems [8] [7] [15].

This view technology is one of the enablers of our approach. Each local database is a view on the global universal storage. Each modification to local data can thus also mapped to a modification in the global universal storage. Each global change that has an impact on the local data can also be mapped to local changes as long as it is necessary.

The main advantage of our approach is the opportunity to flexibly and agile support collaboration among n parties by 1-n data support from the global database to the local databases within this collaboration. So, any party requires only a hock facility to the global universal storage instead of a n-times-n support. Due this simplification we develop also services that are requested by many local partners and by many collaborating partners. A typical service of this kind is a geographic service that can be used by any of partners in the same way as the other parties use it.

In short, our approach is based on a universal storage schema and methods for connecting storage and local environments. This will be further discussed in Section 3. Section 2 is thus used to discuss the problem itself and already existing approaches. Section 4 discusses whether our approach handles the challenges from Section 2 in a proper manner. Finally Section 5 shows an application case and Section 6 concludes the paper.

2 Data Management in Collaborative Research Projects

This section will take a closer look at data management, collaboration in general and their characteristic challenges in the context of research projects. Additionally existing approaches to overcome these challenges are presented and evaluated concerning their applicability.

2.1 Data Management

'*Data management*' is a buzz phrase for a set of activities and tasks connected to the usage of data in general. This includes all steps during the data life cycle from creation to final disposal. An exemplary definition is given by [5] and [18]:

> *Data management is a group of activities relating to the planning, development, implementation and administration of systems for the acquisition, storage, security, retrieval, dissemination, archiving and disposal of data.*

In general almost everything connected to administration or management of data can be captured by the term *'data management'*, but it is possible to identify some key activities. According to [5] and [18] these key activities of data management include:

- Data policy development,
- Data ownership,
- Metadata compilation,
- Data lifecycle control,
- Data quality, and
- Data access and dissemination

This list could easily be expanded (for example with the terms *'data integration'* and *'data comparability'*), but ultimately the key activities are dependent of the context in which data management is applied.
It is the context-giving (research) project which determines the kind and the range of needed data management. Some factors for this project based determination are:

- amount of collected/used/processed data
- type and structure of data
- heterogeneity/homogeneity of data
- intended usage of data
- involved people (quantity, background, ...)
- regulations (e.g. by project head or law)

To be able to use those factors for the determination of the necessary data management activities for the context-giving (research) project, it is essential to have in mind what exactly shall be fulfilled by those activities. In other words, one needs to know what the general intention behind those activities is and what purpose data management has at all. Unfortunately the question for the goal of data management is normally not answered directly in common definitions. But the purpose can be derived from guidelines and principles of good data management like the set published by the Data Advisory Group (DAG) of the UK Environmental Observation Framework ([3]).
The goal of data management is basically to support users in their use of data. According to their specific intentions or tasks, they shall be able to get the data they need in a form and a way which suits their workflows.

As this paper deals with users in the context of research projects, where many different requirements of (multiple) users can occur, we will use a user-centred and purpose-orientated definition of data management:

> *Data management in projects is comprised of the activities which are executed to support (all) users in their usage and handling of data.*

2.2 Collaboration

The factor *'involved people'* of the determinants for the kind and the range of needed data management includes the intended rate of collaboration between the participants. Collaboration is basically a sophisticated term for 'working together' and is an essential part of many (research) projects. To illustrate this term a bit more, we refer to the works of Thalheim and Schewe [16] [19]. According to their *'3C approach'* collaboration consists of three main dimensions:

1. **Communication**: deals with the act of exchange of messages, information and data.
2. **Coordination**: is about the managing activities relevant for collaboration. For example specifying the ways for joint use of resources or the allocation of tasks, belong to this dimension.
3. **Cooperation**: is the actual act of working together and being productive.

In this paper we will focus on the aspect communication which is a requirement for coordination and demanded by cooperation.

2.3 Challenges in (interdisciplinary) Research Projects

As already mentioned, the main factor for the kind and range of needed data management is the project itself. Too complex activities for simple intentions can result in refusal by participants, while too simple regulations could supply only insufficient support.

In this section we will examine the specific challenges and problems data management has to deal with in (medium sized) interdisciplinary research projects in the context of collaboration/communication. We refer to single projects and their internal challenges and not to the task of connecting multiple independent projects. Although the issues often seem to be similar when addressing more than one project, they appear on a much larger scale and mostly have to deal with already existing infrastructures, vocabularies etc.. In contrast we want to cope with problems on a smaller scale in single projects. These tasks concern installing (new) infrastructures or building (new) joint vocabularies, which support individual researchers in their ideas and requirements for data usage. Furthermore, single projects do have joint research questions, which set a collective frame and ease the implementation of certain processes.[3]

A typical list of challenges includes:

---

[3] For example people might be more willing to execute additional work, like noting additional information, which are needed by someone else, if they personally know this other person or if it supports their own project wide research question.

1. **Heterogeneous Data**: Especially in interdisciplinary projects the data of different participants is very likely to be diverse in multiple dimensions. Heterogeneous domains resulting from the different disciplines of interdisciplinary projects and their differences in structure and form of data have to be considered. Variations in quality, granularity or robustness[4] have to be taken in consideration as well. Additionally different ways of data acquisition[5] and different levels of access rights have to be kept in mind too. Further possible origins of heterogeneity could be the integration of external data, like data from previous projects, or the distinction between raw data, working data, processed data and results.

2. **Evolving Projects**: Research is a non-linear process and although it is mostly directed towards reaching a more or less specified goal, often the final way and outcome cannot be completely estimated the beginning. Sometimes even the whole intention of a project might change during this process and shift it to completely new directions. These changes are not always drastic, but they are very likely to happen.
As a result the data, which is collected, produced, processed or used, changes too. For example some data could become obsolete, some could need to be more precise and some could be completely new in the context of the project.
Consequently the final results are often the outcome of a (long) series of intermediate stages (with different data structures) and are not generated directly.

3. **Viewpoints**: This point is linked to *'heterogeneous data'*. As participants of research projects have different aspects and foci in their work and therefore not only need to use different data but also need to configure/group data differently, they might have unequal *viewpoints* concerning the same 'real-life' objects.
Those requirements in data representation and selection can be demanded by groups[6] or by a single person. Some may even require more than one viewpoint, depending on their research interests.

4. **Collaboration**: Data which is produced, collected or otherwise acquired will potentially not only be used by the person who introduced it, but also by other participants. Eventually some project partners do not even have own data acquisition activities and totally rely on the data supplied by other participants.
The data which is acquired under a specific *viewpoint* probably cannot be

---

[4] measured in validity, reliability and objectivity
[5] e.g. primary or secondary data source
[6] e.g. given by the discipline of the participants

used directly by other project partners. It needs to be processed to fit their viewpoints. For example this might be necessary to avoid misunderstandings or misconceptions which originate from different meanings of terms in different (academic) disciplines.

5. **Additional Issues**: Besides the previous points, data management in interdisciplinary research projects also has to deal with the usual challenges of data management in general. All challenges, which are not specific for research projects, but are always present in data management attempts, are present in this context too.
A typical example is *'usability'*: Of course in interdisciplinary research projects the solutions and ways of data management should be in a form which (easily) enables participants to contribute, as well as in 'normal' projects.

2.4 Existing Approaches

To overcome some of the just mentioned challenges, or at least to support the handling of those, managing data is commonly organised by ordering and storing the data according to a specific concept. Especially the ways of dealing with several different viewpoints and the connected (heterogeneous) data are important. Also the different data models of the viewpoints have to be addressed. Existing approaches have for instance been described in [14] or [10].

The four most general and common approaches are:

1. **Full Integration**: The full integration approach basically tries to integrate all viewpoints into one general model. Minor differences can be solved by the use of views or by participants ignoring the disused parts. This approach works best for single topic projects, projects with participants from the same discipline or in general projects with few or similar viewpoints. In these cases this approach provides a good, but probably unique, solution for the single project.
Though for a bit larger (interdisciplinary) research projects the following statement of Thalheim and Tropmann-Frick is applicable:
Disciplines have however also their own foundation, their own background, their own culture and their own way of model use. Therefore, it is infeasible to develop a holistic model for everybody in a research team. Models should remain in their local setting and should not be integrated into general global model that is commonly agreed and used [21].
Another problem of research projects is their evolution (see the previous subsection). Each change in personal research can result in the necessity to adjust the general model to fit the new needs of some project members. This process easily leads to complex and not maintainable structures, as

due to new requirements new structures are added, but the already existing structures cannot be deleted as long as someone uses them and thus must persist.

For this reason a special form of this approach is based on the usage of generic databases, which are easier to maintain and manage, and support evolution (see [9]). These databases usually provide functionalities to simulate the relations, which would be physically implemented in the standard form of this approach.

2. **Local Models/Data**: In contrast to developing a full integration model the local approach does not try to integrate the different viewpoints at all. In this approach for every viewpoint (or for each group of similar viewpoints) an own model is used locally and no direct effort is put into linking the different local data models.

   These local models are independent from each other, are therefore quite easy to create and are adapted to the requirements of their users. In addition they can be changed independently without affecting other participants' models, if during the process of research the necessity of a change occurs.

   Difficulties are raised when communication (as part of collaboration) between different models shall be performed. As the contained data must be fitted to the partner (model), this might be quite challenging.

   A possible solution is the usage of *model capsules* presented in [21], which basically add exchange sub-models[7] to each local model. These exchange models are derived from the main model of the local viewpoint and are designed to fit the partners' viewpoints. Though many exchange models might be necessary in a scenario, where many different viewpoints exist and much collaboration/communication shall be performed.

3. **Mapping**: A hybrid strategy would be to maintain both, local and global models, at the same time. This approach originates from the field of data integration, where the local models are represented by already existing (active) data sources/databases which shall be combined. On the other side the global structure with its own (data-)model is represented by some kind of mediator or middleware, which translates queries to the global model into queries to corresponding local models and afterwards joins the results.[8] To enable this translation a mapping between local and global models is necessary, this can, for example, be implemented via the *'local-as-view'* or the *'global-as-view'* method (see [13] or [1]). In [12] Melnik states the importance of metadata for this task.

   In general this strategy has the same challenge as the *'full integration'*-

---

[7] for import and/or export
[8] typical challenges of this field are for instance addressed in [11], [13], [10] and [1]

approach, namely to create and maintain a global model which is suitable to handle all involved local ones. Otherwise it has to limit the accessible information to a certain subset of the actually available information. It is therefore mostly used to combine similar datasources, like multiple databases holding content of the same topic.

4. **Data Library/Catalogue**: This approach uses local models too. But it is focussed on a step in the working progress, which appears before the challenges of importing data from other project partners into the own local model arise. It deals with questions like:

    Is there any data already present in this project which I could use?
    or
    Do participants of other disciplines have something that might be usable for me?

    To enable cross-disciplinary data exchange it is necessary for all participants to be aware of the existing data in the project. Therefore this approach relies on advanced metadata management and enables searching the project datastore on the data set level (in contrast to data tuple level). It can be compared with a library system or catalogue, which is used to locate objects of interest. However this approach does not directly support the exchange of data between different local models.[9]

2.5 Lessons Learned

Research projects likely assemble a variety of researchers, each with individual foci in his or her work. These foci do not only originate from different academic disciplines, but also might occur within one field of research. This results in a variety of data and a heterogeneous project data pool, which needs to be managed. A crucial step to do so, is to create an organised storage.

As *it is infeasible to develop a holistic model for everybody in a research team* (see [21]), it appears to be reasonable to use local and separate (data)models for each research focus. In such a setting the evolution in single local settings, would also just need to be handled in this specific setting, without affecting other environments. The main disadvantage of such an approach would be the need to install singleton solutions for data exchange between every pair of researchers, who would want to share (some of) their data with each other. A centralised solution to coordinate the exchange is not directly present in this approach. Additionally (database-)queries on the complete data pool of the project are not possible and it might be hard for participants to figure out what data is present exactly in other foci.

---

[9] This approach is used by several research projects in Cologne: [24], [22], [23], [6] and can also be used to connect multiple projects (like suggested by the Research Data Alliance: 'https://www.rd-alliance.org/')

Building a metadata-based catalogue to describe the existing data and eventually how to access it, helps in discovering potentially useful datasets, while preserving the local environments, but neither does it directly support data exchange between local environments nor does it enable project-wide (database-)queries.

Although developing a holistic model for everybody in a research team, which suits everybodys' requirements for working with the contained data, is unrealistic to achieve, we believe that it is possible to design a model for just storing the data from the different foci. In the next section we will describe an approach, which on the one hand preserves local environments for working with data and on the other hand introduces an abstract storage based on [9]. In such a way in each research focus data can be processed in a suitable manner, while in a joint storage a single project wide dataset is created. In addition the activities of data exchange between researchers are moderated by the storage system itself.

3 Universal Storage and Holistic Local View Management

This section will present a novel approach, which is especially designed to address the challenges of (interdisciplinary) research projects (presented in Section 2.3). The central idea is the separation of data storage and data computation (/usage). The approach preserves individual viewpoints and local computation environments, while using a universal storage concept for communication between different local settings and creating a project wide dataset. The storage concept enhanced, for this setting, the work about generic databases of [9], while the idea of using it as a universal communication-tool is a generalisation of the exchange submodel approach in [21].

In Section 3.1 the basic idea will be explained more detailed. In Section 3.2 the universal storage concept will be described. The generation of export and import interfaces for the universal storage will be addressed in Section 3.3 and finally in Section 3.4 the (holistic) local view management will be the subject.

3.1 General Concept

The idea of the approach could be described by the slogan:

'global as storage and local as view'

As already mentioned a universal storage concept, inspired by [9], will be introduced. This concept is based on a quite abstract model and for that reason capable to store any data in the same (database) schema. In particular it is able to deal with all data from all participants of a research project, which allows to install a global (centralised) storage based on this universal model.

As the way of storing data in the universal storage rarely[10] fits the required (individual) forms for usage or processing by the participants, their local viewpoints are preserved. The generation, evaluation and in general the computation of data takes place in the local environments of each researcher. In these environments the used data has the individually required form and therefore the risks of refusal or disuse of the overall system are minimised.

Finally, to enable the overall system to work at all, it is necessary to connect the universal storage and the individual viewpoints. To do so an import/export-engine is used similar to mediators in the *'mapping'*-approach. It allows to map the local viewpoints to sets of (conceptual) views in the universal schema. Via those views data from the local environments can be imported and inserted into the global storage or extracted and exported. These conceptual views correspond approximately to the exchange (sub)models in [21].

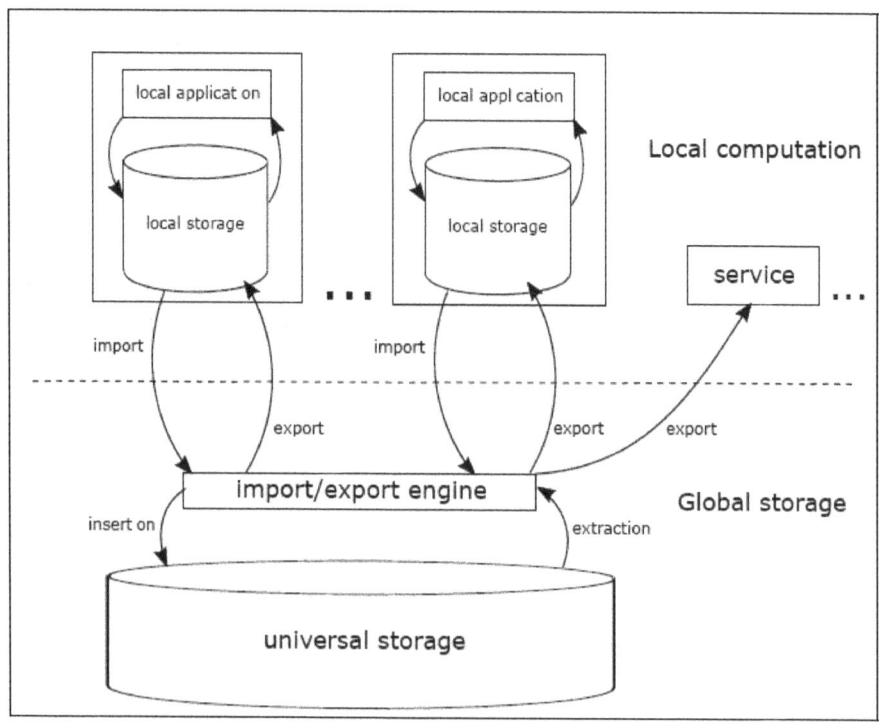

Figure 1. general concept

---

[10] probably never

As the way of storing information in the universal storage is not viewpoint specific, the exchange of data between different researchers can be done by importing their offered data into the global storage and exporting the desired data into their local environment via the (already existing) conceptual views. There is no need to create multiple individual exchange methods for every pair of researchers who want to exchange data, as all of them can just use the universal model as their exchange model. This eases the collaboration, as every researcher just has to deal with his or her own environment and eventually the interface of the global storage, but not with the local environments of other project participants.

Similar to that, it is easy to offer export services based on views, which allow to access explicitly specified subsets of the globally available data. This can for example be useful, if data shall be shared with people outside of the project.

The general concept is illustrated in Figure 1. In Section 5 information about an exemplary implementation of this concept is presented as well as some technical details of that prototype. Prior to that Section 4 recaps and concludes how the challenges mentioned in Section 2.3 are handled by this concept. But beforehand this section addresses more details of our approach.

3.2 Universal Storage

As visible in Figure 1, the global storage based on the universal model is the basis of the overall system. To be able to fulfil or enable its tasks and support for activities, it needs to provide some main abilities :

1. the ability to store a broad variety of heterogeneous data
2. the ability to connect different data sets (from different researchers/sources)
3. the ability to deal with evolving projects (/ evolving viewpoints)
4. the ability to derive conceptual views based on local viewpoints

In order to successfully achieve this, the model is quite an abstract one. Its core is based on the generic database schema from Jannaschks' work [9], in which he presented the idea to directly store information, which usually would be part of a data dictionary. This procedure creates an easier manageable generalisation of the RDF approach[11].

The enhanced model is quite robust concerning heterogeneous data and evolution and should be usable in different projects, needing only minor project-specific adjustments.

To structure the universal data model abstract central units, here called *information objects*, are used. They carry the same function as the central units in star or snowflake schemas (see [20]), as they represent the joint context for

---
[11] for example described in [2]

different (sub)sets of information. For this paper these information objects are any entities, which shall be described further and therefore can be associated with storable/stored data. In short, each information object refers to a real life object, concept or other entity and in the database all stored information concerning this real life instance is directly connected with the corresponding information object.

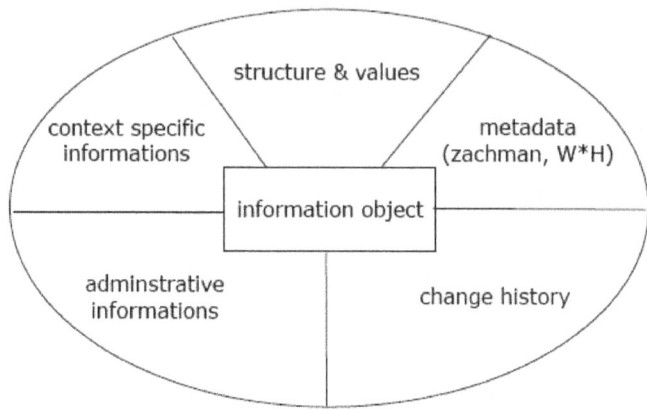

Figure 2. dimensions of data in universal storage

The associated data itself can be split into five different dimensions (see Figure 2). Each dimension captures a different kind of information and can be found in a different part of the model schema.

The central metadata dimension deals with the questions from the W*H and the zachman framework, which for example include information concerning *why*, *how* and *when* a date was created (see [4] and [25]). The Metainformation concerning change history[12] or administrative regulations, like access rights, is treated differently and can be found in own schema parts.

Another dimension is intended for project specific additions, like individual components for data of special interest and with desired separate individual handling. This 'extra dimension' can also be the basis for unusual project-specific (export-) services.

But the, in the context of this paper, most relevant dimension deals with the actual main content.[13] In the schema part of this dimension it is possible to store all kinds of information, without major adjustments of the model itself. This is done by the separation of content into values and structure, which will

---

[12] also called 'provenance'
[13] This is the one based on [9]

be the topic of the next subsection. Due to its relevance the corresponding part of the schema will be referred as the *core schema*.

Core Schema

The higher-order entity relationship model (HERM) for the part of the universal model which is responsible for storing the actual main content is presented in Figure 3. The model is designed to provide further information concerning the *information objects*, which are the defined entities of interest for a given context. As they represent 'real world' objects of investigation or activities, the first and most basic question the model can answer is :

What is this information object?

Each information object must be assigned to a type, which determines the possible structures and additional information, which can (and partly must) be added to the object. The types themselves can be ordered in a hierarchy.
For each type can be configured what kinds of information can be associated with information objects of this type. Instances of information kinds which have not been explicitly enabled for the type of an information object cannot be connected to it.[14] In particular these possible information kinds are split into three categories :

1. **Attributes**: Attributes are the most basic kind of additional information which can be instantiated for information objects (if enabled for its type). They represent atomic characteristics or properties and describe the information object itself.
   In short : in this context attributes are conform with attributes in (H)ERM, XML etc.

2. **Relations**: Relations describe the possible relationships with other instances of types[15]. References, foreign keys and similar concepts can be represented with this possibility. (To further describe the relations, attributes can be enabled for them too.)

3. **Non-Relational-Informationtype (nri)**: The last possible information kind is the 'non-relational information'. It represents everything else, which cannot be described as an atomic attribute or relation and shall nevertheless be stored.
   Typical information of this kind would be binary objects like images or complete documents.

As in different domains the same term can refer to different things, for each information kind a description can be archived to clarify which meaning is meant

---

[14] This should be ensured by database triggers.

[15] to be more precise: information objects assigned to these types

by which kind. This shall help to avoid misapprehensions.

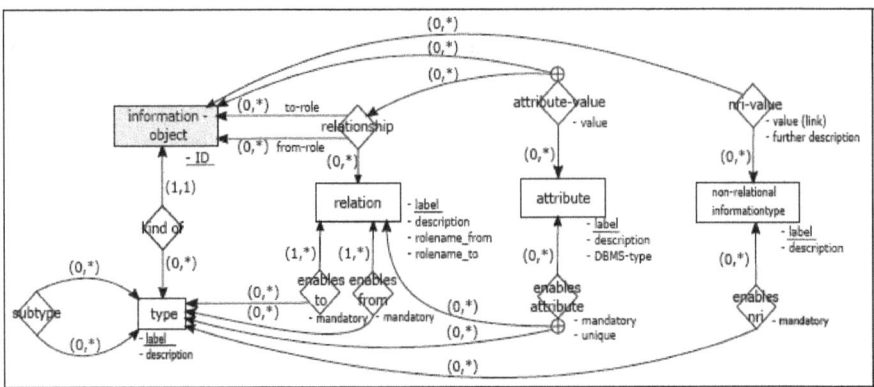

Figure 3. universal storage core schema

For existing information objects the values of enabled information are stored in triples. These are similar to the triples used in the resource description framework (RDF) in semantic web context. In general the stored information can always be read as statements of the form:

- For the information object with ID $= x$, the attribute y has the value z.
- For the information object with ID $= x$, the non-relational informationtype y has the value z.
  Or, if just a reference[16] is stored:
  For the information object with ID $= x$, the non-relational informationtype y can be found in z.
- The information object with ID $= x$ is in relation y to the information object with id $= z$.

In this model the values are stored atomically and not clustered in (tuple-wise or dataset-wise) bulks. This eases the generation of conceptual (import/export) views, described in Section 3.3.

It has to be kept in mind, that not every enabled information needs to be instantiated for each information object. Main reason for this is, that not all attributes fit to every viewpoint. So instead of always creating null-values for unused fields, these attributes can just be ignored. It is imaginable that researchers from different domains examine the same (information) object, but use, due to their different domains, distinct sets of attributes to describe it. To

---

[16] for example to a fileserver

connect their research they just have to refer to the same information object id. (Though it is possible to enforce the usage of very important information, like natural keys to ensure the identification of information objects, via the mandatory flag.)

Obviously (local) changes caused by project evolution can be tranformed to demands like:

- additional relations/attributes/nris become necessary
- some relations/attributes/nris become obsolete
- new types are needed
- old types are obsolete

Those changes can easily be managed by adding or enabling new types/relations/attributes/nris and adjusting the local conceptual views. The universal model itself has no need to be changed.

3.3 Schema Extraction

As already mentioned, it is expected that only few (probably none) of the researchers involved in such a project, will have the intention to structure their data, as it is structured in the universal model. The derivation of conceptual views from the universal schema, which can be mapped with the individual local schemas of the researchers, is for that reason crucial.

Initial Views
As a first step the construction of *'initial views'*, which collect all information objects of the same types with their corresponding values of all enabled attributes in relations, is quite easy to accomplish due to the structure of the universal model. In (relational/SQL-based) databases this would just be a series of (left outer) joins between the information-object relation and the attribute-value relation. The creation of *'initial views'* for each kind of relation and each non-relational information type is analogous

As the construction of those first views is based on the structure information (e.g. What is enabled for which type?, Which types can take which role in this relation? etc.), this can be done automated. Input-triggers, which map changes in those initial views to the corresponding relations, can be generated automated as well. [17] [18]

Together the initial views and their triggers form a basic interface for the storage part of the system.

---

[17] Due to the possible hierarchy of types, each information object can be present in multiple of those 'initial views'.

[18] These operations are similar to the PIVOT/UNPIVOT functions Terwilliger introduced in his thesis [17]

Derivation of Local Views

The initial views offered by the storage are presenting existing data in a more user comprehensible way, but in general they don't fit the local schemas. Nevertheless these views can be used to create a mapping between local and storage schema. Via projection, renaming, selection or similar operations on top of the initial ones, more layers of views can be established in order to get a fitting set of derived views for each viewpoint and local schema.

As long as the ids of the corresponding information objects remain part of the higher layer views, the triggers of the initial views are still able to deal with changes in these views and can be used to communicate with the storage.[19] [20]

Suitable Structures

```
1   FOR all reasonable units of the local schema DO {
2   IF(Is there already an existing suitable structure?){
3   do nothing;
4   }
5   ELSIF(Can an existing structure be reasonably enhanced to fit?[21]){
6   enhance structure;
7   rebuild corresponding initial view and trigger;
8   }
9   ELSIF(Can the unit be seen as super/sub-type of an existing type?){
10  create new type;
11  insert type into hierarchy;
12  if necessary: enable additional structures for new type;
13  build initial view and trigger for new type;
14  }
15  ELSE {
16  create completely new structures;
17  if possible and necessary: enable new structures for suitable types;
18  if necessary: (re)build corresponding initial views and triggers;
19  }
20  }
21  contruct local views based on fitting structures;
22  delete structure elements without any corresponding data
23  AND without presence in any local view;
```

Listing 1.1. proposed process of local view derivation

---

[19] Operations like aggregation would prohibit this too, but, as presented in the first part of this section, computation should be done in the local environments, which includes the computation of aggregates et al..

[20] In contrast to the full integration approach, where the local viewpoints are directly integrated into the global schema, the local viewpoints are integrated as viewsets on the abstract schema, but can be addressed in a similar way.

[21] e.g. enable additional attributes

A precondition for the construction of fitting sets of views is the existence of suitable types, attributes, relation kinds and non-relational information types in the core schema (with suiting enablements). When new viewpoints or new researchers shall be integrated or when, due to project evolution, some viewpoints changed, it might become necessary to not only adjust the local views, but to edit these structure information as well.

This process is about managing instances, while the (universal) model itself stays unchanged. Nevertheless it is important to be careful when editing, in order to preserve the integrity and usability of the system. Actions which could result in errors or problems are for example the creation of types (or attributes, or relations) which already exist under another name, but refer to the same concept and the overhasty deletion of an enablement which is still used in other viewpoints. Listing 1.1 shows a recommended procedure for the creation of a mapping between storage and local views.

Remark :
In order to import secondary data sources, their schema can be mapped to the storage schema in an analogue manner to generate an import interface.

3.4 Holistic Local View Management

After in the Sections 3.2 and 3.3 it was described how (almost any) data can be stored in the universal storage and how the abstract schema can be connected to the local ones, this Section will focus on the usage of data and the local environments. As already presented the computation and storage of data is separated in this approach and the local environments are the designated place for computation activities.

In local schemas data can be structured in a way which is best for individual processing activities and therefore ease or accelerate[22] the computation and individual work. Another advantage of this separation is, that the joint storage is not flooded with short term temporary data.

The keyword 'temporary data' leads us to the main question of this section :

> *Which data is exchanged between the various viewpoints and the storage?*

In the previous Section it was explained how local schemas can be mapped with the storage, however a viewpoint is not solely describable by its schema. The examined data itself is at least as important as simply its structure. For example it makes a great difference whether research questions are addressed on a regional or a supra-regional scale, or whether all available data is scanned

---

[22] The speed benefit is intensified by the fact, that each researcher him/herself can choose the machine for computation matching his/her own requirements.

or just a subset, which fulfils specific requirements. In general there might exist viewpoints (in the same project), which use the same local data structure and even might have similar research questions, but are independent because they use different[23] (sub)sets of data. Especially in bigger research projects it is not unusual to apply the same basic research question in multiple dimensions, like different regions, different time periods, different social groups or different platforms, in order to compare the results in a second step.

parameter	examplary requirement
content	- Only data from the excavation in Vráble.
	- Only data where the value of attribute y is greater than z.
administrative metadata	- Only data inserted by person x.
	- Only data created after my last export.
	- Only data I am allowed to use.[24]
qualitative metadata	- Only age determination data based on ^{14}C analyses.
	- Only data of a granularity better than c.

Table 1: exemplary filter parameters for exporting data

For this reason the conceptual views managed by the import/export-engine (see Figure 1) do not only map the universal schema to the local schemas, but apply data filters as well. When exporting to local environments these filters are responsible for the extraction of the proper dataset for the target viewpoint.

In Table 1 exemplary parameters for filtering are collected. Obviously metadata can be important for creating those filters and to find the right data for the different viewpoints. Which filters exactly should be used is dependent on the requirements of the individual research questions and the administrative regulations given by the project as well. The same applies to the frequency in which the exports should be repeated.

The importance of metadata implies, that it should be collected and imported into the storage as well. When importing main data from the local environments, it is desirable to also add the corresponding metadata. In some cases it is necessary to do this explicitly, but for some other parts the import/export-

---

[23] not necessarily disjoint
[24] should automatically be integrated in the filters

engine is able to determine the metadata itself.
Examples for metadata, which can be created/derived automatically are :

- administrative data like:
  - *When was the data inserted into the storage?* → system timestamp
  - *Who inserted the data?* → system login role
- metadata based on viewpoint like:
  - *The researchers of this viewpoint are all working solely in the Vráble excavation.*
  - *The researchers of this viewpoint are solely using secondary data sources.*
- metadata based on agreed default values or standards like:
  - *Standard age determinition method is the ^{14}C-method, if nothing else is specified, this method is assumed.*

Additional tasks of the import/export-engine arise from administrative regulations. An example of such tasks could be checking whether a person, who tries to change something in the storage, has the administrative rights to perform this change. While adding new datasets should in general be unproblematic, the updating or deletion of already existing datasets or the addition of meta-information might be prohibited, if the person who tries to do this is not the original creator.

This corresponds to the question in exporting *'Who is allowed to see/use which data?'* and the general question *'Who is allowed to perform which actions?'*. This is for example dependent on the chosen collaboration style[25] and should be decided a priori by the head of the project.

4 Contribution to State of the Art

The previous sections described the main challenges of data management in interdisciplinary and collaborative research projects (Section 2.3) and presented our approach (Section 3). Before concluding this paper we will now analyse to what extent the introduced challenges can be handled by the presented approach. To do so we will show for each of our main ideas explicitly its integration into our approach and its contributions towards the management of the challenges.

Separation of Data Storage and Data Computation
This separation of concerns is the basic idea of our approach. The universal storage system is in charge of addressing the challenges of maintaining heterogeneous and evolving data structures and supports general management activities. In contrast to that the local environments[26] are in charge of processing their data.

---
[25] see [16]
[26] e.g. local database engines

Due to this major principle, each project member is enabled to configure her-/his own computation environment adapted to his/her specific requirements and adapted to the used data. There is no central computation system in our approach. This is a very positive aspect, as such a system would typically be either adapted to all requirements of all members and therefore complex and hard to use, or based on a compromise and therefore not fit the individual requirements best.[27]

Despite of this separation data exchange and collaboration among project members can be managed via the storage. This makes the approach applicable for collaborative projects.

Viewpoints of Users on their own Data

Each user has an own viewpoint on data that reflects individual data representations/structures and a number of (sub-)sets of the complete available data. The data structures are typically designed to match the data to the viewpoint in a way that supports the effective execution of often used queries or computations. In general the structures and data match the local environments, the local research questions and the local technical setup. Our approach respects these viewpoints based on local configurations of each project member and incorporates them as essential parts. (As mentioned before, they are the locations for computation and processing activities.)

The incorporation of the individual viewpoints increases the acceptance and decreases antipathies for the complete system. The expertise of the researches concerning their working processes and handling of data in context of their research question is explicitly integrated into the overall system. Independent from the disciplines and research foci of other participants, everyone can continue in using her/his individual setup. Anyway collaboration among participants is still possible in a standardised way. Precisely collaboration and data exchange between project members is possible in a satisfying form just because everybody is responsible for computation himself/herself. This results from the fact that the universal storage (see next Section) must not offer explicit computation support for various individual research questions. It only needs to represent the pure data and data structures as views on the universal schema.

Universal Data Storage

The universal storage is the technological key element of our approach. Fortunately it does not have to support processing and computation activities for the manifold research questions itself, because these activities take place in the local setups. But still its tasks are extensive. It has to provide and deliver the data for those computation activities in a form that matches the requesting local viewpoint. And it is also the central part where heterogeneous data,

---

[27] Though systems based on a compromise are normally easier to use.

schema evolutions and viewpoints have to be managed.

The heterogeneity of the data structures - and thus the heterogeneity of the corresponding data - is handled by the usage of an abstract database schema that is capable of storing any kind of structured (relational) data. This is achieved by storing the structures of data separated from the data values. Each data tuple in a local environment can thus be mapped to an *information object* in the universal storage which reflects the information about the structure and the values of the original tuple. It thus is possible to store the data of each viewpoint in the same database schema in a way that allows to create view schemata which match the structures of the original local viewpoints.[28]

The evolution of local schemata and local data structures can in this system easily be transferred to the universal storage by adapting the stored structure information. There is no need to create new tables or to convert old datasets to the new structure. The decomposed way of storing the values allows to construct structures in a modular manner.

This modular and decomposed method of storing is also in use when dealing with multiple similar viewpoints. For viewpoints which concern the same data but differ in a few details - like additional attributes in one viewpoint -, just their corresponding structure information have to reflect this difference, instead of finding a compromising joint structure.

Collaboration View Schemata for Collaborative Data Exploitation

As the previous section elaborated it is possible to create view schemata on top of the universal storage that match the local viewpoints. This allows to integrate the data from the local viewpoints into the universal storage by using this view schemata as interfaces. To share data among viewpoints - thus among researchers - collaborating parties who use partly the same structures and views can use these existing views directly to access other participants data. But also for participants who do not share exactly the same structures and views collaboration and data exchange is possible. Therefor just another view schemata needs to be created on top of the view schema of the desired data or directly on top of the abstract core schema. Due to the modular and decomposed structure of the core schema, it is in general no problem to create views and view schemata that combine all available data in the storage in desired ways.

These views can not only be used to integrate data from the local environments into the global storage and vice versa, but also to supply data in an adequate form for services like GIS-applications or to make a specific subset of the data available for the public.

Maintenance of Local Data at the Local Level

This aspect was not in direct focus of this paper. Although the maintenance of

---

[28] using the stored structure information and other meta-information

structures (due to evolution) has been mentioned in Section 3, the maintenance of the data values was not. Anyway it is in general possible to perform the classic CRUD operations on all values in the universal storage via the view schemata. This enables project members not only to insert and extract data from the storage, but also updates and deletions are possible.[29]

Similar to the decision to incorporate the local environments, we also want to benefit from the expertise and discipline specific expert knowledge of the collaborating researchers by enabling them to maintain the data in the storage. We are convinced that this local maintenance strategy performed by domain experts is more efficient than any global cross-disciplinary quality standard.[30]

Although the content is mainly maintained by the researchers, we support the integrity of the global storage by enforcing the inserted data to be conform with the defined structures[31] and enable the tracking (and eventually rewinding) of changes with a provenance system (see next section).

Global Versioning of Data

Like the previous one, the global versioning aspect has not been in focus of this paper too. Nonetheless in Section 3.1 (and Figure 2) the 'change history' was mentioned as one of the five dimensions of data in the central storage. This dimension is responsible for global provenance tracking, versioning and logging of significant activities in the storage.

The need to keep old data as well after new data is introduced, the wish to reproduce the steps that lead to the current state of a data set, and the possibility to rewind to previous states, are no exclusive requirements for research projects. But as the same datasets might be used and edited by various different project members, this context implies that the creation of a single, joint, and global system for this tasks is a better approach than local and redundant change histories with the need to synchronise with each other.

Due to the decomposed and modular structures of our global storage we are able to address changes in each single attribute, relation et cetera on this small scale. This is very useful when tracking or logging changes which do not affect the whole datum.

Remark :

In short, our main ideas can be combined into one approach and thus support the handling of typical challenges in interdisciplinary and collaborative research

---

[29] Which participant is allowed to edit which data should be determined for each project a priori, for example by choosing a collaboration style (see [16]).

[30] Of course one could employ one expert per discipline to check all data of his/her colleagues, but this does not sound like a fun job.

[31] We check for example if attribute values have the correct data type, if all elements marked as mandatory are present, and if only information have been instantiated that are enabled for the type of the information object.

projects. Not all aspects could be addressed in full detail in this paper, but a general overview over was given. The next and last Section will now conclude this paper and shortly present the context giving project for this work.

5 Realisation of the Approach

Our approach is currently applied in the context of the collaborative research centre 1266 *'Scales of Transformation - Human-Environmental Interaction in Prehistoric and Archaic Societies'*[32] funded by the DFG[33]. This project deals with phases and processes of transformations in early times.[34] It sets the frame for joint research of participants from diverse disciplines spanning from prehistory and archaeology over biology and geography to geophysics, philosophy and ecosystem research. Collaboration and data exchange is crucial in this context. Some of the researchers are almost completely dependent on data from others to be able to develop and evaluate their theories, as they do not perform any own field work.

We are currently installing a prototype system based on the approach presented in this paper. This system fully relies on open-source technologies. It uses *PostgreSQL*[35] databases for the universal storage (see Figure 3) and an adapted *GeoNode*[36] content management system as user interface and controlling entity for the systems logic. GeoNode itself runs on the Python-based web framework *Django*[37] and is designed to handle spatial data.[38] The central *import/export engine* introduced in Section 3.1 is realised via a combination of Python modules in the GeoNode system and user defined functions in the database.

The technical realisation of our approach is intended to be a central service for each of the local research projects. The prototype system is currently in a validation and data enrichment phase. We are including data from selected project partners according to the process described in Section 3.3 and enhance the system step by step.

Once this first version of the (prototype) system is installed entirely, the members of our project will be enabled to integrate and manage their data regardless of its domain and structure. The localisation of useful data from colleagues and respectively the process of sharing data will me manageable via basic grant-revoke methods for access-rights. The integrated viewpoints thereby allow each researcher to access all data available for him/her in a form which fits to their

---

[32] https://www.sfb1266.uni-kiel.de/en
[33] http://www.dfg.de/en/index.jsp
[34] An example for such examined transformations is the *'neolithic revolution'*.
[35] https://www.postgresql.org/
[36] http://geonode.org/
[37] https://www.djangoproject.com/
[38] This ability is important for other aspects of our project.

research questions and preferences. As these integrated viewpoints are computed dynamically, sharing the newest data revisions will be easily feasible too. In addition the deleted data and previous versions will be stored as well, to enable reproducible research.

In later versions of the system more features will be added. E.g. web map services and other special forms of visualisation are planned. Also the already available features will become more ergonomic and enhanced as they will be improved based on the users feedback.

The context of CRC 1266 is feasible to test our approach due to the joint project wide research question which requires interdisciplinary collaboration of the subprojects. We eagerly anticipate the potentials for improvement, that will likely be revealed by the application of our approach in this project. Especially possibilities for further (partial) automation or support for the mapping of local environments with the storage are of interest hereby.

6 Conclusion

Our approach combines the flexibility of local environments - which are probably best adapted to perform individual tasks - with the strength of a universal storage to connect those research foci. This enables the creation of a project wide dataset, while avoiding the problem to install and maintain a project wide (working-) data model. Queries over the full set of available data are enabled. Additionally it is not necessary to integrate multiple individual solutions for pairwise schema mappings, as the global project storage also fulfils the function of a global exchange model and platform.

The storage itself is, like its inspiration from [9], robust concerning (local) schema evolution and heterogeneous data. Its core schema additionally supports non relational information, type hierarchies and references to joint entities of interest. The explicit addition of further dimensions[39] makes the approach better applicable in 'real-life' contexts. As it bases on already existing technologies its application is feasible.

The presented data management strategy is quite simple to maintain and evolution safe. Regardless of their discipline all participants of a research project may join the central database together with their data. In the same way it is no problem to integrate new members or datasets from research partners.

As mentioned the recommended community of practice for our approach are medium sized (interdisciplinary and collaborative) research projects. The advantages when using it in very small or very large projects are probably limited. On the one hand will this approach be inapplicable large projects that deal with big data, as the storage concept to store each (local) tuple in a decomposed way inside of few relations will probably fail due to the capacity of the used

---

[39] e.g. metadata

database⁴⁰. On the other hand, for small projects with only few viewpoints, limited domains and little evolution, our approach might be over the top. In other words, our approach is designed to handle broad varieties of evolving data structures and allows collaboration and individual data processing, while expecting an only moderate amount of data. In contexts where the data structures are fixed and simple this approach would be needlessly complex. In contexts of very huge data amounts it should be used neither.

But besides these restrictions, our strategy should as well be generalisable for other kinds of data management projects outside of the research area.

6.1 Acknowledgement

This research was performed and funded in the framework of the CRC 1266 'Scales of Transformation - Human-Environmental Interaction in Prehistoric and Archaic Societies' which is supported by the DFG. We thank both institutions for enabling this work.

We also appreciate the advices, constructive proposals, and detailed comments made by our reviewers.

References

1. Jens Bleiholder and Felix Naumann. Data fusion. *ACM Computing Surveys (CSUR)*, 41(1):1, 2009.
2. Jeremy J Carroll and Graham Klyne. Resource description framework (RDF): Concepts and abstract syntax. 2004.
3. Data Advisory Group (DAG). *The principles of good data and information management*. UK Environmental Observation Framework.
4. Ajantha Dahanayake and Bernhard Thalheim. W*H: the conceptual model for services. In *Correct Software in Web Applications and Web Services*, pages 145–176. Springer, 2015.
5. IGGI (Intra governmental Group on Geographic Information). *The Principles of Good Data Management*, 2005.
6. Dirk Hoffmeister and Constanze Curdt. Research data management services for a multidisciplinary, collaborative research project: design and implementation of the tr32db project database.
7. Fynn Holst. *Konzeptuelle Sichten*. Christian-Albrechts University of Kiel, 2015. (Master Thesis).
8. Hannu Jaakkola and Bernhard Thalheim. Multicultural adaptive systems. In *Information Modelling and Knowledge Bases XXVI, 24th International Conference on Information Modelling and Knowledge Bases (EJC 2014), Kiel, Germany, June 3-6, 2014*, pages 172–191, 2014.
9. Kai Jannaschk, Claas Anders Rathje, Bernhard Thalheim, and Frank Förster. A generic database schema for CIDOC-CRM data management. In *ADBIS 2011,*

---

⁴⁰ or at least lack desired performance

Research Communications, Proceedings II of the 15th East-European Conference on Advances in Databases and Information Systems, September 20-23, 2011, Vienna, Austria, pages 127–136, 2011.
10. Ulf Leser and Felix Naumann. *Informationsintegration - Architekturen und Methoden zur Integration verteilter und heterogener Datenquellen*. dpunkt.verlag, 2007.
11. Peter C. Lockemann, Ulrike Kölsch, Arne Koschel, Ralf Kramer, Ralf Nikolai, Mechtild Wallrath, and Hans-Dirk Walter. The network as a global database: Challenges of interoperability, proactivity, interactiveness, legacy. In *VLDB'97, Proceedings of 23rd International Conference on Very Large Data Bases, August 25-29, 1997, Athens, Greece*, pages 567–574, 1997.
12. Sergey Melnik. *Generic Model Management: Concepts and Algorithms*, volume 2967 of *Lecture Notes in Computer Science*. Springer, 2004.
13. Felix Naumann, Johann-Christoph Freytag, and Ulf Leser. Completeness of integrated information sources. *Information Systems*, 29(7):583–615, 2004.
14. Erhard Rahm. *Mehrrechner-Datenbanksysteme - Grundlagen der verteilten und parallelen Datenbankverarbeitung*. Addison-Wesley, 1994.
15. Klaus-Dieter Schewe and Bernhard Thalheim. Design and development of web information systems. (unpublished, forthcoming Springer book).
16. Klaus-Dieter Schewe and Bernhard Thalheim. Development of collaboration frameworks for web information systems. In *20th Int. Joint Conf. on Artifical Intelligence, Section EMC07 (Evolutionary models of collaboration)*, pages 27–32, 2007.
17. James Felger Terwilliger. *Graphical user interfaces as updatable views*. Portland State University, 2009.
18. Bernhard Thalheim. Data management: Challenges and solutions a critical reflection of data management practices in research projects. (unpublished).
19. Bernhard Thalheim. Informationssystem-entwicklung. *BTU Cottbus, Computer Science Institute, Technical Report I-15-2003*, 2003.
20. Bernhard Thalheim. Engineering database component ware. In *Trends in Enterprise Application Architecture, 2nd International Conference, TEAA 2006, Berlin, Germany, November 29 - December 1, 2006, Revised Selected Papers*, pages 1–15, 2006.
21. Bernhard Thalheim and Marina Tropmann-Frick. Model capsules for research and engineering networks. In *East European Conference on Advances in Databases and Information Systems*, pages 202–214. Springer, 2016.
22. Christian Willmes and Georg Bareth. A data integration concept for an interdisciplinary research database. In *Proceedings of the Young Researchers forum on Geographic Information Science-GI Zeitgeist, ifgiPrints*, volume 44, pages 67–72, 2012.
23. Christian Willmes and et al. An open science approach to gis-based paleoenvironment data. *ISPRS Annals of Photogrammetry, Remote Sensing and Spatial Information Sciences*, pages 159–164, 2016.
24. Christian Willmes, Daniel Kürner, and Georg Bareth. Building research data management infrastructure using open source software. *Transactions in GIS*, 18(4):496–509, 2014.
25. John A Zachman. A framework for information systems architecture. *IBM systems journal*, 26(3):276–292, 1987.

Part III
Rigorous Methods

Cyberphysical Systems:
A Behind-the-Scenes Foundational View

Richard Banach[1] and Wen Su[2]*

[1] School of Computer Science, University of Manchester,
Oxford Road, Manchester, M13 9PL, U.K.
banach@cs.man.ac.uk
[2] School of Computer Engineering and Science,
Shanghai University, Shangda Road, Shanghai, China.
wsu@shu.edu.cn

Abstract. Hybrid and cyberphysical systems pose significant challenges for formal development approaches based on pure discrete events. In this essay, after a brief look at the CPS landscape, the foundations of CPS systems are examined from the ground up, with a particular view to aspects rooted in the continuous part of the CPS spectrum. We take a journey starting from the foundations, through a number of ways of addressing the continuous mathematics aspects, to phenomena latent only in the world of physical descriptions, such as the onset of instability due to passing through bifurcation points in the problem parameter space. We argue that such phenomena, that can plague CPS design when optimising for performance metrics, can only be understood by sufficient engagement with the continuous world.

1 Introduction

In today's world of cheap processors, memory, sensors and controllers, the enthusiasm for hybrid [8] and cyberphysical [12] systems (CPS) is veritably exploding. This is increasingly fueling the cost-effectiveness of a smart-everywhere approach to services and systems. New initiatives pour forth at a seemingly ever-increasing rate, in many domains, e.g. health, transport, city infrastructure, communication etc., and their many subdomains.

The presence of control as first class citizen in these systems leads to the impingement of discrete techniques from the computing sphere on the one hand, onto a plethora of techniques from continuous mathematics and the physical systems sphere on the other, lending a highly multi-disciplinary nature to this discipline. It is fair to say, that more than in almost any other multi-disciplinary area, the fundamental role that mathematics plays in all the disciplines that impinge here, means that these disciplines can interact in a deep way, rather than merely providing a distinct view on each other, or offering complementary but still separate families of techniques.

---

* Wen Su was supported by the National Natural Science Foundation of China (No. 61602293), and the Science and Technology Commission of the Shanghai Municipality (No. 15YF1403900).

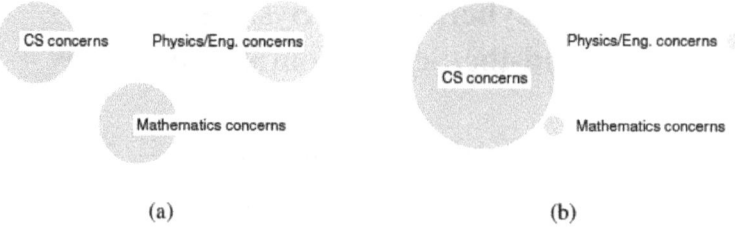

Fig. 1. CPS and computer science: (a) how it should be, with all disciplines involved exerting comparable influence; (b) how it actually is, with overwhelming weight placed on computing perspectives, paying scant regard to the other disciplines.

It is often claimed that completely new formalisms will be needed to reason about systems of this kind, a view that is a little puzzling considering that every component or aspect of such systems comes with a well understood mathematical framework that captures the predictability of its behaviour in engineering contexts. The presence of these, and the *a priori* consistency of mathematics, thus tends to suggest that the underlying mathematical dialogue has not been pursued in sufficient depth. Since Klaus-Dieter Schewe has long had a strong interest in the foundational aspects of cyberphysical systems, it is a pleasure to discuss some of these issues, from a particularly personal perspective, in this festschrift essay.

These days, most design and development of cyberphysical systems is very much rooted in the integration of, and cooperation between, existing tools and techniques from different areas of computer science and different branches of engineering and technology. Overwhelmingly, especially on the computer science side, such tools and techniques are focused on discrete descriptions of system behaviour, and usually pay scant regard to the continuous aspects of physical behaviour. Fig. 1 gives an illustration of the uneven focus.

Unsurprisingly, such approaches are frequently fraught with problems of compatibility and of unpredictable interworking. This arises from a lack of attention to the different semantic foundations of the contributing formalisms, and a lack of precision with which they view issues which are fundamentally continuous. Regarding the latter, frequently, the formalisms in question are unable to speak at all, or can say very little. Since continuous phenomena can display extraordinary subtlety, such a dislocation is evidently undesirable.

In this essay, we look at these issues from the perspective of rigorous model-based system development and verification, but taking a keener interest in the more problematic areas rooted in the continuous world. We will find that we can point to many things which, although perplexing from a conventional discrete/computational perspective, become much clearer when enough notice is taken of what continuous mathematics can tell us. We infer that if we are suitably cognisant of the insights available from *all* the disciplines that contribute to CPS, then most of the foundational problems for CPS melt away, even if the practical problems of constructing large real-world systems both optimally and verifiably, assuredly do not.

The remaining sections of the paper are as follows. In section 2 we briefly survey the most visible features of the CPS world as perceived within the research community. After a few comments, in Section 3 we start from the foundations, reviewing the elements that underpin the mathematical foundations of the theories that contribute to CPS descriptions. The foundational view is important, since a consistent picture must operate across *all* the contributing disciplines, and must connect with the world of discrete mathematics that operates in the computing sphere. This journey through a number of mathematical subdisciplines culminates in Section 4 with the prospects for mechanically supported verification, based on Collins's groundbreaking Cylindrical Algebraic Decomposition, and the possibilities for adapting non-semialgebraic descriptions by using suitable approximations. In Section 5 we use this basis to review a number of phenomena rooted in the continuous world, whose implications are less obvious from a purely discrete perspective. They include: differential-algebraic equations, control issues such as stability and the effects of multi-system descriptions, technical issues in control, delay differential equations, and bifurcations. The section continues by discussing numerical approaches and sampling and quantization issues. Section 6 summarises and concludes.

2 The CPS Landscape

Nowadays, computing devices get ever smaller, more distributed and interconnected, both to each other, and to the physical environment. This enables the construction of systems with a bewildering variety of architectures, required performance characteristics, and interplay with the real world. A very major role is played by simulation in the design of such cyberphysical systems, with popular software suites like SIMULINK [22], 20-sim [2], Modelica [24], much to the fore. Simulation and experimentation are certainly the most appealing ways to realise such systems, since they are so accessible and easily usable, with a relatively modest investment in preparation.

Somewhat more rigorous than pure simulation and experimentation are approaches based on the control aspects of the CPS system. A large literature has grown up around the exploration of appropriate stable control regimes for particular CPS configuration styles and application regimes. Most of this work appears within the wider control systems literature.

As ever though, simulation and experimentation in principle cannot achieve the level of assurance that verification can give (provided, of course, that the models being verified can justify the faith placed in them). Here, the self-evident undecidability of any language expressive enough to describe an interesting set of CPS systems impinges directly on what can be verified and how. The hybrid automaton paradigm (qualified in various ways, as needed) is the default descriptive mechanism in this space. In [8] there is a good survey of well established systems for this, and overwhelmingly, these tend to focus on linear behaviour, because of the tractability of the fragments of arithmetic that are involved.

Another approach takes the analytical descriptions of non-trivial hybrid and cyberphysical systems at face value, and, reflecting centuries-old practice in applied mathematics and physics, engages with them symbolically. The aim is to formalise and to mechanise what can be done via such techniques. Among these approaches we can cite Hybrid CSP [19,32], KeYmaera [26,1] and Hybrid Event-B [5,6].

Invariably, the above sketch is an oversimplification, and there are a large number of variations on these themes to be found in the literature, e.g. [28]. We will have more to say about the connections between some of these styles of approach below.

3 Starting from the Foundations

Cyberphysical systems, by their very name, involve physics (and thus its practical application, engineering). Immediately this implies the involvement of mathematics. They also involve computing, and this too implies the involvement of mathematics. Ideally, all three contributing disciplines, namely physics, mathematics and computing, would play an equally significant role in the development of the subject. However, what is overwhelmingly seen is a very heavy emphasis on the computing aspects, as shown in Fig. 1. The texts [18,3] rather bear this out.

Interestingly, the two mentions of mathematics above refer to very different areas of the subject. In the physical sciences, the mathematics is predominantly continuous, dealing with real valued quantities changing according to physical laws, often expressed via differential equations. Extremely rapid variation in these quantities is idealised as impulsive change, resulting in discontinuities in the real valued behaviour. In the computing sphere, the mathematics is overwhelmingly discrete, with instantaneous change of state being the normal paradigm. The discrete mathematics is overwhelmingly concerned with properties and behaviours expressible via discrete, very often finite, sets.

The discrete sets consist of elements that have no internal structure, and usually, few relationships between them. This contrasts with the world of reals, in which, although the reals are also treated as having no internal structure, we have an enormously rich selection of properties at the disposal of the utiliser, this being due to the fruits of the mathematical analysis that has been created over the last couple of centuries.

3.1 Insights from the Contributing Disciplines

Accepting the broad sweep of issues just mentioned, brings a number of points into prominence:

– Cyberphysical systems are concerned with physical quantities. In physics all quantities are functions of time t, and time is real. Time is also not manipulable (in classical physics). Thus, physics deals with the way that various quantities change over time, but time itself is not one of them.[1]

---

[1] Formally, this means that time is read-only, and that physical quantities must be defined for all applicable time points (if one interprets normal physical discourse from a formal point of

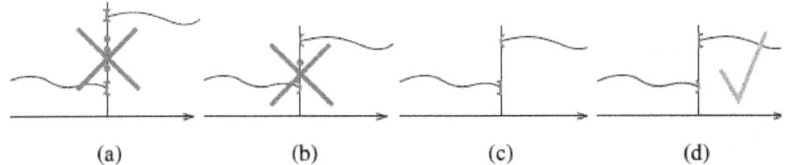

Fig. 2. Piecewise continuous functions of time, and intervals: (a) multivalued functions of time are simply unphysical; (b) individual values at isolated times have no physical significance; (c) left-open right-closed intervals compose nicely, but yield executions without a definite starting point (since there is no earliest point of a left-open interval); (c) left-closed right-open intervals compose nicely and yield a definite starting point.

- Turning to the logical/foundational aspects of the reals, we realise that any real-valued expression, once its parameters are fixed, can only refer to an isolated real number: there is never a 'next real', nor indeed a 'previous real'.
- The computing world re-emphasises the need for discontinuous change. Together with the previous point, a CPS formalism must thus be capable of expressing isolated discontinuous changes in value.
- Typical engineering and physical models require the use of differential equations. So any CPS formalism needs to encompass those.
- Moreover, the normal calculational problem solving techniques used in applied mathematics need to work, otherwise any putative formalism would struggle to achieve anything useful.

3.2 Consequences

We regard the preceding observations as a kind of requirements list that sets out some conditions that a CPS formalism must meet, and we now examine where this leads.[2]

We start with discontinuities. If discontinuities are isolated, then in between, functions of time (that describe values of variables) must be continuous. So we are dealing with piecewise continuous functions, which must therefore be continuous on intervals that can potentially be open or closed at either end. Additionally, in formalisms like the duration calculus [33], multiple state transitions are allowed to take place at a single moment of time. The latter would lead to state functions on intervals that abut at points where the function may be multi-valued — we can dismiss such functions immediately as being unphysical.[3]

---

view). Moreover, formally, each physical quantity is identified with a particular free variable, because physical discourse is typically much more open-ended than typical formal theories allow.

[2] Many existing CPS formalisms deal with these issues in various *ad hoc* ways. In this esay we take their mathematical consequences at face value, and see where this takes us.

[3] We are not saying that the duration calculus is not useful, merely that it is not useful for the task at hand.

If a right-open interval is followed by a left-open one, then there is a single real value in the middle at which a function that is continuous on those intervals may be defined differently. Such individual point values also have no physical significance since only the integrals of functions over regions (whether large or small) can have an impact physically. So we can disregard this configuration of intervals.

So we are left with left-closed right-open intervals, or alternatively left-open right-closed intervals. Both kinds abut nicely with others of the same kind, making bigger ones. However, left-open right-closed intervals raise an exception at the initial time of an execution, since there is 'no earliest moment' of the execution, and a single point does not define a left-open right-closed interval. Thus, left-closed right-open intervals, permitting a definite starting point for an execution, which can then continue over a succession of such intervals, emerge as the winning candidate. Fig. 2 summarises this line of argument.

Next, differential equations (ODEs). Immediately, the discontinuities create a technical issue. If we can anticipate in advance where the discontinuities in functions occur, we can arrange for them to fall exactly at the boundaries of our left-closed right-open intervals. But *if we cannot*, then the right hand side of an ODE may contain discontinuous functions, and it itself may be discontinuous, making the derivative on the left hand side badly defined at such a point.

Here, the instinct of pure mathematicians to imaginatively generalise previously established notions comes to our aid. These days, ODEs are studied assuming their right hand sides are *measurable* over time. Isolated points of discontinuity do not spoil this property, allowing us to use this, the Carathéodory interpretation of ODEs [29], for cases where there might be unanticipated discontinuities on the right.

Of course, if there are no unanticipated discontinuities, we do not need the additional sophistication. However we *do* need a criterion like Lipschitz continuity of the right hand side of an ODE [29], to avoid cases like the non-Lipschitz $\mathcal{D} x = x^2 + 1$, whose solution $x(t) = \tan(t)$ explodes at $t = \pi/2$.

Differentiability goes along with *absolute* continuity (rather than unqualified continuity) [31,27]. A function f is absolutely continuous (AC), iff the fundamental theorem of calculus works, iff an increment of f (over an interval) is the integral of its derivative (which exists almost everywhere, over the same interval). This, in tandem with the preceding observations, suggests the world of piecewise AC real functions as a suitable semantic universe for grounding the semantics of CPS formalisms.

As a diversion, Fig. 3 shows a non-absolutely continuous function, the famous Cantor ternary function. It is defined to be flat (with value $\frac{1}{2}$) on the middle third of the unit interval, to be flat (with values $\frac{1}{4}$ and $\frac{3}{4}$) on the middle thirds of the previous outer thirds ... and so on recursively. All the middle thirds add up to length 1, yet there are enough points left over in the 'Cantor dust' that remains to have the cardinality of the entire real line. Though continuous, and flat almost everywhere, this function cannot be the integral of its derivative.

This grounding in a class of functions of time, specifically the piecewise AC real functions of time, permits viewing the two kinds of update needed in CPS systems (discrete and continuous) from a remarkably similar perspective. A discrete update is a pair of valuations of the variables of the system (the before-valuation and the after-

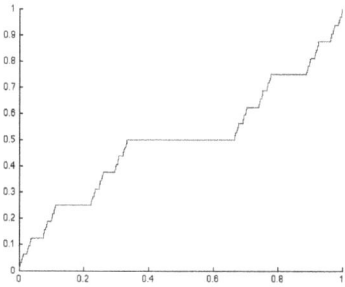

Fig. 3. The Cantor ternary function, a function which is continuous but not absolutely continuous. On all the 'middle thirds' (whose total length is 1) it is flat, so its derivative is zero almost everywhere. Yet it increases in value from 0 to 1, so it is not the integral of its derivative.

valuation) pinned to a particular moment in time (the time of the discontinuity). A continuous update is a time indexed family of pairs of valuations of the variables, with the before-valuation being the valuation at the start of the continuous behaviour and the after-valuation being the valuation at any choice of time subsequently. Viewing time as merely indexing, makes it the only aspect that is different.

Although piecewise AC real functions solve the semantic foundations issue, they permit behaviours that are extremely poorly behaved, if judged by the standards of day-to-day applied mathematics calculations. As an example, consider the function $f(t) = \exp[-\frac{1}{t^2}]$. It is zero at $t = 0$, and so flat there that all its derivatives at $t = 0$ are also zero. This means that the Taylor series derived from these derivatives defines the zero function, quite different from $f(t)$. This implies that Taylor series in general are unreliable when one is dealing with real functions.

We are rescued by a quotation from P. A. M. Dirac: "*A number theory is beautiful, but the complex number theory is more beautiful.*"[4] Its message is that although there are many kinds of 'numbers' explored within mathematics, when we consider complex numbers specifically, and in particular complex analytic functions, a vast array of uniquely powerful properties suddenly burst into vivid relief [17,11]. Here are a few.

Firstly, Taylor's theorem works. A function which is complex differentiable is automatically complex analytic, i.e. Taylor's theorem defines it uniquely. Secondly, there are few awkward issues to worry about: poles of finite order (e.g. $1/(z-a)^k$ for integral k); branch points (e.g. $1/(z-a)^k$ for fractional k); essential singularities. Thirdly, we have unique analytic continuation: knowing complex analytic $f(z)$ precisely in any region, no matter how small, determines $f(z)$ everywhere where it is defined — an incredibly powerful result for practical day-to-day calculations. Fourthly, the authors' own particular favourite, Picard's Great Theorem: *Every analytic function assumes every complex value, with possibly one exception, infinitely often in any neighborhood of an essential singularity.*

---

[4] Note the exquisite use of the indefinite and definite articles in this quotation. Unfortunately we are not aware of the original source of the quotation.

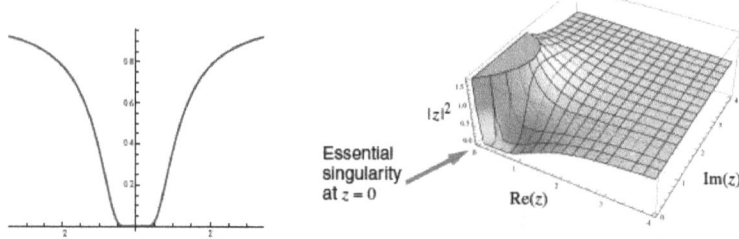

Fig. 4. The real function $f(t) = \exp[-\frac{1}{t^2}]$ on the left. On the right, the magnitude squared of the complex function $f(z) = \exp[-\frac{1}{z^2}]$ which extends it. Picard's Great Theorem implies that $f(z)$ cannot be depicted close to the origin, because of the essential singularity there.

Viewing t as the real part of complex z in our above example, $z = 0$ is an essential singularity of the function $\exp[-\frac{1}{z^2}]$, and this is the source of its wierd behaviour *vis à vis* Taylor's theorem. If we restrict our view of this function to just the real axis, we see nothing untoward, aside from the unnerving degree of flatness; but look one iota away from the real axis in any neighbourhood of the origin, and it's bedlam. Fig. 4 shows both the behaviour on the real line, and the complex behaviour away from the real line. Thus, in general, in order to recover good calculational properties, we need to restrict the piecewise AC real functions, to those which arise as piecewise complex analytic functions which are real on the real line.

4 Proof and Verification

Complex analytic functions (even if we just focus on the real part) are fine, but to get the best out of them often requires human ingenuity. This provides food for applied mathematics enterprises in universities the world over. In the computer science world, the desire for automation dictates that we prefer to invest human ingenuity in deriving powerful algorithms that solve broad classes of problems automatically, rather than focusing on individual problems. This sets up a tension between the applied mathematics and computer science perspectives — a tension between *calculation* and *proof*. Let us look briefly at a couple of examples.

Take the case of Propositional Logic (PL). In this case, there are only Boolean values. There is essentially *no* calculation, all manipulation being equivalent to proof. By the time we move to First Order Logic (FOL), calculation starts to play a role: there are constant and function symbols and their interpretations, and expressions formed from them denoting values. Yet handling these is still very generic: Hintikka sets, Herbrand universes and the like, and they lead to the known generic semi-decidability results, etc. Moving on to languages that are expressive enough to plausibly represent CPS systems, generic model theoretic and proof theoretic techniques have long taken a back seat. Whether one can solve a particular CPS system, or prove some property of it, depends almost entirely on the specific constants and function symbols (and on their standard interpretations) that occur in it, and what one knows about them.

Thus, at low levels of formal language expressivity, inference and decidability form the focus, and, to the extent that particular formal languages allow, decidable language fragments are typically defined in terms of connectives that occur, numbers of variables, permitted numbers and alternations of quantifiers, etc. At high levels of expressivity, inference is determined by in-depth investigation of special cases, and typical 'decidable language fragments' (although the phrase is seldom used in these contexts) are often defined by parameterisations of these special cases, making their scope conceptually much narrower than is usual in mathematical logic.

In the traditional applied mathematics sphere, aspects that would usually be attributed to 'logic', were, in the old days, simply done by hand in the meta-level discourse. These days, as automation increases the size and complexity of problems that are tackled, there is a benefit in using automation to manage such aspects, error-prone as they can be. They include: case analysis, completeness of coverage, bound variable scopes, Skolem constant management, SMT-like calls to calculational oracles.

Referring back to the foundations of the CPS world, we must confront the vast gap between the plain set theory of discrete state change on the one hand, and on the other, sophisticated phenomena like essential singularities (that must be avoided if we are to have any hope of calculating anything). Of course the route from simple set theory, via naturals, integers, rationals, reals to complexes, is well known and can be formalised in various ways. The authors have come across a number of such endeavours over the years. When it is attempted for real, it always goes the same way, described as follows.

4.1 Formalising from the Foundations

At the start there is great enthusiasm (the work, if funded, has just been approved). Much enthusiastic hacking of foundations takes place. There are many interleavings of quantifiers to deal with, but morale is high, and progress is made. Work continues, and after a year or two the foothills of applied mathematics slowly start to become visible. As a result of all the valiant struggles with the foundations in the early days of the project, by now, all the foundational issues have been surmounted, and what remains is just hard work.

At this point it is legitimate to ask just how much hard work is at stake. The NIST Handbook of Mathematical Functions [25], a standard bible of results for theoretical physicists, applied mathematicians and their ilk, amounts to almost 1000 pages. The typical foundational effort that has just been described seldom covers more than 50 pages of [25]. Even accepting that 500 or so of those 1000 pages might be regarded as somewhat esoteric for everyday applications, the achievements of the typical formalisation of applied mathematics do not amount to a serious resource for general purpose applied mathematics problem solving.

Going back to our typical foundational endeavour, by the time the point described has been reached, there being no further foundational issues to chew on, the earlier enthusiasm of the foundations enthusiasts has become severely depleted. The endeavour quietly dies.

Of course, there are tools that confront the needs of applied mathematics, as depicted in [25], head on. They are the computer algebra tools such as Mathematica [21], Maple [20], MATLAB [22], etc. Typically, they work at a much higher level of abstraction than the foundations-led outline above. In effect they encode the cases that can be solved, and put great effort into powerful pattern matching routines, so that solvable cases can be discerned as often as possible. The commercial basis of most such tools enables the large amount of work indicated above to be undertaken in a uniformly compatible way, and they are used extensively in real engineering design. These days, despite not starting from the kind of foundations we discussed, the residual risk in the mathematical core of established tools of this kind is vanishingly small, compared with other risks in the design processes during which they are used, in the construction of the systems we rely on every day.

To summarise the above, given some fixed, specific result in mathematics, the complexity of proving it increases dramatically as one descends into increasingly deep foundations in order to insist on basing the proof at that level of axiomatisation. There is a rather splendid parable about this, which the reader may enjoy, following paragraph 7011 in Andrews's admirable book [4].

4.2 Towards Verification

In the computing world, the added value that a formal approach brings (above and beyond the capabilities of conventional development), amounts to the ability to prove properties of a systems model, in a mechanically verified way. The properties in question are predominantly safety properties. In the context of safety properties, choice is always interpreted demonically, since safety demands adequate behaviour under all possible circumstances. However, in the CPS world, control problems are very much to the fore, and in the control world, the *controllability* property is key. Roughly speaking it says: *for every permitted way of setting up the system and system goal,* there exists *a control input that steers the system to the goal.* The emphasis on existential quantification is deliberate, it tends to imply angelic choice, and is unavoidable.[5] However, as often happens when there is an assumed progress property and an angelic property contingent on it, the latter can be wrapped into a safety property (containing the existential quantification) of the assumed progress property, permitting a focus on safety properties in verification after all. (This typically happens for data refinement properties.) The inescapable progress of time, outside the control of the human user, provides a very useful progress property in the context of CPS systems.

A major contributor to the possibility of verification in the CPS arena is Collins's groundbreaking *Cylindrical Algebraic Decomposition* (CAD) [9,7]. This is concerned with semi-algebraic constraints, which are quantified Boolean combinations of formulae of the form $P(x_1 \ldots x_n) \geq 0$, where P is a polynomial expression with real (in practice rational) coefficients.[6] It was Tarski who observed that the disposition of real roots of a real polynomial could be discerned using an extrapolation of classical techniques: Sturm sequences, polynomial GCD calculations, and other results, that used only the coefficients of the polynomial, and which only needed to be manipulated

---

[5] No conveyance, be it a plane, car, or other vehicle that allows its driver to determine its travel parameters and destination, can prevent its driver crashing it, if the driver so chooses.

[6] Thus, polynomial equations and proper inequalities are subsumed.

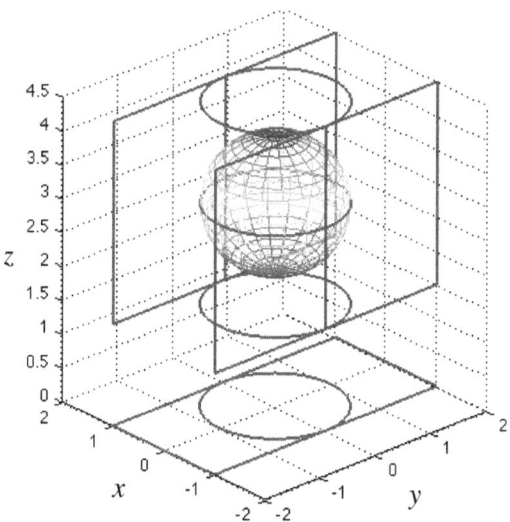

Fig. 5. Three dimensional space, recursively divided into cylindrical regions appropriate to the CAD description of a single sphere of unit radius. Projecting on z produces a circle in a plane; then projecting on y gives an interval bounded by two points, within a line. The two points define two lines in the plane, and these then define two planes in three dimensional space. In general, the decomposition of an $n-1$ dimensional region gives rise to cylinders in n dimensions when the projection is reversed.

symbolically. This led eventually to a decision procedure for semialgebraic constraints that had non-elementary complexity. Collins's great achievement was to reduce this to doubly exponential complexity in the number of variables — making it practical for a moderate number of variables. Collins's result came about by paying close attention to detail, and systematically organising the elimination of variables one at a time until the single variable techniques could be applied, after which the eliminated variables are reintroduced and characterised, again one at a time. We illustrate this fascinating process in Figs. 5, 6, 7.

We take the often used single sphere of unit radius as a running example, specifically, its interior, given by the inequality $P(x, y, z) < 1$ (the negation of the obvious improper inequality), where $P(x, y, z) = x^2 + y^2 + (z - 3)^2$. Fig. 5 overviews the process. In the projection phase of CAD, a rather complicated procedure, but one that is nevertheless symbolically computable in terms of semi-algebraic constraints, implicitly identifies the regions of the (x, y) plane where the disposition of the roots of $P(x, y, z) = 0$ (when viewed as a polynomial in z with coefficients in (x, y)) is invariant. In practice this defines the circle $x^2 + y^2 = 1$. This is repeated to project out y, leaving $x^2 = 1$. This is now solved giving $x = \pm 1$.[7]

---

[7] In general, algebraic numbers, also symbolically computable, are needed to find the roots of an arbitrary real polynomial.

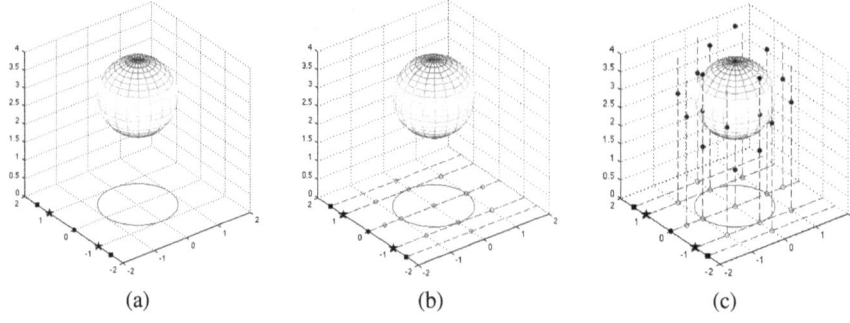

Fig. 6. Illustrating the CAD extension process. Witness points identifying: the region $x < -1$, the point $x = -1$, the region $-1 < x < 1$, the point $x = 1$, the region $1 < x$ are all shown in (a). In (b), this is extended to the (x, y) plane, identifying regions and points of the plane where the projected constraints have invariant truth value. In (c) this is extended to three dimensional space, systematically dividing it into regions in which the original family of constraints (i.e. $x^2 + y^2 + (z-3)^2 < 1$) have constant truth value.

The CAD process now moves into the expansion phase, illustrated in Fig. 6. The solution points for x identify regions of the x axis and their boundary points. Witness points are chosen in the interiors of the intervals (squares and dot in Fig. 6.(a)); the boundary points are also highlighted (stars in Fig. 6.(a)). These values are substituted into the previously derived semi-algebraic constraints in the (x, y) plane giving lines parallel to the y axis. These lines are subdivided into intervals (with their boundary points) according to how they intersect the solution set of the (x, y) plane's semi-algebraic constraints — and witness points are again found (the small circles in the plane $z = 0$ in Fig. 6.(b)). The procedure is repeated: the (x, y) plane witness point values are substituted into the preceding set of semi-algebraic constraints, giving lines parallel to the z axis. Again, witness points are chosen on these lines in the same manner. This gives the dots depicted in three dimensions in Fig. 6.(c). The essential fact throughout this procedure is that the intervals inside which the witness points are chosen have the property that the truth values of the whole family of semi-algebraic constraints (at the requisite level of projection) are invariant within the region containing the witness point. So it is sufficient to evaluate the constraints at each witness point, and to make suitable logical combinations of the answers, in order to know the truth value of the whole family throughout the whole region.

The procedure as a whole builds a tree, shown for our example in Fig. 7. For instance, the middle quintuplet in the bottom row of Fig. 7 (labelled (x, y, z)) corresponds to the vertical line through $x = y = 0$ in Fig. 6.(c), and represents, respectively: the semi-infinite interval below the sphere, the intersection of the line with the lower hemisphere, the interior of the sphere, the intersection of the line with the upper

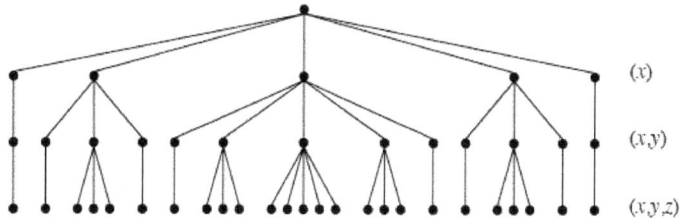

Fig. 7. The tree of witness points for the extension phase of the CAD procedure in the three dimensional sphere example. Descending one level in the tree corresponds to one 'unprojection' in the projection phase.

hemisphere, the semi-infinite interval above the sphere. It is clear that the decomposition along the variable axes generates considerable complexity —after all, there are only three regions of interest in our example: the interior, exterior, and surface of the sphere— yet the tree in Fig. 7 has many nodes. Nevertheless, the procedure is regular and scalable (modulo feasibility considerations), making it widely applicable in practice.

Unfortunately, a decision procedure for semialgebraic constraints does not directly solve the problem of automation of reasoning in CPS systems because, aside from a few rare exceptions, the solutions of CPS systems are not (made out of) polynomials. However, all is not lost. Consider a solution to a CPS system of the form $f(t) = \exp[-\lambda t]$, for which the safety invariant $t \geq 0 \Rightarrow f(t) \leq A$ needs to be established. As a Taylor series, $f(t) = \sum_{k=0}^{\infty} (-\lambda t)^k / k!$. Suppose we did not know that $f(t) = \exp[-\lambda t]$ was monotonically decreasing (making it sufficient to check the value at $t = 0$ to establish the safety invariant). For N big enough, successive terms of the Taylor series are monotonically decreasing in magnitude and alternating in sign. Thus, if N is big enough and odd, successive pairs of terms make a net negative contribution to f so that we can write $f(t) = \text{poly}(t) - |\text{corr}(t)|$, and discarding the correction $\text{corr}(t)$ gives a safe overestimate for f via the polnomial $\text{poly}(t)$. Such an approach allows the safety invariant to be proved within an interval using CAD, if all goes well. (Note that this depends on the safety invariant being given by an inequality, rather than requiring something more precise.)

The example just given was discussed in detail because the general principles apply widely. Useful solutions to (the continuous portions of) CPS systems are asymptotically stable, which means they decrease in magnitude over time. This also means they are given by series whose terms decrease in magnitude and alternate in sign (perhaps in groups). So a technique of judicious grouping of terms can manipulate the series into a polynomial plus a correction that acts in favour of the safety inequality that needs to be established. The MetiTarski tool [23] embodies ideas of this kind.

5 Gremlins Rooted in the Foundations

The preceding sections took us on a journey that spanned the chasm between the simple discrete set theory of familiar computational state change, and a number of formulations in the continuous world that emphasised existence of semantics, or calculational utility, or potential for automation. Any foundational approach to CPS needs to be able to speak somehow to all these issues, whether explicitly or implicitly.

In the computing world, there is a tendency to the view that once a formal framework has been set up, then 'the rest is programming'. This is largely a consequence of the fact that in the world of discrete set theory and bits, the lack of structure leads to a dearth of generic results. If an arrangement of sets and bits leads to one set of circumstances, it is usually not too hard to find another arrangement of sets and bits which leads to the exact opposite circumstances. In the continuous world this is not the case. The subtle constructions that lead from the discrete foundations to the rich continuous world that we have indicated lead to various non-trivial phenomena that apply to broadly applicable classes of system. In this section we consider a number of topics that are connected with problem solving for CPS systems that arise in this way.

5.1 The Influence of Control Problems

The nature of CPS systems, with their interplay between the physical world and computer control, means that control systems *per se* form a large part of the remit of CPS systems. Control theory has been intensively studied since early in the 20th century, and is by now a large and mature discipline. The point of this brief paragraph is to flag up that while many of the topics in this section are not necessarily directly couched in control terms, their relevance to CPS systems is inevitably because of their impact on control systems.

5.2 Differential-Algebraic Equations

Most formal systems for CPS tend to be monolithic, meaning that the whole of the system under consideration is included in a single model, which is then subjected to whatever analysis is envisaged. There is a good reason for this of course. Having a single model makes the maximum information available to the analysis, maximising its potential power. We illustrate this more concretely below.

However, non-trivial practical systems are made of collections of separate components. This means their global properties emerge indirectly. One very mundane aspect of this is that the separate components of the system are directly connected to each other. Consequently, where a CPS formalism handles the continuous world exclusively via ODEs, this can make for some awkwardness, because the relations expressing direct connections are algebraic; e.g. the voltage and current across an electrical connection between two components are equal, rather than being related via an ODE. Thus the continuous world of complex CPS systems is best described by systems of differential-algebraic equations (DAEs) [16], which combine algebraic relations with ODEs, rather than by just using ODEs.

DAEs evidently offer more possibilities than ODEs alone. Clearly, taking at face value arbitrary algebraic relations involving variables and their derivatives as a specification mechanism forces the adoption of a purely simulation/numerical approach, since there is no hope of an analytic solution in such a setup. This prevents proof of properties in the usual model based way. Restricting to linear DAE specifications gives greater leeway for proof based approaches, but still permits a much wider range of behaviours than we see with just linear ODEs. We illustrate what can happen with some very simple examples. In the following, x, y are state variables, f_1, f_2 are inhomogeneous terms (all of these being potentially time dependent), and $||$ is simultaneous definition. We will assume that there are no other constraints on the state variables than the ones we write.

Among the possibilities that may arise, we have the following: inconsistency, e.g., $x := y || y := x+1$; unique solution with forced initial value, e.g., $x := 1$; unique solution with arbitrary initial value, e.g., $\mathcal{D}\, x = 1$; solution with constrained initial values and given by an unspecified arbitrary function, e.g., $x := y+1$; solution with arbitrary initial values and given by an unspecified arbitrary function, e.g., $\mathcal{D}\, x = y$; solution with forced initial values and involving constrained inhomogeneous functions, e.g., $\mathcal{D}\, x = f_1 || x := f_2$.

In the case of linear systems with constant coefficients (such as all our examples) the Kroneker canonical form of the so-called matrix pencil of coefficients of the system covers all the possibilities that arise. DAE systems are sufficiently complicated that even just relaxing the constant coefficients constraint to allow the coefficients to vary over time is sufficient to materially alter the collection of possibilities available for linear systems.

5.3 Stability Considerations, Multiple Machines

As mentioned above, there is a visible tension between the compelling verification impulse towards monolithic descriptions, yielding global information and maximum power for inference on the one hand, and on the other hand, the pragmatic engineering impulse towards partitioned descriptions, yielding the maximum potential for separate working and thus (optimistically) shorter time to market, but permitting reduced information in the context of each individual component. We give a small illustrative example.

Suppose we have an integrated system containing two variables $x(t)$ and $y(t)$ subject to the dynamical equations $\mathcal{D}\, x = y$ and $\mathcal{D}\, y = -x$. In the context of global information, the solution of this system is familiar: $x(t) = \sin(t)$ and $y(t) = \cos(t)$. These behaviours for $x(t)$ and $y(t)$ are bounded and are (marginally) stable, realising the bounds $|x(t)| \leq 1$ and $|y(t)| \leq 1$.

Now, in the interests of separate working, suppose we are obliged to put $x(t)$ and $y(t)$ into separate constructs, with only partial information available to each about the other. Specifically, let us suppose that in the $x(t)$ construct we only know $|y(t)| \leq 1$ and in the $y(t)$ construct we only know $|x(t)| \leq 1$. Then the locally known versions of the dynamical equations become $\mathcal{D}|x| \leq 1$ and $\mathcal{D}|y| \leq 1$. The worst case of these is

$|x(t)| \le t$ and $|y(t)| \le t$. Evidently these behaviours are not stable, so that the loss of information attributable to system partitioning has destroyed certainty about stability.

We can contrast the preceding situation with that in which we have stronger information about stability. Thus instead of the preceding integrated system, suppose that the integrated system's $x(t)$ and $y(t)$ variables have the dynamical equations $\mathcal{D}\,x = y\,\mathrm{e}^{-\lambda t}$ and $\mathcal{D}\,y = -x\,\mathrm{e}^{-\lambda t}$. Now, the integrated solution is $x(t) = \sin(\frac{1-\mathrm{e}^{-\lambda t}}{\lambda})$ and $y(t) = \cos(\frac{1-\mathrm{e}^{-\lambda t}}{\lambda})$. Again we have bounded and stable behaviours, realising the bounds $|x(t)| \le 1$ and $|y(t)| \le 1$. In the partitioned system, if we have the same loss of information about $x(t)$ and $y(t)$ as we had before and must replace occurrences of $x(t)$ and $y(t)$ with the bounds, then the best locally known versions of the dynamical equations become $\mathcal{D}\,|x| \le \mathrm{e}^{-\lambda t}$ and $\mathcal{D}\,|y| \le \mathrm{e}^{-\lambda t}$. Now, the worst case becomes $|x(t)| \le \frac{1-\mathrm{e}^{-\lambda t}}{\lambda}$ and $|y(t)| \le \frac{1-\mathrm{e}^{-\lambda t}}{\lambda}$. This is still stable behaviour.

From the formal verification standpoint, the essential aspect of this version of events is that, even with the degraded knowledge attributable to the partitioning, if the worst case behaviour is nevertheless stable, safety invariants about the behaviours of $x(t)$ and $y(t)$ may still be provable, even if they are suboptimal compared with the global information case.

In section 3.2 we observed the analogy between discrete and continuous updates. This analogy extends to stability considerations. In the discrete world, we often prove termination of a sequential process by proving that each of its steps strictly decreases a *variant expression* which takes values in a well-founded set. The lower bound provided by well-foundedness prevents the sequential process from proceeding indefinitely. By contrast, in the continuous world, we often prove stability by proving that the continuous process (over time) strictly decreases a *Liapunov expression* which takes values in a portion of the reals that is bounded below. As the continuous process forces the Liapunov expression nearer and nearer the lower bound, the dynamics is typically increasingly confined to an asymptotic region, yielding asymptotic stability. Viewed in this light, the termination of the sequential process may also be seen as a kind of stability criterion, since, having terminated, the sequential process is unable to change the value any more.

5.4 Technical Issues in Control

One consequence of the relative longevity and maturity of control theory is that a number of different approaches have been developed to the mathematical analysis of control problems over the years. Given the necessity of relating CPS formulations to foundational issues that we have explored above, the impact of a foundational perspective on different control approaches merits examination.

Most control engineering, as taught in engineering curricula, takes place in the frequency domain, using the Laplace transform (in the case of continuous control) or z transform (for discretized control). Practically useful results are readily obtained, and mathematically, the results in this domain are derived using an L^2 notion of convergence (convergence using mean square error, in less technical language).

An alternative approach, made more popular with the availability of modern simulation based tools, seeks to solve control problems directly in state space. In this domain, some results are also derived using the L^2 formulation, while others are derived in L^∞ (convergence using maximum pointwise deviation). While the L^2 results in the frequency domain and in the state space domain can be related via Plancherel's theorem, they speak about integrated square error rather than pointwise deviation. This makes the results derivable in the two kinds of approach incompatible. A good property derived in either domain does not carry any implication of the analogous property holding in the other domain — in fact quite the opposite is usually the case: a good property derived in one domain spawns a counterexample in the other.

Thus, the various domains of control theory (and others we have not mentioned) tend to exist in separate mathematical silos to some degree, owing to the detailed differences in the rigorous formulations that define them. This being so, it is nevertheless notable that these subtle issue have rather little impact on the day-to-day practice of control engineering. Nevertheless, they *do* have impact on an integration of control issues with model based formal methods techniques, because the latter readily exhibit a sensitivity to the kinds of mathematical subtleties mentioned — and day-to-day control engineering does not. This in turn, is attributable to the fact, as we mentioned above, that model based formal methods are rooted in simple set theory, and so any integration with other concerns has to be sound when based on such set theoretical considerations.

5.5 Delay Differential Equations

The ODEs that we have focused on hitherto provide an excellent framework for describing fundamental physical processes at the microscopic level. However, in system engineering we inevitably deal with finite macroscopic components, and their size and other characteristics mean that it is often the case that there is a delay between the inputs that a component is subject to, and the outputs it can deliver in response. This is especially true if the component in question involves a communication network. Delay differential equations (DDEs) can provide a useful description of some such systems [15,10].

To see the intriguing phenomena that DDEs can lead to, let us consider the simplest possible example: $\mathcal{D} x = -K x(t - \tau)$. This says that the derivative of x is proportional to a value of x that is τ time units old. In analysing the behaviour of such an equation, is is always most useful to start by linearising, and looking at the stability of small, exponentially varying deviations. In the case of our equation, we start with the simplest possible solution, a steady state solution. In a steady state solution x does not vary with time, and the significance of the delay disappears. It is easy to see that $x^{\blacklozenge}(t) = 0$ is a steady state solution. If we now add a small exponential perturbation $A \mathrm{e}^{\lambda t}$ to $x^{\blacklozenge}$, we can analyse the tuples of K, A, λ values that yield solutions, and examine the stability (or otherwise) of such solutions. Substituting $A \mathrm{e}^{\lambda t}$ into the equation and simplifying yields $\lambda = -K \mathrm{e}^{-\lambda \tau}$. Unfortunately, even in this simplest of problems, the equation is a transcendental equation, and the analysis of the character of its solutions is not trivial.

With K real and positive, if λ is real, it must be negative (because $\mathrm{e}^{-\lambda \tau}$ is then always positive). This is good news, because it implies that $x^{\blacklozenge}$ is a stable solution.

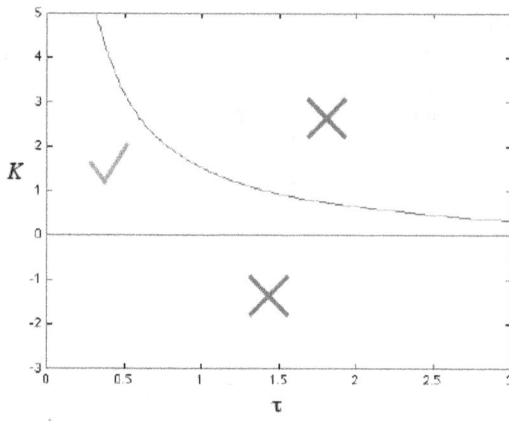

Fig. 8. Regions of solution stability and instability for the simple DDE $\mathcal{D} x = -K x(t - \tau)$.

Unfortunately, as we vary the three parameters, we find that there are also solutions to $\lambda = -K e^{-\lambda \tau}$ with non-real λ, and these lead to oscillatory behaviour around $x^{\blacklozenge}$.

Fig. 8 shows the general characteristics of the parameter space. Evidently, $K < 0$ forces real λ to be positive, yielding instability. Also, for $K > 0$ and τ small enough, we find negative λ and stability. But as τ increases (with K fixed), we eventually cross into another unstable region. What we have outlined in embryonic form is the onset of a steady state bifurcation, in which a seemingly innocuous system, upon being subject to a change in the values of its static parameters, suddenly destabilises and exhibits oscillatory behaviour.

5.6 Bifurcations

Varying parameters to optimise resource utilisation or other performance metrics is grist to the mill in engineering design. Equally, the sudden and unexpected onset of instability and oscillatory behaviour as parameters are varied, is the bane of the engineer's life. Such behaviour is particularly perplexing to the engineer steeped in discrete methods, since there is no possible way to discover the possibility of onset of instability of the kind described, by looking at the system from a purely discrete perspective. Only by non-trivially engaging with the continuous mathematics of the system can one hope to discern the root cause of instabilities of this kind.

The steady state bifurcation we outlined above is merely the simplest example of sudden change in the global characteristics of the solution space of a system as its parameters are varied. Another commonly occurring kind of bifurcation is the *Hopf bifurcation* [13,14].[8] In this, varying the system's static parameters causes a pair of characteristic roots of the stability equation to cross the imaginary axis with non-zero

---

[8] Historically, Poincaré and Andronov also studied this phenomenon, before Hopf's account made it more widely known.

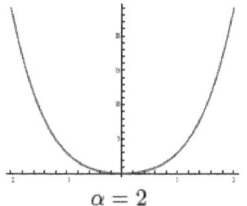 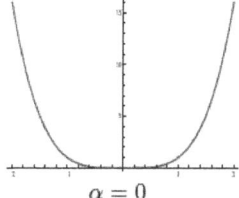 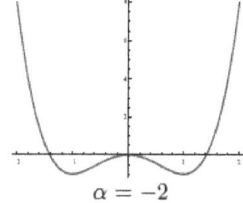

Fig. 9. Values of the schematic 'energy function' $E = r^4 + \alpha r^2$, where $r^2 = (x^2 + y^2)$ for various values of α.

velocity. As in the preceding case, a previously stable (but not steady state) solution to the system suddenly loses stability and starts to behave in an oscillatory manner.

Figs. 9 and 10 illustrate how a Hopf bifurcation works. We imagine a system containing two dynamical variables x, y (amongst others). For convenience, we suppose that there is circular symmetry in x, y so that we can use $r^2 = (x^2 + y^2)$ and get some simplification. We suppose that there is an 'energy function'[9] for x, y that looks like $E = r^4 + \alpha r^2$ where α is some system parameter that is subject to optimisation. We also assume that the system is dissipative, so that, left to its own devices, the dynamical trajectory would seek a point of minimum 'x, y energy'.

For α positive, the 'energy function' E looks like Fig. 9.(a); the minimum is comfortably at $r = 0$. When $\alpha = 0$ the 'energy function' E looks like Fig. 9.(b); the minimum is still at $r = 0$, but the neighbourhood of $r = 0$ is flatter. But when α is negative, e.g. $\alpha = -2$, the 'energy function' E looks like Fig. 9.(c); the minimum is no longer at $r = 0$ but at a non-zero value.

Typical system trajectories for α values of $+2, 0, -2$ are shown in Fig. 10. In Fig. 10.(a), for $\alpha = 2$, a typical trajectory rapidly sinks to the 'energy' minimum $r = 0$; the system is stable. In Fig. 10.(b), for $\alpha = 0$, a typical trajectory still sinks to the 'energy' minimum $r = 0$, but more slowly because of the absence of the quadratic term in E; the system is still stable because of the quartic term. But for $\alpha = -2$ the quadratic term is negative, and close to $r = 0$ it overcomes the quartic term. The 'energy' minimum jumps away from $r = 0$. There is now a circular limit cycle of minimum 'energy' in the x, y plane. Fig. 10.(c) shows how typical trajectories flee from $r = 0$. For either variable, x or y, the observed behaviour is oscillatory once the dynamics is tracing out the limit cycle.

In fact the 'energy' minimum jumps away from $r = 0$ infinitely fast as a function of α as α crosses the value 0 in a negative direction. Fig. 10.(d) shows the shape of the minimum 'energy' manifold as α varies. The paraboloid on its side in Fig. 10.(d) witnesses the sharp departure from $r = 0$ at the critical point $\alpha = 0$, at which the

---

[9] Assuming the existence of an 'energy function' is not essential here but it helps to illuminate the example. The 'energy function' does not have to literally be a form of energy — many energy analogues have been identified over the years that share some of the mathematical properties of genuine energy, without actually being forms of energy.

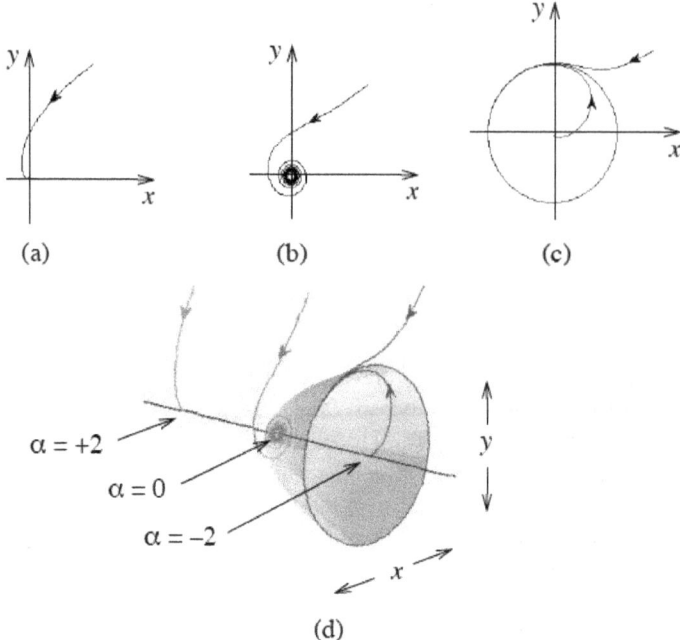

Fig. 10. Typical trajectories for the dissipative 'energy functions' of Fig. 9. In (a) $\alpha = +2$; in (b) $\alpha = 0$; in (c) $\alpha = -2$ and trajectories flee from $r = 0$. The minimum 'energy' manifold consists of the non-negative α axis together with the paraboloid on its side for negative α.

minimum energy manifold switches from a single zero value to the circle of radius $\sqrt{-\alpha/2}$. The various example trajectories are superposed on it.

Of course these examples are just the best known instances of change in the global characteristics of the solution space, and they provide an inroad into a wealth of increasingly complicated phenomena that can be increasingly difficult to discern from the standard way in which problems are posed, in terms of differential and algebraic equations. In all cases though, an approach based purely on discrete techniques is doomed to never make contact with the underlying causes of the loss of stability, and to leave designers who rely solely on such techniques perplexed by the instabilities that they encounter.

5.7 Numerical Approaches

From the perspective of gaining the greatest possible insight into the behaviour of a system, being able to attack it analytically (as we have done in the examples above) is evidently the ideal. But this ideal approach is available in an acutely small number of cases. The case of linear systems is familiar and provides the basis for most exact results in the kinds of case we have discussed, but beyond linear systems, the pickings for analytically based approaches are sparse indeed. Unfortunately, most realistic descriptions of physical systems contain nonlinearities, immediately putting them beyond the reach of analytical approaches.

Under the circumstances, numerical approaches come to the fore. Here we can distinguish two kinds of problem tackled numerically. The first kind concerns the calculation of numerical values for situations where there is an underlying analytical model. A typical such case would be a linear equation, $\mathcal{D}x = Ax$ say, for which we have an analytic solution e^{At}, and we need a specific value, e.g. at $t = 5$, so we need to calculate e^{5A}. For such problems, there are typically power series (e.g. the familiar one for e^t), which we can use to home in on the value needed. It is important to note that in such cases there is a large amount of group theoretic machinery doing a lot of the heavy lifting behind the scenes, which allows us to focus on just the value $t = 5$. Such numerical problems are relatively easy. (Evidence for the implicit presence of the group theory is given by such familiar things as the multiplication laws for the exponential, e.g. $e^a e^b = e^{a+b}$, or the addition laws for trigonometric or hyperbolic functions, and so on.)

The second, much tougher kind, are situations in which there is no discernible group theoretic machinery around (aside from very generic results, e.g. concerning the flow semigroup of a dynamical system, that tend not to lead to calculational techniques). Then, if the dynamics starts at $t = 0$ and we are interested in what happens at $t = 5$, there is no longer a convenient formula to have recourse to, which captures the initial conditions at $t = 0$, and into which we can just plug the value $t = 5$ to get the answer. Now, we must grasp the differential equation (or other dynamical system) and must integrate it by brute force, inching along until we reach $t = 5$. This is much more difficult for the following reason. Let $\mathcal{D}x = \phi(x)$ be our ODE. Assuming we know the value of x at $t = t_0$, we are interested in the value at $t = t_0 + h$. For this we will need (sooner or later) the values of $\mathcal{D}x$ at $t = t_0$ and $t = t_0 + h$, the latter of which depends on the value of $\phi(x)$ at $t = t_0 + h$, which depends on the value of x at $t = t_0 + h$, which is what we are trying to find! An enormous literature has arisen around this issue, because of both its technical challenge and its enormous practical utility [13,14]. The fact that there are readily available existence theorems that assure us that all these things are well defined [29] only adds to our chagrin, since they do not easily translate into efficient numerical algorithms.

A further, related point concerns the kind of results that *are* available regarding this kind of numerical integration of differential equations. Typically, they state that in the limit of the step size h going to zero, such and such an algorithmic procedure converges to values which are on the solution trajectory of the ODE. However, of more interest to the engineer is a result which quantifies the closeness of the approximation to the solution, in terms of the step size needed to achieve it. Such results are more rare, especially since the estimates used in proving the convergence tend to be suboptimal, to improve the perspicacity of the proof.

Given the somewhat pessimistic drift of the last paragraphs, we can reasonably ask whether there is much scope for the kind of analytically based approaches that we have, rather implicitly, been advocating. More specifically, how much can proof and verification based approaches contribute to the development of CPS systems? The answer is twofold. Firstly, there is the bespoke option. Although many systems are not solvable analytically, mathematical ingenuity can often rigorously elicit specific facts about the solution space of the problem, which can help improve what can be deduced numerically. Such work goes case by case, and helps sustain applied mathematics departments in universities the world over, and will continue to do so for the forseeable future. Secondly, there is the fact that engineers need to be able to predict how the artifacts they design will behave. If linear systems give the only route to routine predictability, then engineers will, overwhelmingly, tend to use linear techniques in their designs, using them in designs built out of linear pieces, combined in suitable ways to give overall behaviour which is not linear in the large. The first option offers a significant challenge to verification approaches since each problem will demand its own verification strategy, but the second is much more tractable, since it just requires the flexible combination of pieces which can be handled analytically.

5.8 Sampling, Aliasing, Quantization

Although we have focused rather heavily on phenomena latent in the continuous world in the preceding sections, in practice, system development inevitably descends to the world of discrete implementations. In this world, the smoothly changing phenomena of the continuous world dissolve into jumpy broken-up phenomena, whose behaviour is not always a simple retraction of their continuous counterparts. Of course, the desire in performing a discretization step is for the result *to be exactly* a simple retraction of the continuous version, but, as often happens, the desire for the design to optimise certain performance measures may make the discretization step cross a boundary between relatively faithful reflection of an earlier continuous design and a more chaotic regime.

The study of the correspondence between continuous and discretized worlds is perhaps most highly developed in the signal processing world. Real signals are continuous, whereas the processing of them is done almost exclusively in the digital sphere nowadays. The digitization process involves, firstly, the choice of a suitable sampling rate, and at each sampling point, the conversion of the value of the continuous signal to one of a finite number of discrete values: quantization. The converse applies when a digital signal has to be put back into the continuous world.

Much has been learned about these processes over recent decades [30]. The rule of thumb *sine qua non* in this sphere, is the Nyquist Sampling Theorem. This states that provided the sampling rate is at least twice the highest frequency present in the signal to be processed, discretization will not introduce false harmonics into a reconstructed signal. In fact, the onset of false harmonics in a reproduced signal can be seen as a kind of bifurcation in the underlying detailed dynamics. Thus, as a parameter (the sampling rate) is steadily decreased, a threshold is crossed and the dynamics bifurcates to allow the presence of not only the correct frequencies, but of the false ones too. The false frequencies give rise to what is known as aliasing. These phenomena become apparent when we precisely model what is going on in the discretization process in the continuous sphere. Of course, what works in the time domain (i.e. as regards sampling rate) works equally well in the value domain. Thus, too coarse a quantization strategy can be as damaging to a discretization process as too low a sampling rate.

Inevitably, what holds in the signal processing world applies directly to the control problem world of CPS, since both the input and output of a control process are themselves signals. The subtlety here is that, in essence, all the results derived in the signal processing world regarding sampling and quantization are derived in the L^2 sphere. This contrasts with the model based and verification approach to CPS systems which is much more concerned with results in the L^∞ sphere. This is because the model based approach works with the current state (at the current moment in time) and fidelity to some desired notion of acceptable behaviour is based on the pointwise deviation between actual and desired state, as time proceeds.

As pointed out in Section 5.4, the L^2 and L^∞ worlds are, strictly speaking, mathematically incompatible. Therefore, fully understanding the complex phenomenon of the discretization process, with its many opportunities for bifurcation arising from the additional presence of the discretization parameters remains a considerable challenge.

6 Summary and Conclusions

The kind of mathematics that might conceivably have an impact on the formulation and behaviour of CPS has been in development for at least 400 years. During this time an enormous amount of relevant knowledge has been accumulated, from the applied mathematics that strives to accurately quantify the behaviour of components and systems, to the pure mathematics that underpins the foundations of differential equations and connects these to the kind of discrete formulations familiar in the computational world.

In this essay we started by arguing for a world of piecewise absolutely continuous real functions of time, since these include both the absolutely continuous functions within which the modern theory of differential equations resides, and the discontinuous changes needed by computational frameworks and impulsive physical control. This world allows the semantics of typical syntactic frameworks for CPS to be formulated fairly easily. However it offers few guarantees regarding calculation.

Within this world we identified the functions that were (restrictions of) piecewise holomorphic functions that were real on the real line, as offering dramatically better prospects for calculation. However, despite this, they offer limited prospects for automation, since mathematical creativity is often required to get the best out of this world.

To maximise the possibilities for automation, we pointed at the functions characterised by semialgebraic properties, for which relatively more recent advances in algebraic geometry have created decision procedures based on cylindrical algebraic decomposition. Implemented in modern tools, these procedures have led to a vast surge in the mechanically supported design of complex engineering systems, characterised using semialgebraic properties.

It is only fair to point out that none of the preceding involves noise, noise being a property that every physical system exhibits to some degree. Since, in many cases, the noise can be regarded as negligible, constructing a framework that disregards it is not a waste of time. However, in many other cases noise is not negligible, and for such cases, we would need a stochastic extension of the preceding theory. Removing the 'absolutely' qualifier from our first world gives us a playground in which a stochastic calculus extension of the ideas discussed here may find a semantics. For convenience, we regard all this as outside the scope of the present essay.

Around this main thread, a number of other ideas swirl. In Section 5 we pointed out a number of them: DAEs, control issues, DDEs, bifurcations, numerical issues and quantization. Although all of them can make a difference to the description and behaviour of CPS systems, it is perhaps the bifurcations that have the most visible and dramatic impact: a system, hitherto quite well behaved. suddenly loses stability and starts to oscillate wildly, in response to an innocent looking adjustment to some static parameters.

In reality, the above constitutes a rather demanding sweep of theoretical techniques to take on board, and almost all approaches to CPS focus on one or other fairly narrow portion of this spectrum. The narrowness of focus is, of course, unfortunate, since it precludes locating the source of some difficulty that arises in the design of a system in the correct way, if the cause of the difficulty lies in some part of the spectrum unfamiliar to a particular individual.

This phenomenon is particularly prevalent in the computational world. The somewhat understandable inclination from the computational point of view to relegate the non-discrete aspects of CPS to a rather distant world of continuous behaviours, belies their ability to dramatically affect overall system behaviour in a manner that is essentially impossible for a discrete system formulation to engage with. Particularly when discussing bifurcations, we have indicated just how dramatic the effects of this can be. If this essay is to serve any useful purpose at all, it would be to help highlight the need for a deeper appreciation of just how wide the spectrum of ideas that impact CPS behaviour actually is, and thus to help stifle poorly judged views about the inadequacy of the theoretical and foundational framework for CPS. The fact of the matter remains, that *all* elements of CPS systems have well established mathematical descriptions that can help explain their behaviour. It remains the responsibility of CPS designers to appreciate the implications of all of them.

References

1. KeYmaera, http://symbolaris.com
2. 20-sim: http://www.20sim.com/
3. Alur, R.: Principles of Cyberphysical Systems. MIT Press (2015)
4. Andrews, P.: An Introduction to Mathematical Logic and Type Theory: To Truth Through Proof. Academic (1986)
5. Banach, R., Butler, M., Qin, S., Verma, N., Zhu, H.: Core Hybrid Event-B I: Single Hybrid Event-B Machines. Sci. Comp. Prog. 105, 92–123 (2015)
6. Banach, R., Butler, M., Qin, S., Zhu, H.: Core Hybrid Event-B II: Multiple Cooperating Hybrid Event-B Machines. Sci. Comp. Prog. 139, 1–35 (2017)
7. Basu, S., Pollack, R., Roy, M.F.: Algorithms in Real Algebraic Geometry. Springer (2006)

8. Carloni, L., Passerone, R., Pinto, A., Sangiovanni-Vincentelli, A.: Languages and Tools for Hybrid Systems Design. Foundations and Trends in Electronic Design Automation 1, 1–193 (2006)
9. Caviness, B., Johnson (eds.), J.: Quantifier Elimination and Cylindrical Algebraic Decomposition. Springer (1998)
10. Erneux, T.: Applied Delay Differential Equations. Springer (2009)
11. Gamelin, T.: Complex Analysis. Springer (2001)
12. Geisberger, E., Broy (eds.), M.: Living in a Networked World. Integrated Research Agenda Cyber-Physical Systems (agendaCPS) (2015), http://www.acatech.de/fileadmin/user_upload/Baumstruktur_nach_Website/Acatech/root/de/Publikationen/Projektberichte/acaetch_STUDIE_agendaCPS_eng_WEB.pdf
13. Hairer, E., Norsett, S., Wanner, G.: Solving Ordinary Differential Equations I Nonstiff Problems. Springer (1993)
14. Hairer, E., Wanner, G.: Solving Ordinary Differential Equations II Stiff and Differential-Algebraic Problems. Springer (1996)
15. Hale, J., Verduyn Lunel, S.: Introduction to Functional Differential Equations. Springer (1993)
16. Kunkel, P., Mehrmann, V.: Differential-Algebraic Equations: Analysis and Numerical Solution. European Mathematical Society (2006)
17. Lang, S.: Complex Analysis. Springer (2008)
18. Lee, E., Shesha, S.: Introduction to Embedded Systems: A Cyberphysical Systems Approach. LeeShesha.org, 2nd. edn. (2015)
19. Liu, J., Lv, J., Quan, Z., Zhao, H., Zhou, C., Zou, L.: A Calculus for Hybrid CSP. In: Ueda (ed.) Proc. APLAS-10. vol. 6461, pp. 1–15. Springer, LNCS (2010)
20. Maple: http://www.maplesoft.com
21. Mathematica: http://www.wolfram.com
22. MATLAB and SIMULINK: http://www.mathworks.com
23. MetiTarski: https://www.cl.cam.ac.uk/~lp15/papers/Arith/
24. Modelica: https://www.modelica.org/
25. Olver, F., Lozier, D., Boisvert, R., Clark, C.: NIST Handbook of Mathematical Functions. Cambridge University Press (2010)
26. Platzer, A.: Logical Analysis of Hybrid Systems: Proving Theorems for Complex Dynamics. Springer (2010)
27. Royden, H., Fitzpatrick, P.: Real Analysis. Pearson (2010)
28. Su, W., Abrial, J.R., Zhu, H.: Formalising Hybrid Systems with Event-B and the Rodin Platform. Sci. Comp. Prog. 94, 164–202 (2014)
29. Walter, W.: Ordinary Differential Equations. Springer (1998)
30. Widrow, B., Kollar, I.: Quantization Noise: Roundoff Error in Digital Computation, Signal Processing, Control, and Communications. Cambridge University Press (2008)
31. Wikipedia: Absolute continuity.
32. Zhan, N., Wang, S., Zhao, H.: Formal Modelling, Analysis and Verification of Hybrid Systems. In: Proc. UTPFM-13. vol. 8050, pp. 207–281. Springer LNCS (2013)
33. Zhou, C., Hoare, T., Ravn, A.: A Calculus of Durations. Inf. Proc. Lett. 40, 269–276 (1991)

The Role of Validation in Refinement-Based Formal Software Development*

Jean-Pierre Jacquot[1] and Atif Mashkoor[2]

[1] Université de Lorraine & LORIA, Vandœuvre-lès-Nancy, France
`firstname.lastname@loria.fr`
[2] Software Competence Center Hagenberg GmbH, Hagenberg, Austria
`firstname.lastname@scch.at`

Abstract. In this chapter, we consider the issue of validation in the context of formal software development. Although validation is a standard practice in all industrial software development processes, this activity is somehow less well addressed within formal methods. As the needs for formal languages, tools and environments are increasing in producing real-life software, the validation issue must be addressed. In this chapter, we discuss what the place of validation within formal methods, what specific issues there are associated with formal methods as far as validation is concerned, and what tools can be used in this regard. We then present a few examples of the usefulness of validation from the case studies we have developed. The chapter is concluded with a few open research problems associated with validation and future work.

1 Introduction

This chapter discusses the role of *validation* in the context of formal software development. We borrow the definition of validation from the US Food and Drug Administration (FDA) principles of software validation[3] which state that "validation is a confirmation by examination and provision of objective evidence that software specifications conform to user needs and intended uses, and that the particular requirements implemented through software can be consistently fulfilled." In this chapter, we focus on model-oriented and state-based (refinement-based) methods such as B [1], Event-B [3] or Z [25]. Such methods are based on two principles:

1. the specification of a system or software can be cast as a mathematical model expressing invariant properties on a state, and

---

* The work of Atif Mashkoor is supported by the Austrian Ministry for Transport, Innovation and Technology, the Federal Ministry of Science, Research and Economy, and the Province of Upper Austria in the frame of the COMET center SCCH.
[3] https://www.fda.gov/downloads/MedicalDevices/.../ucm085371.pdf

2. a correct implementation can be derived by a chain of refinements of the formal specification adhering to a strict set of correctness rules.

It thus follows that, by using such a method, developers are guaranteed to produce software which is provably consistent with the specification. They have "built the software right." However, nothing guarantees that they have "built the right software." So, the verification part of the engineering effort is covered, but not the validation part. We believe that software developers will be more likely to use formal methods if they take into account this latter part.

The issue of validation within formal methods has come up only recently for two main reasons, one good and one bad. The good reason is that the research community had to focus first on hard theoretical problems, e.g., how to automate most of the verification process. Without a high level of automation, formal methods are not usable in practice. Now, the power of theories and tools allows us to tackle real-life problems, see for example [7–10]. Provers know how to discharge complex formulas, model-checkers can deal with huge spaces, and formalisms allow developers to simply express complex properties. While there is still room for much improvement, it seems safe to say that we have passed the threshold where the methods could reasonably be deployed. The bad reason is the belief that validation is not necessary: we "just" need to get all the right properties; i.e., we just need to focus on getting a complete specification from the requirements analysis. We should also mention a probable last reason: validation requires human judgment, and so is outside the formal world.

While requirements engineering is indeed a key element of any successful development, it is unrealistic to expect this activity to produce a complete set of expectations, constraints, and assumptions before the development begins: requirements evolve, decisions during development must be made, assumptions are overlooked, and not all requirements can be expressed in a specific formal method.

Our position is that validation must be part of all stages of a development conducted with refinement-based formal methods: from the initial specification, to the implementation, passing through each strategic refinement [17, 19].

The chapter is organized as follows. We first discuss what validation is and why we must use it in the context of formal software development. Then, we present the important issues associated with validation. Then, we discuss the kind of tools, existing and to be invented, necessary for validation. Last, we give some examples where validation uncovered critical problems in case studies which were already verified.

2 The Place of Validation in Formal Methods

The term "validation" is understood, slightly, differently in science and in engineering. In the former, validation consists of checking that the predictions of a model conform with reality. In a sense, validation measures the explicative power of a model. In the latter, validation consists of checking that the produced artifact is consistent with its users' expectations. As formal methods are an engineering technique based on mathematical modeling, "validation" concerns both meanings.

In this section, we analyze validation in relation to different phases of a software development project. We discuss what can be validated and why it is important to do it.

2.1 Initial model

The starting point of any development using a formal method is a formal specification of software to produce. It consists of a set of mathematical and logical formulas. We are then in the classic scientific validation issue:

- do the formulas express properties we want to account for?
- do the formulas lead to the correct predictions?
- do the formulas cover all the properties we want to account for?

A major difference between models in software engineering and models in other engineering is that in the latter models are often more "numerical" and in the former models are more "logical." Numerical models are, in principle, easier to validate since computing with them produces numbers which can be checked against existing data or measured on physical objects. Logical models are, in principle, more complicated to validate since we need to draw all possible inferences from the formulas.

For models with few formulas, one can expect domain specialists to assess the validity of the specification through thorough reading. For larger models, the task requires the use of tools to explore the possible inferences.

2.2 During development

In refinement-based methods, a development step is a refinement: an evolution of the model which must meet very strong consistency relations between before and after the step. Those constraints have the form of theorems, or proof obligations, which must be discharged. There is then a solid safety net for the model's evolution. However, validation is necessary for several reasons: extension of the model, fixing-up of formulas, developer choices, emerging behaviors, and assessing progress.

Extension of the model Some methods, such as Event-B, allow developers to introduce new properties during refinement. On the plus side, this allows to build incrementally the model by introducing gradually its different features. The notion of observation level[4] is an attempt to organize rigorously such

---

[4] An observation level is a focus on a specific part of a formal model for a fine-grained analysis [15, 16].

introduction. The new formulas must of course be validated in the same spirit as for the initial model, but we must also check that they interact well with the part of the model that has already been specified.

Fixing-up formulas The general direction of refinement is to go from non-deterministic to more deterministic behaviors, and from abstract data models to implementable data structures. The proof obligations guarantee that the refined formulas are consistent, but not that they are a valid model. For instance, in Event-B, the guard of a refined event must be stronger than the guard of the corresponding abstract event. That can be cast as a proof obligation which must be discharged. But, upon success, we must also check that the new guard is not "too strong," thus discarding wanted behaviors from the model.

Developer choices Refinements, most often, require the developer to make some choices. Regularly, even good and precise specifications do not offer any clue to guide the choices. The problem is that all choices are not equal! For instance, the complete specification of the withdrawal function of an ATM is an attainable goal. There are three inputs (card, PIN-code and amount), three outputs (card, cash, account modification) and a limited set of conditions (even integrating hardware problems). But, at some time during the development, the two operations of giving back the card and the cash must be ordered. Interestingly, the first generation of ATM gave first the cash, then the card. Now, the order is reversed: one must retrieve the card to get the cash. The most probable explanation for this switch of behavior is that humans are forgetful and left many cards behind. This is the kind of assumption that is likely to be omitted in a formal model. The point is not that the problem would have been caught during the development, but that, without validation, it is nearly impossible to spot.

Emerging behaviors As refinements proceed, the model is growing in size and details. It contains more and more formulas which may interact in unexpected ways. This growth has the potential to introduce emerging properties in the model, some maybe good, some maybe neutral, and some maybe bad. Of course, it is better to avoid the bad ones as soon as possible.

Assessing development progress A (probably undesirable) feature of formal methods is their opacity: potential users and stakeholders do not possess, generally, the mathematical background necessary to read and understand thoroughly the models. This can lead to uncomfortable situations because stakeholders cannot be included in the development process [11]. This has several consequences:

- problems with requirements and unwanted behaviors are not caught before the development is finished,
- nobody, except the developers, can assess whether the development is going in a good direction,
- without continuous monitoring of the development progress, it will be more difficult to convince managers of the benefits of formal methods.

2.3 Implementation

Using formal methods does not forgo the necessity to test the final product. However, the kind of tests that are required later on are like acceptance tests where it is determined that the product meets users needs and unlike other tests which (try to) assess that the invariant properties are preserved.

Nonfunctional properties Nonfunctional properties, such as response time or maintainability, are equally important requirements. Even when they can be formalized, such as a bound on response time, they are often outside the scope of formal methods. Meeting nonfunctional properties is also often dependent on the hardware and the environment where the implementation runs, which may not be exactly known or modeled during development.

Usability For systems which interact with humans, e.g., ATMs, it may be impossible to predict all the difficulties that may arise. This should be the true role of beta-version software: validating that the system improves the users' ability to fulfill their task.

2.4 Non modeled properties

Irrespective of the considered formal method, there are some properties that are outside its expressive power. For instance, it is very difficult to model general time, causality, or reachability properties in Event-B. So, we can expect, for any real-life software, to have requirements which cannot be formalized. It is then important to test for those requirements. The tools and techniques to test are the same as those used to validate the models through execution. So, these tests can be included in the validation procedure.

As soon as a model takes care of requirements outside the expressive power of a particular method, specific tests must be setup. Further down the development, such tests must be run at every step in the spirit of non-regression testing.

3 The Issues with Validation

The validation of models is not a new idea. It is indeed an important part of traditional software development processes. However, formal methods, until recently, mostly ignored validation. We think the context of formal methods induces some specific issues about the validation process. This section discusses those.

3.1 Requirements

Whatever development method is used, the validation of a piece of software is only worth the quality of the requirements for that element of a system. Precise requirements are the goalposts against which the final product must be measured. Hence it is absolutely important to have a detailed, well written, complete, and unambiguous statement of requirements to start with.

In the context of formal methods, there are two issues with requirements. The first issue lies with the formalization of requirements themselves. A critical question is the coverage of requirements: does the final formal model takes all of them into account? To answer such a question, we need to be able to trace requirements back and forth to the formal model. There is then a need to have a good level of compatibility between the modeling languages used for requirements and for models. This is, for instance, what the ProR tool[5] proposes for Event-B.

The second issue lies with the evolution of requirements. We are not convinced that a better requirement elicitation process would make this issue disappear. Of course, better requirements must always be sought out, but they will still evolve for two main reasons: implicit assumptions and environment modifications. Software which interacts with open environments, notably humans, must rely on assumptions about this environment (ranges of value, behaviors, etc.), many of which are not formalized. Yet, once in a while, we discover that the formal models allow behaviors that are undesired. Also, the notion of environment of a system is a point of view of the system. But there are other points of view, from which the new system is a modification of the environment. By its mere existence, the new system may shift the expectations for its users or create new expectations.

The problem is actually an open research question: how to manage modifications in a formal model? Until now, there is no other option than starting again the development.

3.2 Human judgment

Validation implies that someone makes a judgment on the model. This means that users must assess whether the model meets requirements. There are two

---

[5] http://www.eclipse.org/rmf/pror/

ways to make such an assessment: one is to "understand" the model enough to explain its relation to requirements, the other is to run the model and analyze its output.

The former technique is adequate for models which consist of few formulas (invariant properties, events, automatons, etc.). As the size of the model grows, the interaction between the logical formulas becomes more and more difficult to analyze. For instance, we have often observed that we tend to write stronger than necessary guards for events, therefore forbidding otherwise admissible situations. Such errors are difficult to catch through reading only.

Larger models, in terms of the number of elements and formulas, are better validated by analyzing the space of legal states they specify. The analysis revolves around two kinds of questions: Which states are reachable? Are all the ways to reach a state admissible to the user or the environment? The first question concerns the coverage of requirements by the model; the second question concerns the problem of inadequate behaviors. This can be achieved by techniques akin to those used in software testing. So, we can adapt existing techniques to reduce safely the amount of executions to explore the state space.

3.3 Abstraction and nondeterminism

In using a test-based approach to validation, we are confronted with a major difficulty: how to execute the model? While it is always possible to run a state-based model by computing manually a few traces, a true validation requires to compute many more. We need an automated execution mechanism.

An executable program is a formal model. However, it is way too detailed to allow for formal verifications in general. There lies the reason which motivated the introduction of formal methods. Abstract formal models are allowed to use data which may not be implemented as such, e.g., operations with non-constructive definitions, and nondeterministic execution models, so proofs can be conducted.

Execution tools must provide solutions for two different problems. The first is a way to implement data. Since efficiency can be put aside (up to a reasonable degree), we can use canonical definitions and libraries. The second is the exploration of the state space. Both nonconstructive and nondeterministic definitions increase the size of the state space to explore. There is a threshold below which automated exploration is possible. Over this limit, we must resort to human intervention to guide the exploration.

When using an execution tool, we must be confident that the results and behaviors we observe are indeed specified by the model. The tool must be correct. Fully automated tools can be trusted more easily than tools where user's intervention is required. In both cases, we need some insurance that tools correctly implement the operational semantics of the formalism. In the last case, we also need the insurance that elements provided by users are consistent with the model. We have addressed this issue with the notion of simulation "fidelity" which is looser than the strict behavioral equivalence relation, but ensures that observations made on the executable model can be trusted as observations on the original one [19].

4 The Tools

The actual use of formal methods is highly dependent on sophisticated tools. Formal models are too complex to be manipulated "by hand," and the operations on them are highly complex, relying on elaborate theories and extensive exploration. Verification, either through theorem proving or model checking, is not possible without tools; validation is not different. We present here a typology of tools we have found useful.

4.1 Translators

A development using formal methods and refinement will eventually produce a program written in a programming language. So, at some point, the mathematical and logical model must be translated. Then, standard testing techniques can be used to validate the final program. Sometimes, when the model is deterministic and the data has some canonical implementation, translators can be used during the development. Of course, this will generally happen when the development is near its end.

The main advantage of using a translator in a validation procedure is that the executable model is very close to the final program. Assuming the translator is proven (or at least verified), the observations made while executing the program correspond exactly to the model. Execution can be trusted without further consideration. The environment in which the model is executed is close to the environment for the final product. So, the kind of non-functional requirements that depend on the actual environment can begin to be assessed.

Of course, the main limitation is that the model must be close to implementation. This limits the use of translators to the very last stages of the development. We should strive to have the major problems with requirements been discovered and overcome before those stages.

4.2 Model checkers

The validation of a software model is to explore the state space defined by the model. Model checkers are built for that exact purpose and, indeed, can be used during the development. ProB [12], a model checker for B adapted to Event-B, has been extended by an interface control and shows the execution of the model.

The main advantage of model checkers is their solid formal foundation. When they provide an answer, we are assured that it is a true consequence of the model. Also, their exhaustive exploration of the state space insures a full coverage of the model's behaviors.

The main limitation of model checkers is the combinatorial explosion of the state space exploration. Nondeterminism fuels the explosion, however, even deterministic systems can lead to models with a state space too large for practical application of model checking. For instance, in the landing gear case study proposed for ABZ'2014 [6], most published studies acknowledge that the use of model checkers becomes impossible when a certain level of details has been reached. Yet, the system is purely deterministic.

4.3 Animators

Animators are tools which implement the operational semantics of the formalism and are able to interpret directly the formal code. The nondeterministic features are treated by enumerating all the possible local values and picking randomly the actions to do when necessary. The enumeration can use either smart constraint solving techniques (ProB for instance), or brute force enumeration (AnimB[6] for instance).

The main advantage of animators is their close proximity to the formal model. They interpret directly the code over a "virtual machine" which is relatively easy to verify.

Animators have the same limitations as model checkers. Since they need to enumerate values, those need to belong to relatively small sets. It is possible to develop heuristics to transform nonanimatable models into animatable ones, e.g., as proposed in [14, 18]. It can be done safely but at some cost: transformations generate proof obligations which must be proven, and the resulting model is only a sub-model of the original model. Some behaviors and data state may be lost, but no spurious ones are introduced.

4.4 Simulators

To make validation accessible at all times during a development, there is a need to execute very abstract and nondeterministic models on which all the preceding tools fail. The idea then is to rely on automatic interpretation on most of the model and ask the user to provide explicit definitions for nonconstructive constructs and choice restriction for the nondeterministic constructs. The rationale is that developers must begin with very abstract models, but they know in which direction the development will proceed to become implementable. Of course, the hand-coded parts provided by the users constraint and restrict the space that can be explored, but we can expect that the cut-out parts of the state space are actually irrelevant for the final software. For instance, when modeling some movements in a 3D space, we may need some distance function

---
[6] http://www.animb.org

which is not explicitly modeled; for practical validation, "implementing" it as the standard Cartesian distance is effective.

The main advantage of the simulation technique [26] is that models at all levels of abstraction can be validated, at least partially.

There are two limitations to the technique:

- the explored state is smaller than the specified space; we must trust users to provide solutions that explore the space which is important in practice,
- the user's additions must be consistent with the model. This is a complex issue. For Event-B, for instance, we have developed a notion of "fidelity" which captures this property [28]. It entails the generation of proof obligations which must be discharged to guarantee the consistency of the simulation.

4.5 Scenario managers

To check whether a given requirement is covered by a system, we need to set up a test situation: getting to an initial state value, firing a predefined sequence of commands or actions, and comparing the expectations against actual results and observations. To validate a model, we need to check all requirements the model is supposed to cover. So, we must have some tools to manage a large collection of scenarios.

Furthermore, for validation to be a continuous process accompanying refinements, we must check that no requirement is dropped off during a refinement. This is the standard issue of non-regression testing that faces the maintenance process and the incremental development methods.

5 Case Studies

In this section, we present a few examples of our developments where the execution of models helped us spot and eventually fix anomalies in our models. All the following examples are taken from our work with Event-B. So before discussing examples, we first introduce this formal method.

5.1 The Event-B method

Event-B is a state-based formal method for systems and software engineering. A model in Event-B is composed of three elements:

1. a state, which is a function mapping names to values. The values are inductively defined as atoms (integers, symbols, booleans), sets of atoms, and set-theoretic constructions of values. There are special notations for typical set constructions such as relations, power sets, total relations, or bijective functions. Syntactically, the state distinguishes between constants and variables.

2. an invariant, which is a logical first-order formula on the state that specifies the legal values of the state of a system. Syntactically, the invariant is a conjunction of smaller formulas.
3. a set of events, which are guarded substitutions on the state. The guard is a conjunction of smaller first-order formulas on variables of the model.

The formal semantics of the model is defined by four properties:

1. all expressions must be well-typed; the typing relation is essentially the "belongs to" set-theoretic relation,
2. there must exist at least a value of the state which satisfies the invariant. Syntactically, there is a special event ($INITIALISATION$) which sets the initial legal state,
3. any event whose guard holds can be fired (the choice is nondeterministic), each event fired from a legal state leads to another legal state: i.e., they maintain the invariant,
4. all events must be feasible, i.e., if the guard is true, there must exist a solution for all substitutions which leads to a legal state.

The model development scheme associated with Event-B is based on formal refinement. Refinement is a relation between models. A model MR is a refinement of MA if: (1) there is a gluing invariant between the state of MR and the state of MA, this gluing invariant is part of the invariant of MR, and (2) all events in MR have their counterparts in MA and have a stronger guard ensuring that a refined event can only be fired if the abstract event can.

In practice, refinements can be classified into four general kinds:

State restriction The invariant or guards of events are strengthened so the legal state space becomes smaller.
Data refinement Some variables are "replaced" by others closer to an implementable data structure.
Property introduction New variables and related invariants are introduced to model a property.
Behavior refinement New events are introduced to model new behaviors. Technically, new events must be formal refinements of the $SKIP$ (do-nothing) event.

The syntax and semantics are designed so that, for each model, a set of independent proof obligations are generated. The model is verified when all proof obligations are discharged. The major advantage of refinement-based methods is that the proof of correctness of the implementation is broken into many small and relatively simple proofs which are spread out all over the development process.

5.2 The platooning case study

The platooning case study [27] was intended to assess the possibility of using Event-B to prove an existing algorithm of platooning by redeveloping it. The algorithm controls the movement of autonomous vehicles forming a platoon by using only perception of the preceding vehicle in the formation. The safety property of interest is non-collision of vehicles participating in the platoon. The development was conducted in two phases: the first phase considered only a 1-D version of the algorithm which only controls speed, the second phase considered a 2-D version which adds direction control. The examples below concern only the 1-D version.

The validation was done by using the animator Brama [23] to which a graphical interface was added. It was performed at the end of the development, on a model which included an event to set the target speed of the platoon which was not part of the specification. This unguarded event does not modify the state space or the essential behaviors of the model.

The animation uncovered two interesting problems with the model.

Oscillations During the animation, it appeared that when the platoon reached its target speed, although the leading vehicle had a smooth continuous speed, the following vehicles were alternating braking and accelerating with each cycle. Furthermore, acceleration/deceleration values increased with the position in the platoon. This is of course an unintended behavior of the system. This situation comes from the incompleteness of requirements which must include a statement about oscillation. It should be noted that the definition of oscillation in Event-B may be tricky.

Blockages When running a scenario where the platoon comes to a stop, we observed that it could not start again. Analysis of the problem reveals that the platoon algorithm was actually caught in a deadlock. The reason was that Event-B supports only integers while the algorithm uses real numbers. The slight difference in values prevented some guards to be true. The solution was to modify slightly the model to account for the numerical discrepancy. Event-B has no provision to guarantee that models do not deadlock. It is possible to define invariants which express the deadlock freeness property, but they must be introduced explicitly.

5.3 The landing gear case study

The landing gear case study [5] was introduced as part of the ABZ conference to allow comparison of formal systems and development approaches. The task is to design the control system of the landing gears of an airplane. The requirements are given in a natural language document which describes the physical components, the sequence of movements, and the timing constraints of a live system.

A very important property of the control system is to allow maneuver reversibility at all time, i.e., the pilot must have the possibility to stop and reverse the maneuver at any time. Such a property cannot be specified in Event-B. However, as the system can be described as a finite automaton, it is possible to design an exhaustive set of scenarios to check the reversibility.

All solutions of the case study based on Event-B observed the necessity to use a form of execution to check the property; all developers admitted to have reworked their models a few times to get it right.

An interesting observation about the tools is that, even if the system is finite and deterministic, automatic tools, such as ProB, stop to work as the description of the hardware elements gets more detailed.

We used JeB [19]. An important part of the code we had to write consisted in the executable models of the hardware, the sensors in particular.

5.4 The transport domain model case study

The transport model [13] is an experiment to assess the potential expressiveness of Event-B to specify domains. The requirements include the following:

- vehicles move along a network akin to a road system,
- vehicles must not collide,
- two kinds of collisions are considered: rear-endings along a path and intersection crashes,
- vehicles obey the usual kinematics law,
- travel time must be modeled,
- energy consumption must be modeled.

A very important feature of the development is the gradual introduction of the properties.

We used two kind of tools to execute the model: Brama (an animator) and JeB (a simulator). While the use of Brama prompted the design of transformation heuristics (please see [14] for more details) to make models amenable to animation, JeB allowed us to execute all models in the 13 refinement steps. Execution of the model helped us on several grounds.

Context validation Although execution tools are designed to check the behavior of the model, the technology can be used to validate that the static part (the contexts in Event-B) is a valid model of the environmental data of reality. Modeling a small village showed that we need to introduce new kinds of nodes in the network to model the entry and exit points as (infinite) sources and (bottomless) sinks.

Deadlocks identification A difficulty with our model is that it must account for gridlocks, i.e., places in the network where no movement is possible, but must also avoid deadlocks, i.e., situations where a real vehicle could move but the model prevents it. Traditional techniques for deadlocks checking do not work well as the model allows for all vehicles to be parked.

Guard debugging Most of the anomalies seen during the executions were traced back to incorrect guards in events. Most often, the guards were too strong for one of two reasons. The first is "logically" too strong, i.e., some cases were forgotten when writing the guard. The second is "numerically" too strong. As for the platooning case study, the kinematic functions are modeled with integers. So, the formulas must take into account some kind of "imprecision" due to the differences in arithmetics.

Termination checking An important implicit property is that unobstructed vehicles must progress toward their destination. Such a property is difficult to model in Event-B. Execution allows us to assess the progress.

6 Research Issues

In this section, we discuss more precisely two issues which have recently emerged as new research topics.

6.1 Interaction between models

Formal methods are very relevant for software pieces which are part of wider systems where they control hardware elements. Cyber-Physical Systems (CPS) are of this kind, and they are becoming ubiquitous in our everyday environment. The validation of software models then also requires hardware elements to be modeled and participate to the execution strategy. Except for the rare hardware pieces which act as digital state machines, most are analog devices obeying some physical laws. Thus, they are more likely to be modeled with techniques such as differential equations or frameworks like Scilab[7]. We know how to produce executable models of such artifacts, but the interaction between the executable models of software and hardware is an open problem.

The following issues must be addressed:

1. Connecting the executable models. This is a standard issue in software engineering when we need to build systems composed of units running in different environments. In the context of Event-B, several directions are investigated. The European FP7 ADVANCE project[8] explored platforms build on Functional Mock-up Interface (FMI) [21], Rodin [4], and ProB to test CPS system with multi-simulation. Generic platforms, such as MECSYCO [24], provide a basis for connecting JeB generated executable models with models developed in other languages. Preliminary studies we have conducted ahve shown the potential of such approaches.

---

[7] http://www.scilab.org
[8] http://www.advance-ict.eu

2. Equalizing the abstractions. At a given refinement step, the formal model is an abstraction of actual software. The model considers the outside environment from this abstract point of view and must sees it at the same abstraction level. We then need to bring the hardware simulation models up to the same abstraction level. There are two basic possibilities. Either a model of the hardware at the appropriate level is built by the specialists of these artifacts, or, an "abstracting interface" is wrapped aroung a detailed model of the hardware to make it interact at the required abstraction level.
3. Which ever of the preceding two techniques is used, there remains the essential question: can the observation on the simulation be trusted? This boils down to the fidelity of the hardware simulation model to the actual hardware piece.

6.2 Relation between refinement and validation

In formal methods, such as Event-B, the notion of refinement was introduced and formalized to support the incremental verification of the final software. While each refinement is a complete model by itself and its consistency must be proven, the syntactic and semantic structures have been defined so that only proof obligations associated with the newly introduced elements need to be discharged for the proof to be complete. The proof that the final software conforms to its specification progresses along with the proof of each refinement. The verification is then monotonic with respect to refinements. Such a monotonicity would also be desirable for validation. However, much research needs to be done to achieve this goal. Several points must be established:

1. How to identify and associate a particular refinement with the "new" requirements being taken into account?
2. How to associate requirements and test scenarios?
3. How to generate new test scenarios for the refined model from test scenarios for the abstract model? This is related to the issue of non-regression testing: we need to be able to "replay" tests to check that behaviors are valid "refinement" of behaviors validated on abstract models. A definition of scenarios which can integrate a notion of abstraction/refinement is then required.

At present, points 1 and 2 are more a, complex, engineering problem than a research issue. Tools, such as ProR, already allow the elicitation and formalization of requirements in a format compatible with Event-B models in Rodin [22].

Point 3 requires first to have a formal definition of scenarios compatible with refinement. Since the semantics we use to define fidelity, i.e., the relation between formal models and executable models, is based on traces, we can consider scenarios as traces. More precisely, we can define scenarios as constraints on traces such as succession of events (from a given refinement level), succession of states, and any mix.

7 Conclusion and Future Work

Although we did not start from this point of view, we have come to realize that our work on validation led to ideas and tools that are close to the Agile spirit [20], at least the Good ideas as identified by Meyer: short development cycles, extensive "testing," and permanent user validation.

The similarity between the short development cycles (the "runs") of Agile methods and the short refinement steps as advocated in [2] is striking. In both cases, the development proceeds through a sequence of correct models, either formal or executables. Of course, the notion of correctness is very different in each case. By including validation of the formal models, we can get even a closer analogy as users can be involved and assess the adequacy of the requirements all along the development.

Behind the practice of "permanent testing" of Agile methods is the idea that, once established, the correctness of a feature must always be guaranteed. Refinement-based formal methods ensure this property by other means, and, too often, despise tests. However, testing has other roles which are useful in formal methods. The first role concerns the verification of the properties that cannot be modeled within the method. The second role is the validation that the requirements have been met.

The possibility for users to interact with models during the whole development is essential to ensure that, at the end, software will meet the actual, useful, requirements. Requirements' evolution is a hot research topic, particularly in the context of formal methods. At present, we do not know how to refactor a formal development when a requirement changes. So it is very important to detect as early as possible a problem with requirements: either missing, incomplete, or mis-adapted.

The diffusion of Agile methods in industry is a model we should strive for formal methods. As discussed above, by introducing validation into each requirement step, we hope to get the positive arguments to counter those used to put formal methods aside.

References

1. Abrial, J.R.: The B Book. Cambridge University Press (1996)
2. Abrial, J.R.: Formal methods in industry: achievements, problems, future. In: Proceedings of the 28th international conference on Software engineering. pp. 761–768. ICSE '06, ACM, New York, NY, USA (2006)
3. Abrial, J.R.: Modeling in Event-B: System and Software Engineering. Cambridge University Press (2010)
4. Abrial, J.R., Butler, M., Hallerstede, S., Hoang, T., Mehta, F., Voisin, L.: Rodin: an open toolset for modelling and reasoning in Event-B. International Journal on Software Tools for Technology Transfer 12(6), 447–466 (2010)

5. Boniol, F., Wiels, V.: The landing gear system case study. In: ABZ 2014: The Landing Gear Case Study, Communications in Computer and Information Science, vol. 433, pp. 1–18. Springer International Publishing (2014), http://dx.doi.org/10.1007/978-3-319-07512-9_1
6. Boniol, F., Wiels, V., Ameur, Y.A., Schewe, K.D.: ABZ 2014: The Landing Gear Case Study Case Study Track, Held at the 4th International Conference on Abstract State Machines, Alloy, B, TLA, VDM, and Z Toulouse, France, June 2-6, 2014, Proceedings, Communications in Computer and Information Science, vol. 433. Springer (2014)
7. Bormann, J., Lohse, J., Payer, M., Venzl, G.: Model checking in industrial hardware design. In: Proceedings of the 32nd Annual ACM/IEEE Design Automation Conference. pp. 298–303. DAC'95, ACM, New York, NY, USA (1995), http://doi.acm.org/10.1145/217474.217545
8. Butler, R., Caldwell, J., Carreno, V., Holloway, C., Miner, P.S., Di Vito, B.: NASA Langley's research and technology-transfer program in formal methods. In: Computer Assurance, 1995. COMPASS'95. Systems Integrity, Software Safety and Process Security. Proceedings of the Tenth Annual Conference on. pp. 135–149 (Jun 1995)
9. Cimatti, A.: Industrial applications of model checking. In: Cassez, F., Jard, C., Rozoy, B., Ryan, M. (eds.) Modeling and Verification of Parallel Processes, Lecture Notes in Computer Science, vol. 2067, pp. 153–168. Springer Berlin Heidelberg (2001), http://dx.doi.org/10.1007/3-540-45510-8_6
10. Kaufmann, M., Moore, J.: An industrial strength theorem prover for a logic based on Common Lisp. Software Engineering, IEEE Transactions on 23(4), 203–213 (Apr 1997)
11. Kossak, F., Mashkoor, A., Geist, V., Illibauer, C.: Improving the understandability of formal specifications: An experience report. In: Salinesi, C., Weerd, I. (eds.) Requirements Engineering: Foundation for Software Quality, Lecture Notes in Computer Science, vol. 8396, pp. 184–199. Springer International Publishing (2014), http://dx.doi.org/10.1007/978-3-319-05843-6_14
12. Leuschel, M., Butler, M.: ProB: An Automated Analysis Toolset for the B Method. Journal Software Tools for Technology Transfer 10(2), 185–203 (2008)
13. Mashkoor, A.: Formal Domain Engineering: From Specification to Validation. Ph.D. thesis, Université de Lorraine (Jul 2011), http://tel.archives-ouvertes.fr/tel-00614269/en/
14. Mashkoor, A., Jacquot, J.P.: Stepwise validation of formal specifications. In: 18th Asia-Pacific Software Engineering Conference (APSEC'11). pp. 57–64 (2011)
15. Mashkoor, A., Jacquot, J.P.: Utilizing Event-B for Domain Engineering: A Critical Analysis. Requirements Engineering 16(3), 191–207 (2011)
16. Mashkoor, A., Jacquot, J.P.: Observation-Level-Driven Formal Modeling. In: High-Assurance Systems Engineering (HASE), IEEE 16th International Symposium on. pp. 158–165 (2015)
17. Mashkoor, A., Jacquot, J.P.: Validation of formal specifications through transformation and animation. Requirements Engineering pp. 1–19 (2016), http://dx.doi.org/10.1007/s00766-016-0246-6
18. Mashkoor, A., Jacquot, J.P., Souquières, J.: Transformation Heuristics for Formal Requirements Validation by Animation. In: 2nd International Workshop on the Certification of Safety-Critical Software Controlled Systems (SafeCert'09). York, UK (2009)

19. Mashkoor, A., Yang, F., Jacquot, J.P.: Refinement-based Validation of Event-B Specifications. Software & Systems Modeling pp. 1–20 (2016), http://dx.doi.org/10.1007/s10270-016-0514-4
20. Meyer, B.: Agile! The Good, the Hype and the Ugly. Springer Verlag (2014)
21. Savicks, V., Butler, M., Colley, J.: Co-simulating Event-B and Continuous Models via FMI. In: Proceedings of the 2014 Summer Simulation Multiconference. pp. 37:1–37:8. SummerSim'14, Society for Computer Simulation International, San Diego, CA, USA (2014), http://dl.acm.org/citation.cfm?id=2685617.2685654
22. Sayar, I., Souquières, J.: La validation dans le processus de développement. In: Actes du XXXIVème Congrès INFORSID, Grenoble, France, May 31 - June 3, 2016. pp. 67–82 (2016), http://inforsid.fr/actes/2016/INFORSID2016_paper_3.pdf
23. Servat, T.: BRAMA: A New Graphic Animation Tool for B Models. In: B'07: Formal Specification and Development in B. pp. 274–276. Springer-Verlag (2006)
24. Vaubourg, J., Presse, Y., Camus, B., Bourjot, C., Ciarletta, L., Chevrier, V., Tavella, J.P., Morais, H.: Multi-agent Multi-Model Simulation of Smart Grids in the MS4SG Project. In: Demazeau, Y., Decker, K.S., Bajo Pérez, J., de la Prieta, F. (eds.) PAAMS'15. Lecture Notes in Computer Science, vol. 9086, p. 12. Springer, Salamanca, Spain (Jun 2015), https://hal.inria.fr/hal-01171428
25. Woodcock, J., Davies, J.: Using Z: Specification, Refinement, and Proof. Prentice-Hall, Inc., Upper Saddle River, NJ, USA (1996)
26. Yang, F., Jacquot, J.P., Souquieres, J.: JeB: Safe Simulation of Event-B Models in JavaScript. In: Software Engineering Conference (APSEC), 20th Asia-Pacific. vol. 1, pp. 571–576 (2013)
27. Yang, F., Jacquot, J.P.: Scaling up with Event-B: A case study. In: Bobaru, M., Havelund, K., Holzmann, G., Joshi, R. (eds.) NASA Formal Methods, Lecture Notes in Computer Science, vol. 6617, pp. 438–452. Springer Berlin Heidelberg (2011)
28. Yang, F., Jacquot, J.P., Souquières, J.: Proving the fidelity of simulations of Event-B models. 15th IEEE International Symposium on High-Assurance Systems Engineering pp. 89–96 (2014)

Hagenberg Business Process Modelling Method
Towards a Homogeneous Framework for Integrating Process, Actor, Dialogue, and Data Models[*]

Verena Geist, Felix Kossak, Christine Natschläger, Christa Illibauer, Thomas Ziebermayr, and Atif Mashkoor

Software Competence Center Hagenberg GmbH, Hagenberg, Austria
{firstname.lastname}@scch.at

Abstract. Due to different aspects, such as data, functionality, communication, and interaction, modelling information systems is a complex and challenging task. Each aspect in general calls for its own model, which has to be integrated in the overall system model. In this paper, we present the Hagenberg Business Process Modelling (H-BPM) method as a novel approach for designing business process management systems, which – in addition to process modelling – integrates several other important aspects for workflow specification like user interaction, data, and organisational modelling. On top of these aspects, we propose the enhanced Process Platform (eP^2) architecture to illustrate the static view on formal model integration. We apply the notion of Abstract State Machines (ASMs) to formalise the components, their interfaces and behaviour, and to specify the collaboration of the components in a rigorous and reliable way. Formal semantics for executing the integrated models in a business process management system is provided using ASM ground modelling and refinement. We finally demonstrate the suggested method using the example of processing user tasks and show its usability by a model for ordering supplies.
Keywords—H-BPM, Business Process Modelling, Formal Software Architecture, Horizontal Model Integration, Abstract State Machines

1 Introduction

The number of models required for different aspects, such as data, functionality and communication, makes modelling information systems a complex endeavour [28]. In addition, those models are generally related to different abstraction layers, leading to discrepancies when integrating them.

The motivation for the development of a comprehensive modelling method proposing a design for Business Process Management (BPM) systems which

---

[*] The research reported in this paper has been partly supported by the Austrian Ministry for Transport, Innovation and Technology, the Federal Ministry of Science, Research and Economy, and the Province of Upper Austria in the frame of the COMET center SCCH.

comprise much more than just process modelling stems from several of our industrial projects (see [6] and [18]), where business analysts and software developers struggled with redundancies and inconsistencies in system documentation. Process modelling is often considered as a first possible step in BPM, typically including control flow definitions as well as message and event handling (though one could start with user-centric modelling as well, see e.g. [7]); but when it comes to full-scale software support, much more is needed. At this point, if not already before, *actors* (workers) will have to be added to the model and integrated with issues from *user interaction* and *data* modelling.

While different views on business processes represented by different models, see e.g. [14, 18, 11, 13], have been described before, up to our knowledge, these different models have never been integrated. In fact, it is their interplay which makes them useful in practice, and this interplay is nontrivial.

Therefore, in [12] we introduced the Hagenberg Business Process Modelling (H-BPM) method, named after the Upper Austrian village where it has been designed, for seamless modelling of business processes. Thereby, we take a static view regarding software component integration as well as a dynamic, runtime-related view on model integration. As a result, this new modelling method is able to seamlessly integrate different aspects of business process modelling, including actor, dialogue, and data models, on all levels of abstraction. While the comprehensive book [12] contains the complete H-BPM specification (i.e. precise formal descriptions and justifications), in this paper we summarize the key results of our research and, in addition, discuss the general idea of vertical and horizontal model integration, originally suggested by K.D. Schewe in 2010. The main contribution of the paper is to exemplify the comprehensive modelling method by formally specifying the semantics for processing user tasks. In order to show the applicability of the suggested method, we also illustrate our four-step integration approach on the basis of a scenario for ordering supplies.

The semantics of the concepts introduced in this paper is defined using the ASM method [4] that is a system engineering method for developing software and systems seamlessly from requirements capture to their implementation. Within a precise yet simple conceptual framework, the ASM method allows a modelling technique which integrates dynamic (operational) and static (declarative) descriptions, and an analysis technique that combines verification and validation methods at any desired level of detail. Thus, ASMs allow to formulate rigorous specifications while still remaining understandable by a large class of people. Furthermore, we experienced that their employment for specification tasks allows to identify inconsistencies as well as ambiguities and gaps within an initial, natural-language requirements specification at an early stage. This is exactly why we decided to use ASMs for modelling processes semantics.

This paper is organised as follows. In Section 2, we start with a short introduction to ASMs. Section 3 first addresses static component integration, where we present the enhanced Process Platform (eP^2) architecture in which different software components to handle different types of business process models can be integrated to enable compreshensive and smoothly integrated tool support.

The formal specification of the operational semantics is well-suited to be refined towards an implementation. Section 4 (which might be suited also as an introduction to the modelling method for readers who are not familiar with business process modelling) illustrates how different models can play together, using an example scenario. In Section 5, we present related work, and in the final section we conclude with a summary and an outlook to future work.

2 Formal Background

The introduction of H-BPM, including the specification of the eP^2 architecture, is a further step towards the implementation of a framework that utilises integrated models for the implementation of information systems. The application of formal methods for defining the models and describing the structure assures correctness and clarity and also helps to facilitate the implementation of the framework by, at least, partly generated code.

The semantics of the presented concepts is defined using the ASM method. ASM models are state machines, which describe (discrete) states of a system and transitions between those states. They are an extension of finite state machines whose states are defined by arbitrarily complex data structures. ASMs can also simply be seen as "a rather intuitive form of abstract pseudo-code" [4]. *Functions* define the data structure of an ASM. Concrete values of function parameters define a *location*, and concrete values of the functions for all locations define a particular *state* of the automaton.

Besides elementary update *rules*, there is a limited but powerful set of *rule constructors* that defines state transitions by modifying the function values at a finite number of locations: *if-then* for conditional actions, *par* for simultaneous parallel actions, *seq* for sequential actions, *choose* for nondeterminism (existential quantification), and *forall* for unrestricted synchronous parallelism (universal quantification). Basic *functions*, i.e. state-relevant functions, are classified by their changeability and visibility. Functions that never change during any run of the machine are *static*. Those updated by agent actions are *dynamic*, and distinguished between *monitored* (only read by the machine and modified by the environment) and *controlled* (read and written by the machine). *Derived* functions, an important auxiliary element in ASM models, do not change the state of the automaton and are calculated from the values of other functions.

The ASM method concentrates on the dynamics of a system and thereby makes the intentions of a model easier to grasp. This makes ASMs particularly suitable for specifying business processes. For example, the state of a process diagram can be described by the combination of the lifecycle states of the activities (e.g. "Ready", "Active", and "Completed") and the number of tokens at the sequence flows between them. In addition, ASMs are especially suitable for parallel processing; they can model both parallel paths in processes and concurrent token instances running through processes.

Due to the easily understandable nature of the ASM text [15], we believe that the functions and rules defined in this paper are intuitive to understand without a deeper background knowledge of ASMs. For an understanding of the semantic subtleties, we recommend to consult "the ASM Book" by Börger and Stärk [4].

3 The H-BPM Method

The Hagenberg Business Process Modelling (H-BPM) method proposes several extensions to a traditional core business process description language, covering aspects which are often needed in business process modelling and execution (see also [35]). The original idea of horizontal and vertical model integration suggested by K.D. Schewe is shown in Fig. 1.

Since business process models can easily become too large and complex to grasp and communicate them as a whole, hierarchical modelling should be used to treat them in manageable chunks. Hierarchical business process modelling can also be applied in a horizontal direction. So the idea of K.D. Schewe was to just model a part of the aspects in a first step and then to complete the remaining ones in following steps. This can be nominated as *vertical* (hierarchical) and *horizontal* (extensional) refinements. Both horizontal and vertical refinements should be expressed on the fundament of formal semantics, i.e. using the ASM method and, in particular, ASM ground modelling and ASM refinements.

For horizontal refinements, the structure of the Business Process Model and Notation (BPMN) shall be exploited, starting with only activities, control flow and sub processes, afterwards extended with messages and events. Key steps are the addition of data consumption/creation to activities, formalized by dialogues, and deontic constraints for actor modelling as well as view constraints for data modelling. Thereby, also some simplification by investigating whether BPMN constructs can be merged or generalized (and thus simplified) should be achieved (BPMN+-).

Vertical refinement means that (in contrast to additional functionality) more details are provided, i.e. to go from abstract to concrete definitions. So vertical refinement steps include the replacement of high-level business process specification as ASM definitions using formal verification down to executable models. In addition, the target architecture with autonomous actors and rules based on task allocation (worklist) shall be defined.

In this section, we (1) first give an overview of the different aspects in the schema definition, without going too much into details, (2) introduce the enhanced Process Platform (eP^2) architecture, which allows for executable process definitions covering all aspects in the form of integrated models, (3) discuss the four-step-integration approach for modelling all aspects of business processes, and (4) provide an exemplification of our work on formally specifying operational semantics of the processes.

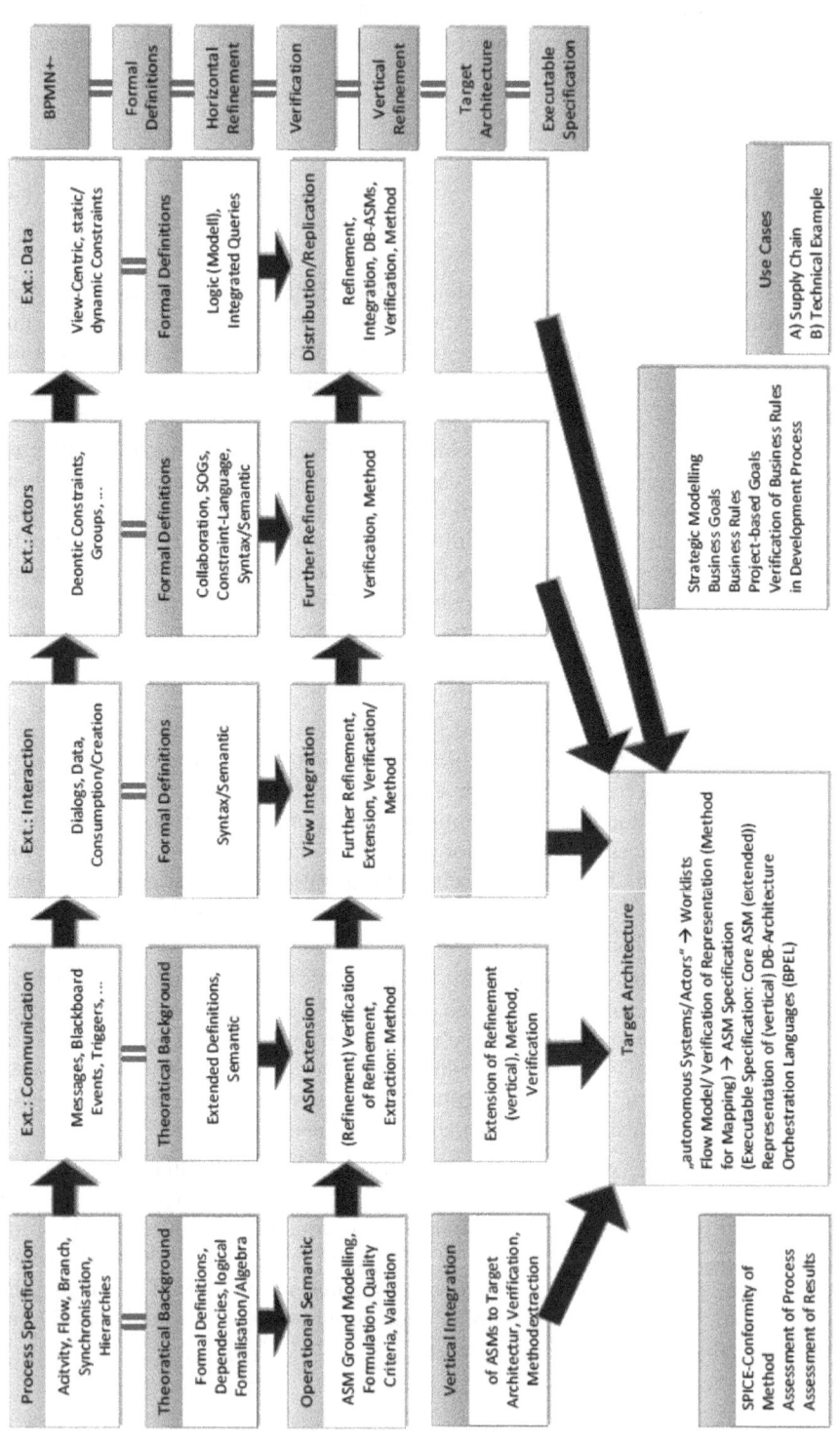

Fig. 1. Idea of H-BPM (suggested by K.D. Schewe)

3.1 Business Process Schema Definition

The *process model* is one possible starting point for business process modelling. In H-BPM, the process model basically follows the syntax of the BPMN 2.0 [34] and can be interpreted by the so-called *Workflow Transition Interpreter*, which is formally specified in [14].

Further preceding research focuses on *deontic logic* to explicitly highlight modalities of tasks in process diagrams in a comfortable and reliable way [21, 17]. In particular, we think that optionality should be expressed directly within the corresponding activity and not through the structure of the process flow [19]. An approach for *actor modelling*, introduced in [18], then builds upon these deontic business process diagrams and allocates different modalities to different actors, depending on role permissions. The layered actor modelling approach comprises several views and gradations that must be considered in model integration. Important concepts of the *process view* are task-based assignment of roles and concatenation of several roles that may execute an activity either together (only in case of non-atomic activities) or alternatively. The *organisational view* comprises an organisation chart with all roles, a role hierarchy, and the individuals. Dependencies between tasks are defined in the *rule view* (e.g. to support patterns like separation of duties or retain familiar). Those rules can also be checked for consistency and derived rules can be generated.

User interaction modelling focuses on the user interface for those users who actually have to perform the tasks defined in a process diagram. Such users, in their daily work, have a totally different view on a process than a business process analyst or higher-level manager and need not be confronted with the whole process diagram. The core of a worker's view is represented by a list of tasks which they are supposed to perform at a given moment. Selecting an item in their worklist, users are led through a set of associated dialogues, consisting of reports and editable forms that give access to the data needed for performing tasks of the process model. The corresponding dialogue model is defined in the notion of typed *workflow charts* [1, 11], which, amongst others, allows to further structure tasks. Thereby, a user can be led through a dynamically assembled succession of dialogues, depending, amongst others, on their own actions and decisions. A workflow chart is specified as a tripartite graph consisting of typed nodes for (i) rendering reports, (ii) providing forms for user input, and (iii) representing task items of the users' worklists. Supporting the worklist paradigm, assignment information, as well as parallelism, this approach provides a flexible process technology based on dialogues. For resolving data anomalies in process models, several approaches have addressed different notions of correctness [3]. In H-BPM, user tasks in the process model are refined by dialogues and data.

However, these diagrams should not be considered in isolation from one another, as they have strong interdependencies. In [29] the interested reader can find a more detailed overview of these most relevant aspects of H-BPM.

3.2 The eP^2 Architecture

The aim of the eP^2 architecture [12] is, on the one hand, to show that the proposed process description language with all its extensions in the form of integrated models is applicable for executable process definitions and, on the other hand, to provide a platform for process execution and user support.

In Fig. 2, we present the architectural overview of the system which is designed to achieve the above-defined objectives. That is, the eP^2 architecture basically includes, besides a central workflow engine, respective components for actor, user interaction, and data management, as well as different interfaces to define inter-component communication. We use the notation of the UML component diagram, which depicts artefacts (e.g. *Process Model*, *Dialogue Model*, etc.) and components (e.g. *Workflow Engine*, *Dialogue Engine*, etc.) as rectangles, interfaces as circles (connected to the provider), and dependencies as directed, dashed lines.

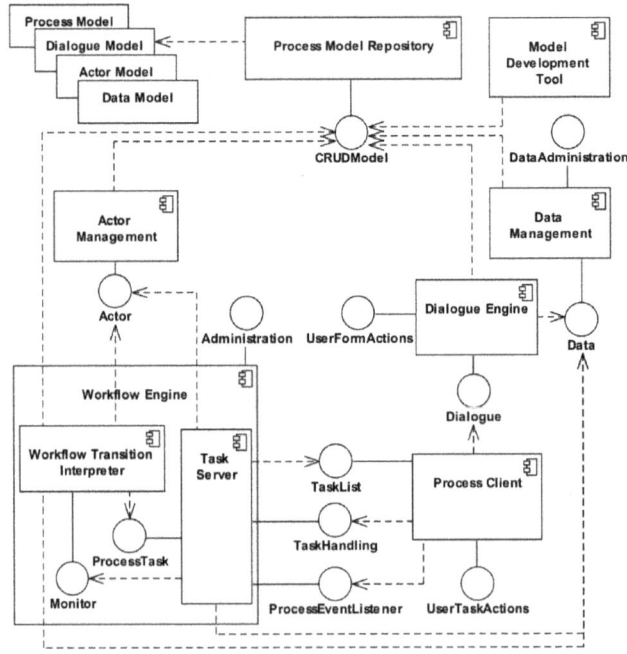

Fig. 2. Architectural Overview of eP^2

The core of this architecture for designing a BPM system is the *Workflow Transition Interpreter*, which is able to interpret H-BPM process models. To facilitate model integration in general and user interaction in particular, we add some interfaces to the ground model of the core *Workflow Transition Interpreter* to incorporate deontic classification and user interaction modelling. The *Task Server*, as a further component of the *Workflow Engine*, is, in particular, responsible for distributing user (and manual) tasks to users via the *Process Client*. This distribution is based on role and assignment information of the *Actor Management*, which also implements the role hierarchy as escalation paths (e.g. for handling situations where no permitted user is currently available). For user tasks, the *Dialogue Engine* displays required data and stores user inputs (using the *Data Management* component). Finally, the *Process Model Repository* provides managed access to the models within the architecture.

Each eP^2 component is specified by an ASM ground model. Its signature includes exported and imported functions and rules (cf. [4]), whereby the rules basically conform to supplied and called operations of the component interfaces. The interactions between those components correspond to synchronously or asynchronously called interface operations, considering contextual constraints and dependencies. Therefore, each component which provides interfaces specifies a rule or a derived function for each operation of those interfaces, in which it reacts to the respective operation call.

3.3 Four-Step-Integration Approach

Based on the architectural overview in Sect. 3.2, we now specify the behaviour of the eP^2 system and show how different aspects of BPM can be integrated.

The suggested approach utilises different levels of abstraction to define the system behaviour in a formal, unambiguous way. The *Workflow Transition Interpreter* (see Fig. 2) plays a central role in controlling the interaction of all eP^2 components. It has been refined from the specification in [14] in order to integrate deontic classification, actor modelling, and user interaction modelling, in particular for user tasks, which, in contrast to a "classical" task, are marked as "obligatory", "optional", "alternative", and/or "forbidden".

These tasks are displayed in a *worklist*, which represents the core of a worker's view. The worklist is the one point where users in a business process select the next step to execute. Each user has their own worklist or rather their own view of the global worklist. The worklist is divided into two areas, i.e. the process and the task area. The process area (start menu) contains all registered business processes that can be started by the user. The entries in the task area are related to dialogues consisting of reports and forms, through which the users are led in order to complete the selected tasks.

In H-BPM, we distinguish between four modelling steps which are required to achieve seamless integration of selected models for defining holistic business processes. All steps are illustrated by a concrete example model in Section 4.

The ***first step*** is dedicated to classical process modelling, including a set of activities, gateways, events, etc., using common modelling languages or notations such as BPMN.

The ***second step*** is to extend the model with *deontic classifications* and required data objects. The core deontic concepts are obligation (O), permission (P), and prohibition (F for "forbidden") extended with an operator for alternative (X). The inserted data objects illustrate the flow of data at a high level of abstraction and will be refined within the next step. The deontic classification allows to remove a multitude of surrounding (exclusive) gateways and alternative empty paths, with the advantage of reducing the structural complexity of the diagram [20] and leading to increased understandability [17].

The ***third step*** includes the specification of user interaction and data using the particularly suitable *workflow chart* notation [11]. A workflow chart is specified as a tripartite graph consisting of typed nodes for (i) rendering reports (*client pages*), (ii) providing forms for user input (*immediate server actions*), and (iii) representing task items of the users' worklists (*deferred server actions*). Basically, the resulting dialogue model describes the same behaviour as the previously defined process model but includes considerably more details on the users' actual work within the defined process. This is because its viewpoint focuses more on users and their interaction with applications by specifying the data and operations available, which will be used to refine user tasks.

The ***fourth step*** is finally designed for integrating the dialogue model with the deontic process model in order to refine user tasks by dialogues and data (as shown in Section 4). This integration is made by groups of client pages and server actions that are assigned to a user task (according to the notion of a *dialogue*, which only includes immediate server actions with unique choice [1]). Different kinds of workflow chart nodes refine the user tasks of the process model based on their types by also integrating the *data model*. The type of a client page describes data that is displayed as a report to the user. The type of an immediate server action represents a user-editable function, i.e. it describes data that the user can enter and submit. The type of a deferred server action is void (as deferred server actions are typically displayed as links in worklists). The types of both the client page and the immediate server action are defined by data views denoting lists of relations. The relations are described by n-ary tuples over data attributes that assign data type and attribute name.

3.4 Specification of Operational Semantics

In the following, we show an exemplification of our approach on formally modelling the semantics of BPMN using the ASM method by means of processing user tasks. As a consequence of several inconsistencies we detected in the BPMN standard, we were not able to stick to the standard in every respect. In some cases, we took the opportunity to clarify open questions with people involved in the standard's committee; in other cases, we had to make additional assumptions ourselves to cover gaps and to remove ambiguities, or we had to decide to regard certain provisions of the standard while ignoring other, contradicting provisions (see [12]).

The *Workflow Transition Interpreter* runs through business process diagrams and passes active task instances via the interface *ProcessTask* to the *Task Server*, which in turn uses the interface *TaskList*, provided by the *Process Client*, to distribute the tasks to appropriate users. When a task is reached by a token (and required data are available), the users currently belonging to the associated role are determined (using the actor model) and deontic classification with possible deontic expressions are evaluated (that is, the task is classified as obligatory, permitted (optional), alternative, or forbidden for the users in question).

In detail, in the rule *StartOperation* [14, Section 4.5.3], for a user task, the rule *ProcessUserTask* is called which was left abstract in the original specification in [14]. (*StartOperation* is engaged when an activity is activated.)

We can specify *ProcessUserTask* now in order to notify the *Task Server* that the task in question is scheduled for processing. Therefore, we have to asynchronously call the operation *processTask* of the *ProcessTask* interface (of the *Task Server* component) with the given parameters. Like ASM agents, the components are associated with the function *self*, which is used to identify the current instance of any component.

In addition, we need to add a rule which computes the data structure of exported functions containing the respective deontic classification for each user. In most cases, the classification will depend on the runtime situation, that is, it will be obligatory, forbidden, etc. under a certain condition. A typical example is that a task is, in general, optional but obligatory if a certain other task has already been performed. Possible inputs are, amongst others, the deontic expression (rules), the executing role, and the history of the encompassing activity instance.

```
rule ProcessUserTask(instance, inputSet, isTransaction) =
  let parameterList = new(eP2.parameters),
  let transactionSubProcess =
      getTransactionSubProcess(instance, isTransaction) in
  seqblock
    add instance to parameterList
    add currentInputSet(instance) to parameterList
    add transactionSubProcess to parameterList
    add deonticClassificationsForUsers(instance) to
        parameterList
    eP2.CallOperationAsync("processTask", parameterList, self,
        eP2.taskServer)
  endseqblock
```

In the following, we will further detail concretisation of the approach during refinement by specifying the processing of a user task as a representative piece of the comprehensive eP^2 architecture. Particularly interesting in this context is the *Process Client*, which is controlled by the *Workflow Transition Interpreter* via the *Task Server* by means of process definitions and reacts to user actions of the interface *UserTaskActions* on the selection of tasks from the users' worklists. In the rule *ReactToCall*, which is invoked for every asynchronously called operation, the *Task Server* propagates the relevant data of the selected user task.

```
rule ReactToCall(calledOperation) =
  let params = params(calledOperation),
  let calledOperationName = operationName(calledOperation) in
    case calledOperationName of
      "processTask" → PropagateUserTask(params)
      ...
```

The rule *PropagateUserTask* handles special needs of transactional and ad-hoc sub-processes and propagates the task instance to all users that may process this task using the rule *PropagateTaskToUsers*, which evaluates the deontic classifications and adds the task to the worklists of the respective users, unless the classification is "Forbidden" (synchronous call). Based on the deontic classification, it must be visualised in the user's worklist whether this task is obligatory, optional, or an alternative to certain other tasks.

```
rule PropagateTaskToUsers(deonticClassifications, instance)
  forall dCForUser ∈ deonticClassifications do
    if classification(dCForUser) != "Forbidden" then
      if forsome client ∈ runningClients holds user(client) =
        user(dCForUser) then
          TaskList(client).addTask(instance, dCForUser)
```

Then the *Workflow Transition Interpreter* waits for feedback from the user. Possible reactions are "task selected" (followed by "task completed") for alternatives, or – as far as this is admissible – "task rejected" for inclusive alternatives and "task skipped" for optional tasks. In the original specification (see [14, Section 4.5.2]), we only waited for *taskCompleted* (in detail, we waited for *exitCondition(instance)* to become *true*, in which case we "exit" the activity in a regular way). Now we must refine the specification to check for the three other monitored functions (which includes introducing two new lifecycle states, "Rejected" and "Skipped", for the lifecycle of a BPMN activity [34]):

- If *taskSelected* ≠ *undef*, we re-classify the alternatives of the given task instance as "Forbidden" and set their lifecycle states to "Rejected". No tokens are produced.
- If *taskRejected* is *true*, we set the lifecycle state of the task to "Rejected". No tokens are produced. Additionally, we re-classify the alternatives of the given task (which might lead to a single remaining alternative being re-classified as "Obligatory").
- If *taskSkipped* is *true*, we set the lifecycle state of the task to "Skipped". In this case, tokens *are* produced.

We, thus, extend the *InstanceOperation* of *ActivityTransition* [14, Section 4.5.2] by three additional checks, querying the status of the task instance in question, in the following way, using the *Monitor* interface:

```
rule InstanceOperation(instance, flowNode) =
parblock
  if lifecycleState(instance, flowNode) = "Ready" then
    GetActive(instance, flowNode)
  if lifecycleState(instance, flowNode) ∈
      activeWaitingLifecycleStates then
  parblock
    CleanUpBeforeExit(instance, flowNode)
    if exitCondition(instance) then
    parblock
      ExitActivity(instance, flowNode)
      lifecycleState(instance, flowNode) :=
          getNewLifecycleState(instance, flowNode)
    endparblock
    if Monitor.taskSelected(instance) then
      ReclassifyAlternatives(instance, flowNode)
    if Monitor.taskRejected(instance) then
      RejectTask(instance, flowNode, true)
    if Monitor.taskSkipped(instance) then
      ExitActivity(instance, flowNode)
  endparblock
endparblock
```

For continuing process execution, further eP^2 components are required. The *Actor Management* is responsible for managing roles and assigning users to these roles and the *Data Management* provides access to the data model.

According to the type concept of workflow charts, the *Data Management* supports generating views based on global data and translation of changes from a local perspective to the global one. More precisely, data is requested for reports (*read* operation) and manipulated in actions (*update*, *insert*, and *delete* operations). Deferred server actions are regarded as side effect free and can be neglected when addressing global consistency. Thus, we capture user interactions by means of operations on data views, i.e. we basically apply operations, causing changes of data, to data views and in turn receive new views (cf. ASM transducers [33]). In this context, beside data availability on activity activation and completion, it must be ensured that the input data specification $\Sigma(i)$ of a user task fulfils the type specification of the first report $\Sigma(r)$ contained in the refining dialogue ($\Sigma(r) \subseteq \Sigma(i)$), where Σ defines the types of data inputs and outputs. In addition, the type specification $\Sigma(a)$ of the final action(s) must fulfil the output data specification $\Sigma(o)$ of the user task ($\Sigma(o) \subseteq \Sigma(a)$).

Returning to our specification on processing a user task, the user's reaction is captured via the *UserTaskActions* interface of the *Process Client*. In the case of *taskSelected*, the rule *SelectedByUserForProcessing* first tries to start the task by checking authorisation of the user and then hands over to the *Dialogue Engine*.

```
rule SelectedByUserForProcessing(instance) =
  let user = user(self) in
```

```
    let isAllowed = TaskHandling.TryStartTask(instance, user) in
      if(isAllowed) then
        if flowNodeType(getFlowNode(taskInstance( instance))) =
          "UserTask" then
          ProcessTaskByAidOfDialogueEngine(instance)
```

The *Dialogue Engine* handles all form-oriented activities that may exist behind the user task, which is refined by at least a *one-step dialogue*, consisting of a (starting) deferred server action, a single, subsequent client page, and its connected (final) server actions. This would mean that every dialogue state belongs to a workflow state, resulting in a high number of elementary tasks with workflow control always returning to the worklist after having submitted a form. Alternatively, a user task can be refined by a *complex dialogue*, including multiple client pages and server actions wired together.

The rule *ProcessDialogue* uses the derived function *enabledNodes* [11] that yields the succeeding, enabled nodes of a given selected node in a workflow chart. For presenting information to the user, the rule *ShowDialogue* is applied, which implies calling the operation *read* of the *Data Management*. The submission of user input is handled by calling the rule *SubmitData*, which implies *create*, *update*, or *delete* operations.

```
rule ProcessDialogue(instance, transactionId) =
  local link := startingDeferredServerAction(
      dialogue(getFlowNode(instance))) in
  local endOfDialogue := false, action, page in
    while !endOfDialogue do
    seqblock
      page := getSinglePage(enabledNodes(link,
          deferredServerActions, clientPages))
      ShowDialogue(page, enabledNodes(page, clientPages,
          immediateServerActions))
      action := selectedNode(self)
      SubmitData(action, transactionId)
      endOfDialogue := isFinalDialogueNode(action)
      link := getSingleLink(enabledNodes(action,
          immediateServerActions, deferredServerActions))
    endseqblock
```

After all defined client pages and server actions have been processed, the *Workflow Transition Interpreter* is notified about the completion of the task and, depending on the specification of the process model, enables new tasks.

4 The Order Placement Process: An Application Scenario

We demonstrate the usability of the model integration approach by means of the *order placement process*, a scenario for ordering supplies, as an illustrative example. We first model the process in BPMN-style, i.e. as a conventional process model. We then add deontic classifications and actors and describe required data by dialogues consisting of typed nodes (reports, forms, and links). Finally, we show how the deontically classified process model can be integrated with the actor modelling approach and workflow charts in H-BPM.

In Fig. 3, we show the classical process model which primarily comprises user tasks. Somebody in the company signals a *demand* for supplies by sending a respective message. Now a purchaser in charge of ordering supplies must first turn this message into a proper *product specification* of the desired goods or services. The purchaser then has to decide whether to use a given *standard supplier* or, alternatively, to issue a *bidding call* to several possible suppliers. In the first case, the purchaser needs to check the *current offer* and general terms of shipment – a rather simple task. A call for bidding, however, requires more work, which we will show in the refined process model in Fig. 6. After a sufficient number of offers has been collected, there is the option to try to improve prices and conditions through *further negotiations* with a few selected trading partners. Finally, a *supplier is chosen* and the process continues in the same way as if a standard supplier had been chosen; the *order is placed* and our simplified process ends.

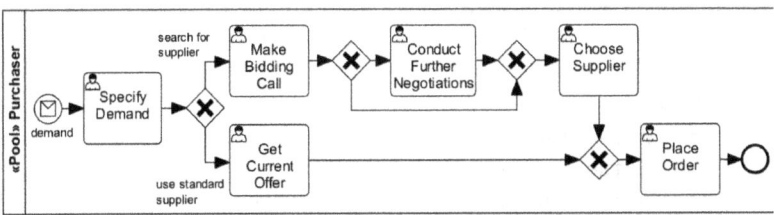

Fig. 3. Order Placement Process: Classical Process Model in BPMN

The next step is to extend the model with *deontic classifications* as shown in Fig. 4 (O for obligation, P for permission, F for prohibition (rarely used in case of multiple deontic classifications to mark activities that are forbidden under some circumstances), and X for alternative). The diagram is also augmented with data objects to illustrate the flow of data.

After the demand for supplies is signalled, the task *specify demand* must be performed within every process instance (i.e. there is no alternative to it) and is, thus, marked as obligatory. It is followed by the two alternative tasks *get current offers* and *make bidding call*. If the purchaser has chosen to search for further suppliers, then the optional task *conduct further negotiations* can be executed, but the completion of task *make bidding call* must be specified as precondition (i.e. this ensures that the upper path has been taken).

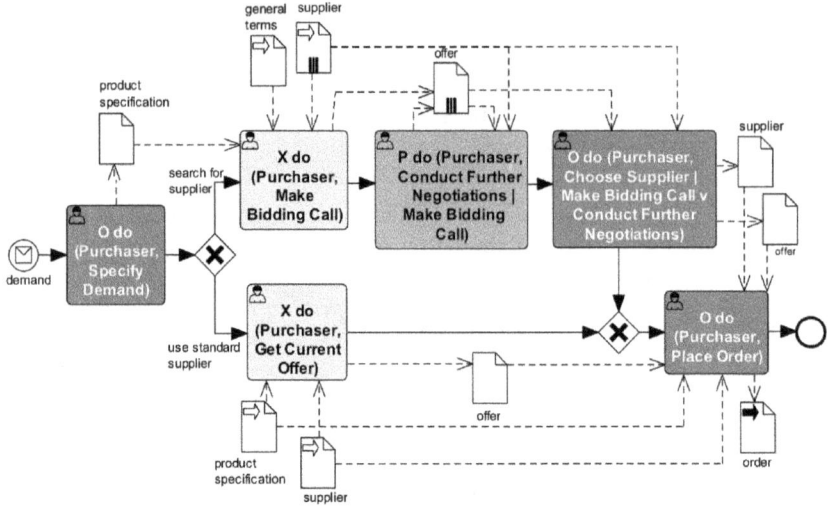

Fig. 4. Order Placement Process: Deontic BPMN Diagram and Data Objects

Based on the deontic classification the surrounding exclusive gateways and the alternative empty path can be removed. Note that all data objects produced by an optional task are also optional, so if a data object is required as input for a subsequent task, then there must be an alternative source, e.g. in our case the original *offers*.

The last task on the upper path is to *choose a supplier*, which is classified as obligatory under the precondition that possible previous tasks are completed. The two alternative paths are then merged and the last, obligatory task *place order* is executed. Finally, we can extend the deontic classification and specify permitted roles for every task. In our case, we assign the role *Purchaser* to each task and remove the corresponding pool shown in the classical process model.

In Fig. 5, we present the order placement process in *workflow chart* notation, which corresponds to the third step of the integration approach. *Client pages*, specified as circles in workflow charts, present information to the user and provide forms for user input. The forms are represented by *immediate server actions*, which are displayed by white rectangles. *Deferred server actions* appear as links to client pages in the worklists of the corresponding users and are specified as rectangles with a gradient fill. Actor information is attached to deferred server actions.

Regarding the order placement process, only one role, i.e. the purchaser, is concerned with ordering supplies, starting from *specifying a demand* to *placing the order*.

The final step is to integrate the dialogue model with the deontic process model. In doing so, user tasks of the process model are refined by dialogues and data as shown in Fig. 6.

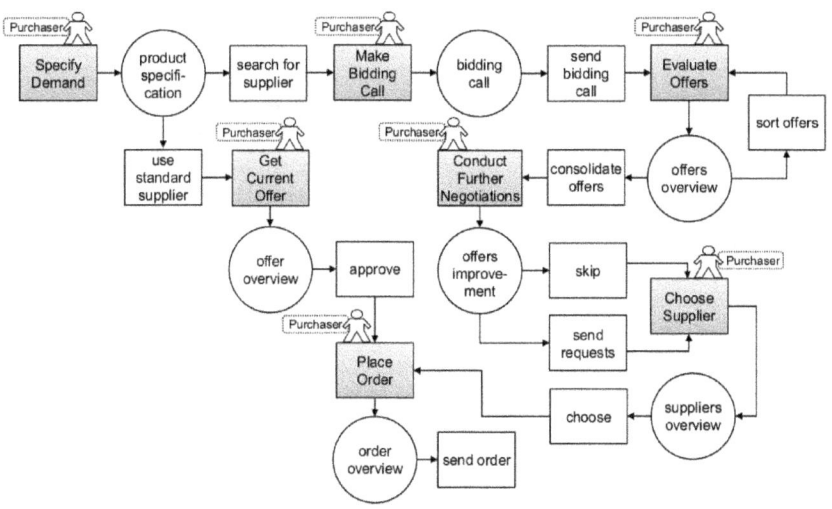

Fig. 5. Order Placement Process: Workflow Chart Diagram

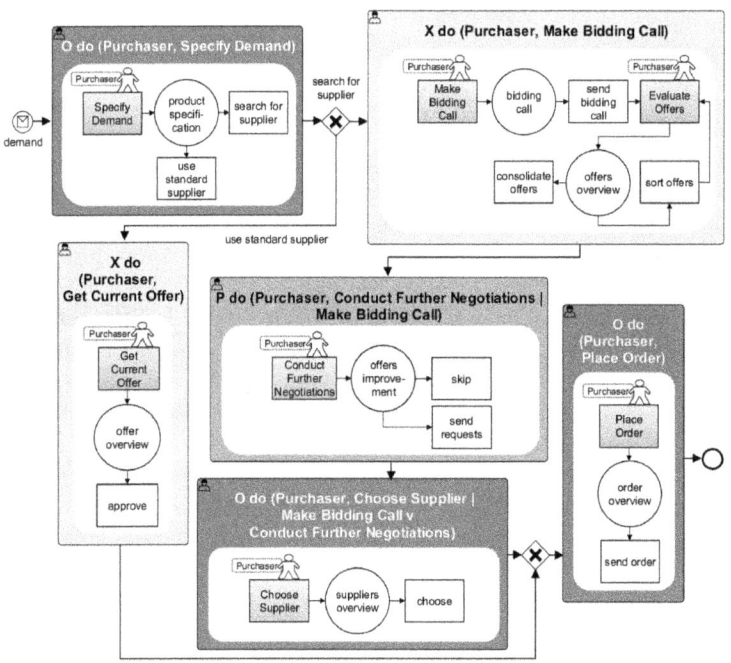

Fig. 6. Order Placement Process: H-BPM Diagram

For example, the task *Place Order* will be refined by the homonymous deferred server action (assigned to the role *Purchaser*), client page *order overview*, and immediate server action *send order*. In the following, an exemplary notation is given for defining the concrete types of these nodes. The types of both the client page and the immediate server action are defined by data views denoting lists of relations. The relation is described by n-ary tuples over data attributes that, in turn, assign data type (e.g. *Integer*) and attribute name (e.g. *quantity*).

client page order_overview: (product_specification, supplier, offer)
immediate server action send_order: (order)

relation product_specification: (product_name:String × description:String × quantity:Integer)

To sum up, the resulting H-BPM diagram comprises all relevant aspects for business process modelling and execution (i.e. control flow, actor, interaction, and data) based on a formally defined integration method. In this way, process modellers and analysts can benefit from this intuitive, integrated view and concentrate on their modelling tasks free from distraction by following the defined steps of the H-BPM method.

5 Related Work

Since the 1980s, different views are used in software and information system architecture as abstractions to comprehend the complexity of such systems. Enterprise architecture frameworks [16] like ARIS, the Zachman Framework, and TOGAF bring systematisation to the plethora of models and views, or more generally to the plethora of related artefacts, which show up during the life-time of systems.

The business process framework *Architecture of Integrated Information Systems* (ARIS) [25] comprises five views. The views are symbolically presented in the form of a house, the so-called ARIS house with the *Organization View* as the roof, the *Data, Control*, and *Function View* as the three pillars, and the *Output View* as the basis of the house. A detailed comparison between ARIS and our approach would not be conducive, because ARIS is a framework for overall enterprise architecture, whereas H-BPM aims at seamless supporting different aspects in the domain of business process modelling. Nevertheless, three views of the ARIS concept are similar to our approach. The *Function View* supports goals and defines processing rules for application software. According to A.-W. Scheer, the designations "function", "process" and "activity" are used synonymously, thus this view is similar to our process model. In addition, the *Data View* comprises the data processing environment including messages, and is therefore similar to our data model. Furthermore, the *Organization View* of ARIS comprises a hierarchical organisational model with organisational units and concrete instances [25], and resembles the homonymous view of our actor model.

The *Zachman Framework* [37] consists of six rows for the viewpoints (*Scope, Enterprise* (or *Business*) *Model, System Model, Technology Model, Detailed Representations*, and *Functioning Enterprise*) and six columns for the aspects (*Data, Function, Network, People, Time,* and *Motivation*). The aspect *Data* includes the logical and physical data model, the data definition, and the actual data; thus it is similar to our data model. The aspect *Function* contains the business process model and resembles our process model. Furthermore, the aspect *People* contains the workflow or organisational model with organisational units and hierarchical dependencies defined between them and correlates to our actor model. However, missing are detailed descriptions of user interactions.

Besides established enterprise architecture frameworks, which address aspects for describing whole enterprises, various approaches propose views suitable for automating business processes, e.g. by Hofmeister et al. [8], Clement et al. [5], or Rozansky and Woods [24]. All of them propose – at least – kinds of (i) a static view (also called *Conceptual Architecture View* or *Module Viewtype*) for describing the static structure of the system, (ii) a dynamic view (also called *Component-Connector Viewtype* or *Operational View*) for describing the runtime level, and (iii) an allocation view (also called *Execution Architecture View* or *Allocation Viewtype*) for describing the mapping of software components to the environment. Additional views of the individual approaches are, e.g., the *Code Architecture View* [8], focusing on the development and deployment, the *Concurrency View* [24], describing the concurrency structure of the system, or the *Context View* [24], where the system is observed from outside. Similarly, the H-BPM approach provides the eP^2 architecture as the static view and then describes the operational semantics using ASMs, thereby focusing on parallel specifications and classified functions to reflect different kinds of visibility.

Beyond this, concrete BPM tools that support enterprise or software architecture frameworks may (explicitly or implicitly) define balancing rules between different kinds of supported artefacts. Such balancing rules target, e.g., the unique naming and unique place of definition of modelling elements or traceability features to materialise the conceptual relationships between the elements of several kinds of artefacts. However, the tools and frameworks do not elaborate domain-specific guidelines for concrete models as we do in H-BPM for actor and user interaction modelling. The approach presented in this work can be characterised as orthogonal to enterprise architecture framework approaches. It concentrates on the definition of process-oriented systems and proposes specific guidelines for this application domain, obviously lacking important enterprise dimensions. The models that are defined and integrated in our approach show up naturally and are provided to reduce complexity. However, the proposed H-BPM method goes beyond the definition of models or views; it is designed as an integrated method for seamless support of different aspects in business process modelling based on different abstraction levels and integration guidelines (as discussed in Section 3). The more concrete guidelines, e.g., the tripartiteness of typed workflow charts or the powerful yet understandable deontic classifications of user tasks that are considered by our approach, can also be re-used or re-elaborated in the realm of other enterprise architecture frameworks.

Regarding the integration of actor modelling concepts in business processes, several evaluations and studies revealed the lack of completeness and clarity regarding this perspective, see e.g. [23, 35]. To address this problem, several approaches have been suggested, such as various extensions to BPMN to express task-based allocation and authorisation constraints using OCL [2, 36] or a rule-based extension of the BPMN metamodel based on the REWERSE Rule Markup Language (R2ML). Also some other business process modelling languages provide task-based definition of resources to integrate actor modelling, e.g. UML Activity Diagrams [22], Event-Driven Process Chains (EPCs) [26], and Yet Another Workflow Language (YAWL) [9]. However, some approaches still use the pool and lane concepts to express the role hierarchy resulting in a mixture of different resource definitions. In addition, it is not possible to specify generic restrictions (e.g. every business travel must be approved by the corresponding manager), dependencies between tasks concerning the executing roles, or to signify general permissions or obligations of users. In H-BPM, we consider such constraints and modalities when integrating the process and actor models and further extend the life-cycle of a classical BPMN activity by introducing two new life-cycle states, i.e. "Rejected" and "Skipped", to be able to handle tasks that are classified as permissible or alternative in the user interface of a *worklist*, which represents a new and innovative concept that is not yet realized in existing tools.

In addition, integrating dialogues into the activity-centred language represents a novel approach for refining business process models, and user tasks in particular, by dialogues and data views. Beyond existing approaches to user interface modelling [31, 32, 10] and integrating user interfaces and databases [30, 27], we specified how typed dialogues can be laid behind (or be graphically embedded in; see Section 4) the deontically classified user tasks of the process model, with designated entry points to the dialogue. As a result, the structure of dialogues becomes changeable, which leads to a natural partition of business logic into services of appropriate granularity. It further allows process designers to flexibly specify which parts of a business process apply to workflow technology and which parts make up the system dialogues, using an intuitive framework with defined modelling steps.

6 Conclusion and Future Work

In this paper, we presented a novel, homogeneous integration framework, which is capable of formally integrating indispensable aspects for business process automation. The proposed extensions to the rigorous semantics for business process diagrams address actor modelling, closely tied to an intelligible way for denoting permissions and obligations in terms of deontic logic, user interaction modelling, and data modelling. All of these models have been described before and their expressiveness and ease of use have been proved independently. However, it is their interplay which makes them useful in practice. By using the proposed H-BPM method, process analysts and modellers are guided via defined steps, providing a complete walk-through for modelling business processes in a trustworthy way.

In this way, the H-BPM method enables seamless modelling of multifaceted aspects of business processes on all levels of abstraction by means of formal refinement. It uses the well-established ASM method, which excels in understandability, in order to provide accurate information for the design of a BPM system. Therefore, we addressed both static component integration as well as a dynamic, runtime-related view on model integration. The formal specification of the semantics, of which we presented the part for processing a *user task*, is well-suited to be refined towards an implementation.

We currently focus on further aspects towards a unified, comprehensive BPM approach. Future research work includes the development of an enhanced communication concept (supporting more differentiated communication types for complex event handling), a deeper investigation of deontic concepts, process adaptivity, and the integration of exception handling mechanisms to handle unexpected situations in a reliable manner in H-BPM. A further task will be to apply the presented approach to a real-world scenario in the form of a comprehensive case study.

Acknowledgment

This publication has been written within the project *AdaBPM*, which is funded by the *Austrian Research Promotion Agency* (FFG) under the project number 842437. Underlying research was conducted within the projects "Vertical Model Integration (VMI)" and "VMI 4.0", which were both supported within the programme "Regionale Wettbewerbsfähigkeit OÖ 2007-2013" by the European Fund for Regional Development as well as the State of Upper Austria.

References

1. Atkinson, C., Draheim, D., Geist, V.: Typed business process specification. In: Proceedings of the 14th IEEE International Enterprise Distributed Object Computing Conference. pp. 69–78. EDOC'10, IEEE Computer Society (2010)
2. Awad, A., Grosskopf, A., Meyer, A., Weske, M.: Enabling resource assignment constraints in BPMN (2009), Hasso Plattner Institute, Potsdam
3. Awad, A., Decker, G., Lohmann, N.: Diagnosing and repairing data anomalies in process models. In: Business Process Management Workshops. pp. 5–16. Springer (2010)
4. Börger, E., Stärk, R.: Abstract State Machines: A Method for High-Level System Design and Analysis. Springer, Berlin, Heidelberg (2003)

5. Clements, P., Bachmann, F., Bass, L., Garlan, D., Ivers, J., Little, R., Nord, R., Stafford, J.: Documenting Software Architectures: Views and Beyond. Addison-Wesley, Boston, MA (2003)
6. Draheim, D., Geist, V., Natschläger, C.: Integrated framework for seamless modeling of business and technical aspects in process-oriented enterprise applications. International Journal of Software Engineering and Knowledge Engineering 22(05), 645–674 (2012)
7. Fleischmann, A., Schmidt, W., Stary, C., Obermeier, S., Börger, E.: Subject-Oriented Business Process Management. Springer, Berlin, Heidelberg (2012)
8. Hofmeister, C., Nord, R., Soni, D.: Applied Software Architecture. Addison-Wesley object technology series, Addison-Wesley (2000)
9. ter Hofstede, A.M., van der Aalst, W.M.P., Adams, M., Russell, N. (eds.): Modern Business Process Automation: YAWL and its Support Environment. Springer, Heidelberg (2010)
10. Kloppmann, M., Koenig, D., Leymann, F., Pfau, G., Rickayzen, A., Riegen, C., Schmidt, P., Trickovic, I.: WS-BPEL Extension for People – BPEL4People. IBM, SAP (2005)
11. Kopetzky, T., Geist, V.: Workflow charts and their precise semantics using Abstract State Machines. In: EMISA. pp. 11–24. Lecture Notes in Informatics, Gesellschaft für Informatik e.V. (2012)
12. Kossak, F., Illibauer, C., Geist, V., Natschläger, C., Kopetzky, T., Ziebermayr, T., Schewe, K.D.: Hagenberg Business Process Modelling Method. Springer, Heidelberg (2016)
13. Kossak, F., Geist, V.: An enhanced communication concept for business processes. In: Kolb, J., Leopold, H., Mendling, J. (eds.) Enterprise Modelling and Information Systems Architectures – Proc. EMISA 2015. Lecture Notes in Informatics, vol. 248, pp. 77–91. Gesellschaft fÃ¼r Informatik (2015)
14. Kossak, F., Illibauer, C., Geist, V., Kubovy, J., Natschläger, C., Ziebermayr, T., Kopetzky, T., Freudenthaler, B., Schewe, K.D.: A Rigorous Semantics for BPMN 2.0 Process Diagrams. Springer, Heidelberg (2015)
15. Kossak, F., Mashkoor, A., Geist, V., Illibauer, C.: Improving the understandability of formal specifications: An experience report. In: Salinesi, C., van de Weerd, I. (eds.) Requirements Engineering: Foundation for Software Quality, Lecture Notes in Computer Science, vol. 8396, pp. 184–199. Springer International Publishing (2014)
16. Minoli, D.: Enterprise Architecture A to Z: Frameworks, Business Process Modeling, Soa, and Infrastructure Technology. Auerbach Publishers Inc. (2008)
17. Natschläger, C.: Deontic BPMN. In: Hameurlain, A., Liddle, S., Schewe, K.D., Zhou, X. (eds.) Database and Expert Systems Applications, Lecture Notes in Computer Science, vol. 6861, pp. 264–278. Springer, Berlin, Heidelberg (2011)
18. Natschläger, C., Geist, V.: A layered approach for actor modelling in business processes. Business Process Management Journal 19, 917–932 (2013)
19. Natschläger, C., Geist, V., Kossak, F., Freudenthaler, B.: Optional activities in process flows. In: Rinderle-Ma, S., Weske, M. (eds.) EMISA 2012 – Der Mensch im Zentrum der Modellierung. pp. 67–80 (2012)
20. Natschläger, C., Kossak, F., Schewe, K.D.: Deontic BPMN: a powerful extension of BPMN with a trusted model transformation. Software & Systems Modeling pp. 1–29 (2013)
21. Natschläger-Carpella, C.: Extending BPMN with Deontic Logic. Logos Verlag Berlin (2012)

22. Object Management Group: OMG Unified Modeling Language (OMG UML), superstructure version 2.4. http://www.omg.org/spec/UML/2.4.1/Superstructure/PDF. Accessed 2015-07-09. (2011)
23. Recker, J.C., Indulska, M., Rosemann, M., Green, P.: How good is BPMN really? insights from theory and practice. In: Proceedings of the 14th European Conference on Information Systems (ECIS 2006). pp. 1582–1593 (2006)
24. Rozanski, N., Woods, E.: Software Systems Architecture: Working with Stakeholders Using Viewpoints and Perspectives. Addison-Wesley (2011)
25. Scheer, A.W.: ARIS – Business Process Modeling. Springer, Heidelberg (2000)
26. Scheer, A., Thomas, O., Adam, O.: Process modeling using event-driven process chains. In: Process-Aware Information Systems: Bridging People and Software through Process Technology, pp. 119–146. John Wiley & Sons, Inc., Hoboken, New Jersey (2005)
27. Schewe, B., Schewe, K.D.: A User-Centered Method for the Development of Data-Intensive Dialogue Systems: an Object-Oriented Approach. In: Proceedings of the IFIP International Working Conference on Information System Concepts: Towards A Consolidation of Views. Chapman & Hall (1995)
28. Schewe, K.D.: Horizontal and vertical business process model integration. In: Database and Expert Systems Applications, Lecture Notes in Computer Science, vol. 8055, pp. 1–3. Springer, Berlin, Heidelberg (2013)
29. Schewe, K.D., Geist, V., Illibauer, C., Kossak, F., Natschläger-Carpella, C., Kopetzky, T., Kubovy, J., Freudenthaler, B., Ziebermayr, T.: Horizontal business process model integration. In: Transactions on Large-Scale Data-and Knowledge-Centered Systems XVIII, pp. 30–52. Springer (2015)
30. Schewe, K.D., Schewe, B.: Integrating Database And Dialogue Design. Knowledge and Information Systems 2(1) (2000)
31. Seffah, A., Vanderdonckt, J., Desmarais, M.: Human-Centered Software Engineering. Springer (2009)
32. da Silva, P.P., Paton, N.: UMLi: The Unified Modeling Language for Interactive Applications. In: UML 2000 - The Unified Modeling Language. Springer (2000)
33. Spielmann, M.: Verification of relational transducers for electronic commerce. Journal of Computer and System Sciences 66(1), 40–65 (2003)
34. The Object Management Group: Business Process Model and Notation (BPMN) 2.0. http://www.omg.org/spec/BPMN/2.0. Accessed 2015-07-09. (2011)
35. Wohed, P., van der Aalst, W.M., Dumas, M., ter Hofstede, A.H., Russell, N.: On the suitability of BPMN for business process modelling. In: Dustdar, S., Fiadeiro, J.L., Sheth, A.P. (eds.) Business Process Management. Lecture Notes in Computer Science, vol. 4102, pp. 161–176. Springer (2006)
36. Wolter, C., Schaad, A.: Modeling of task-based authorization constraints in BPMN. In: Business Process Management: 5th International Conference (BPM 2007), pp. 64–79. Springer, Berlin, Heidelberg (2007)
37. Zachman, J.: A framework for information systems architecture. IBM Systems Journal 26(3), 267–292 (1987)

Closing the gap between the specification and the implementation: the ASMETA way

Paolo Arcaini*[1], Angelo Gargantini[2], and Elvinia Riccobene[3]

[1] *Charles University, Faculty of Mathematics and Physics, Czech Republic,*
arcaini@d3s.mff.cuni.cz

[2] *Department of Economics and Technology Management, Information Technology and Production, Università degli Studi di Bergamo, Italy,*
angelo.gargantini@unibg.it

[3] *Dipartimento di Informatica, Università degli Studi di Milano, Italy,*
elvinia.riccobene@unimi.it

Abstract

In software system development, formal models are used to precisely specify the initial requirements, understand what the system is supposed to do and prove properties of the system. Ideally, formal specifications should be developed long before the system is implemented, and in a model-driven software development the code would be automatically generated from the models. However, quite often either code generation is not available or not reliable enough, or the final specification of a system becomes ready only together with (or even after) its implementation. In all these cases, the conformance relation between the formal specification and the concrete implementation must be checked, in order to assure that the software behaves as expected.

In this paper, we show how the ASMETA framework, based on the use of the Abstract State Machine (ASM) formal method, can be used to assist the developer in the conformance checking of Java code against ASM models. We present a technique to formally link a Java class with its ASM formal specification, and two approaches for checking their conformance: *offline* (i.e., before deployment) using model-based test generation, and *online* (i.e., after deployment) using runtime verification. We also report our experience in applying these two techniques in the context of safety critical systems, where conformance checking of the implementation against its formal specification is highly demanded.

---

*The research reported in this paper has been partially supported by the Czech Science Foundation project number 17-12465S.

1 Introduction

In software system development, the use of formal models is fundamental to assure reliability and robustness of the system to deploy.

Formal models permit the designers to specify what the software is intended to do in a rigorous and precise way, avoiding ambiguities and misunderstanding. They can be used, already at the early stages of the software development, to prove that safety-critical properties are satisfied. Moreover, formal specifications, especially if based on executable computational models, can be exploited for different forms of validation, and in particular to derive test cases in a software independent way.

Model-driven development approaches want the code to be automatically derived from models or obtained by refinement as the last step of a chain of more and more correctly refined models. However, most of the times, this ideal scenario cannot be applied. Sometimes the software is not developed from scratch but it relies on existing code, exploiting component reuse. Very often software is developed by external vendors without any assurance of correctness or traceability w.r.t. a formal model. Or it is implemented by the use of programming techniques and structures that have no correspondence to constructs or elements of formal notations. Furthermore, practitioners find designing models time-consuming and, in practice, code is often developed independently from the model, or simultaneously with the model, or even after only when a quality assurance becomes necessary. In most cases, formal models aim at capturing only the critical behavior of a system, so the implementation is at a different level of abstraction w.r.t. its formal model. Therefore, even if the code has been automatically generated or somehow obtained from its formal specification, it must be modified in order to add all the necessary details, and also later (after deployment) for maintenance reasons.

In all these scenarios, to guarantee software reliability, checking conformance of the running code against models specifying expected and safe behavior is required.

Among the wide range of formal methods, the Abstract State Machines [19] (ASMs) have been used for requirements specification, validation and verifications of different case studies. The method is based on an executable computational model of state machines and allows for an incremental design process based on model refinement. ASMETA (ASM mETAmodeling) [14] is an Eclipse framework for an integrated use of different approaches and techniques developed around ASMs. It helps the developer using ASMs by providing support for model editing, simulation, validation, formal verification, etc. [12].

Although in ASMs a reliable code prototype should be obtained as the final step of correct refinement of models, and although attempts exist that try to generate executable code automatically from models [17], also when using the ASM method the implementation might be developed independently from the model, and therefore, the problem of software conformance checking remains a compelling issue. Thus, we have enriched ASMETA with features to check conformance of executable code against ASM models.

In this paper, we focus on the conformance checking activity of Java code within the ASM/ASMETA-based software development process. We present a technique to formally link a Java class with its ASM formal specification, and two approaches for checking their conformance: an *offline conformance checking*, executable before deployment and exploiting the model-based testing approach, and an *online conformance checking*, executable after deployment and based on the runtime verification. We also report our experience in applying these two techniques in the context of safety critical systems, where conformance checking of the implementation against its formal specification is highly demanded.

In Sect. 2, we briefly frame the conformance checking activity within the ASM-based development process. Then, in Sect. 3, we introduce a classical software engineering example, the one-way traffic light case study, that we use along the paper to illustrate the application of our conformance checking approaches. The choice of a simple and well-know example is intentional. Sect. 4 describes how to perform conformance checking between a Java implementation and its ASM formal specification, and presents the two approaches (*offline* and *online*) for conformance checking. In Sect. 5 we discuss our experience in applying both techniques to a series of non-trivial case studies. Sect. 6 relates our work with other results existing in literature, and Sect. 7 concludes the paper.

2 ASM-based development process

The ASM design formal method is based on the use of Abstract State Machines. They are transition systems that extend the Finite State Machines by replacing unstructured control states by algebraic structures, i.e., domains of objects with functions and predicates defined on them. A *state* represents the instantaneous configuration of the system and transition rules describe the change of state. A *run* is a (finite or infinite) sequence of states $s_0, s_1, \ldots, s_n, \ldots$, where each s_i is obtained by applying the transition rules at s_{i-1}. There exists a classification of functions; in this paper, we only consider *controlled* and *monitored* functions. Controlled functions can only be updated by transition rules and represent the internal *memory* of the ASM. Monitored functions, instead, cannot be updated by transition rules, but only by the *environment*; they represent the inputs of the machine.

The ASM formal method allows an iterative design process, shown in Fig. 1. Tools supporting the process are part of the ASMETA (ASM mETAmodeling)

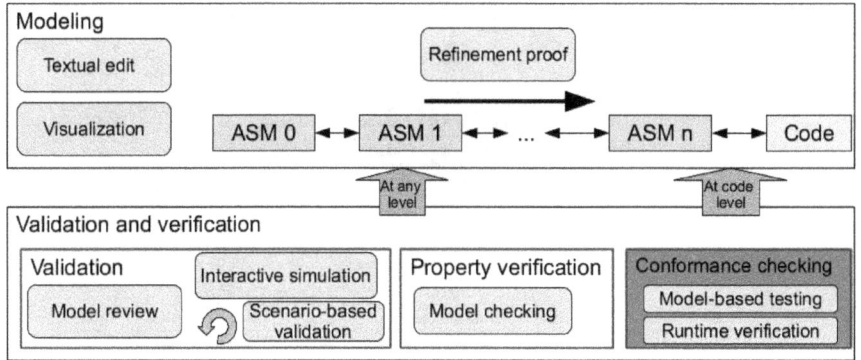

Figure 1: ASM-based process

framework[1] [14], an Eclipse platform that allows for an integrated use of a set of techniques and approaches developed to assist the user along the process.

Modeling is based on model refinement: starting from the *ground model* (ASM 0 in Fig. 1), a high-level specification of the intended system behavior according to the informal requirements, through a sequence of step-wise refined models, further functional requirements and architectural choices can be specified till a complete description of the system. At each refinement step, the refined models should be proved to be correct w.r.t. the abstract one. For a particular (and more common) kind of refinement (*stuttering refinement*) correctness proof can be done automatically [11, 13]. The ASMETA framework is endowed with features for model editing in terms of a concrete syntax [31] and a graphical representation of models [4] (see Fig. 2).

Validation and verification (V&V) are fully integrated into the process, and are possible at any level of abstraction. ASM model validation is possible by means of model simulation [31] and by construction of critical scenarios [20]. Moreover, models can be statically checked if they have sufficient *quality* attributes (e.g., minimality, completeness, consistency) [6]. Formal verification of ASMs is possible by means of model checking [5] *Computation Tree Logic* (CTL) and *Linear Temporal Logic* (LTL) formulas.

Implementation can be either automatically derived from the model [17] or externally provided. In the former case, the conformance w.r.t. the specification should be assured by the translator; in the latter case, conformance checking must be executed. Approaches and functionality of conformance checking supported by ASMETA for Java code will be explained in Sect. 4.

---

[1] http://asmeta.sourceforge.net/

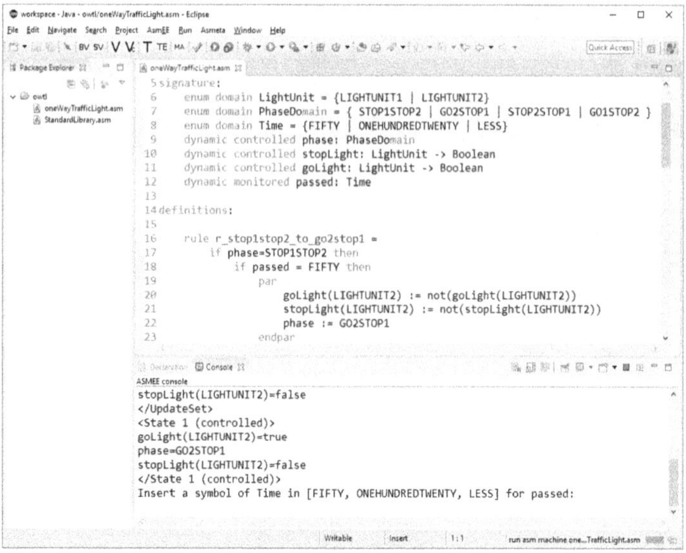

Figure 2: ASMETA in eclipse

3 Running example

As running example, we use the one-way traffic light case study described in [18]. Two traffic lights (LIGHTUNIT1 and LIGHTUNIT2) are placed at the beginning and at the end of an alternated one-way street; both traffic lights are connected to a computer that controls them. Each traffic light is equipped with a *Stop* light (red light) and a *Go* light (green light). The computer turns the lights on and off, following a four phases cycle:
- for 50 seconds both traffic lights show the *Stop* signal;
- for 120 seconds LIGHTUNIT1 shows the *Stop* signal and LIGHTUNIT2 the *Go* signal;
- for 50 seconds both traffic lights show again the *Stop* signal;
- for 120 seconds LIGHTUNIT2 shows the *Stop* signal and LIGHTUNIT1 the *Go* signal.

When a traffic light has the *Go* signal turned on, the cars waiting at that entry of the street can pass through. In the following period, when both units have the *Stop* signal, no car can enter in the street from any entry; in this period, the cars that are driving in one direction have the time to exit the street. In the following period, instead, the cars waiting at the other entry of the street can pass through, and so on.

Fig. 3 shows the semantic visualization [4] of the ASM specification of the example (shown later in Code 1). The semantic visualization shows how the

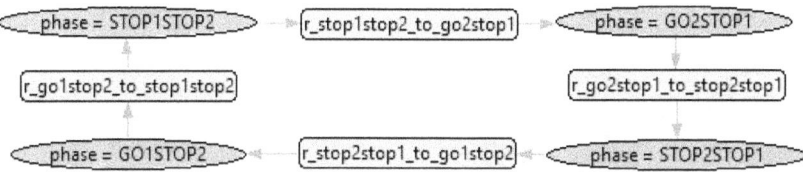

Figure 3: Semantic visualization of the one-way traffic light specification

four macro rules of the ASM specification change the phase of the system.

4 Offline and online conformance checking

We here describe how to perform conformance checking between an implementation in Java and its formal specification in ASMs (the last step of the ASM-based development process described in Sect. 2). We support two approaches for conformance checking:
- an *offline* approach using *model-based testing*,
- an *online* approach using *runtime verification*.

The two approaches are depicted in Fig. 4 and described in Sects. 4.2 and 4.3. In order to show their application, we report in Code 1 an excerpt of the ASM formal specification of the running example. The specification models the time using an enumerative monitored function **passed** saying whether 50, 120, or less than 50 seconds have elapsed since the beginning of the current phase. The system is modeled through functions **stopLight** and **goLight** representing the statuses of the *Go* and *Stop* lights of the two traffic lights; an additional function **phase** identifies the current phase. Code 2 shows the corresponding Java implementation. In the implementation, there is no explicit representation of the phase that is instead modeled by the values of the lights (fields **redLight1**, **redLight2**, **goLight1**, and **goLight2**) and by the field **turn1** saying whether the first or the second traffic light must switch its lights.

Both conformance checking techniques require a way to link elements of the code with corresponding elements in the specification; such a linking is described in Sect. 4.1.

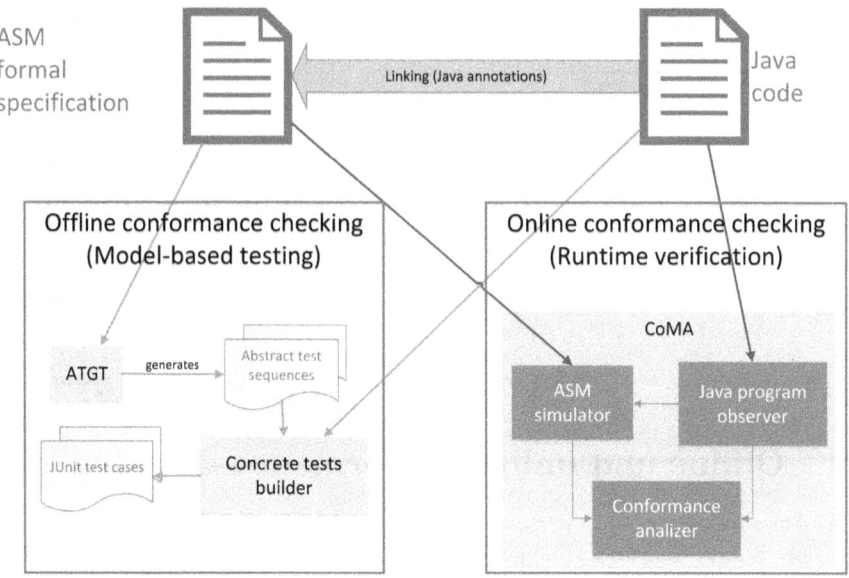

Figure 4: Offline and online conformance checking

4.1 Link and conformance definition between specification and code

For both offline and online conformance checking, we need to define the notion of *conformance* between an implementation and its formal specification. In [8], we proposed an approach to define the conformance relation by using Java annotations[2] originally introduced in [7]. These annotations are used to link a Java class with its corresponding ASM model. There are different kinds of annotations.

- Annotation used to link a Java class to its ASM model:
  - @Asm: it is a class annotation and permits to identify the formal specification (specified using the string attribute asmFile) that corresponds to the annotated class. In the running example, the Java class OWTL.java is linked to the ASM specification oneWayTrafficLight.asm.
- Annotations used to link the class data with the ASM signature:
  - @FieldToFunction/@FieldToLocation: it annotates fields and permits to link a Java field with an ASM function/location. The annotation has a mandatory attribute func for specifying the function

---
[2]A Java annotation is a meta-data tag that permits to add information to code elements (class declarations, method declarations, etc). Annotations are defined similarly as classes.

```
asm oneWayTrafficLight
import StandardLibrary
signature:
    enum domain LightUnit = {LIGHTUNIT1 | LIGHTUNIT2}
    enum domain PhaseDomain = { STOP1STOP2 | GO2STOP1 |
                                STOP2STOP1 | GO1STOP2 }
    enum domain Time = {FIFTY | ONEHUNDREDTWENTY | LESS}
    dynamic controlled phase: PhaseDomain
    dynamic controlled stopLight: LightUnit -> Boolean
    dynamic controlled goLight: LightUnit -> Boolean
    dynamic monitored passed: Time
definitions:
    rule r_stop1stop2_to_go2stop1 =
        if phase=STOP1STOP2 then
            if passed = FIFTY then
                par
                    goLight(LIGHTUNIT2) := not(goLight(LIGHTUNIT2))
                    stopLight(LIGHTUNIT2) := not(stopLight(LIGHTUNIT2))
                    phase := GO2STOP1
                endpar
            endif
        endif

    rule r_go2stop1_to_stop2stop1 = ...
    rule r_stop2stop1_to_go1stop2 = ...

    rule r_go1stop2_to_stop1stop2 =
        if phase=GO1STOP2 then
            if passed = ONEHUNDREDTWENTY then
                par
                    goLight(LIGHTUNIT1) := not(goLight(LIGHTUNIT1))
                    stopLight(LIGHTUNIT1) := not(stopLight(LIGHTUNIT1))
                    phase := STOP1STOP2
                endpar
            endif
        endif

    main rule r_Main =
        par
            r_stop1stop2_to_go2stop1[]
            r_go2stop1_to_stop2stop1[]
            r_stop2stop1_to_go1stop2[]
            r_go1stop2_to_stop1stop2[]
        endpar

default init s0:
    function stopLight($l in LightUnit) = true
    function goLight($l in LightUnit) = false
    function phase = STOP1STOP2
```

Code 1: One-way traffic light – ASM formal specification

```java
import org.asmeta.monitoring.*;

@Asm(asmFile = "oneWayTrafficLight.asm")
public class OWTL {
   @FieldToLocation(func = "stopLight", args={"LIGHTUNIT1"})
   boolean redLight1;
   @FieldToLocation(func = "stopLight", args={"LIGHTUNIT2"})
   boolean redLight2;
   @FieldToLocation(func = "goLight", args={"LIGHTUNIT1"})
   boolean greenLight1;
   @FieldToLocation(func = "goLight", args={"LIGHTUNIT2"})
   boolean greenLight2;
   private boolean turn1;

   @StartMonitoring
   public OWTL() {
      redLight1 = true;
      redLight2 = true;
      greenLight1 = false;
      greenLight2 = false;
      turn1 = false;
   }

   @RunStep
   public void updateLights(@Param(func = "passed") Time passedTime) {
      if((passedTime == Time.FIFTY && redLight1 && redLight2) ||
         (passedTime == Time.ONEHUNDREDTWENTY && greenLight1 != greenLight
         if(turn1) {
            greenLight1 = !greenLight1;
            redLight1 = !redLight1;
         }
         else {
            greenLight2 = !greenLight2;
            redLight2 = !redLight2;
         }
         if (redLight1 && redLight2) {
            turn1 = !turn1;
         }
      }
   }
}
enum Time {FIFTY, ONEHUNDREDTWENTY, LESS;}
```

Code 2: One-way traffic light – Java implementation

name and `args` for specifying the location arguments (if any). In the running example, fields *redLight1*, *redLight2*, *greenLight1*, and *greenLight2* are linked to locations *stopLight(LIGHTUNIT1)*, *stopLight-(LIGHTUNIT2)*, *goLight(LIGHTUNIT1)*, and *goLight(LIGHTUNIT2)*.

- `@MethodToFunction/@MethodToLocation`: it is similar to `@FieldToFunction/@FieldToLocation`, but it annotates methods. The annotated methods are required to be *pure*[3].
- `@Monitored`: it has the same structure and applicability of `@FieldToLocation`, but it must be used to annotate fields that take their values from the environment (e.g., the user input, or streams as files, sockets, etc.), and are not updated by a class method (differently from fields annotated with `@FieldToLocation`). Fields annotated with `@Monitored` can only be linked to ASM monitored locations that, indeed, represent the part of the ASM state determined by the environment.

- Annotations used to link the execution of the Java program with a run of the corresponding model:
  - `@StartMonitoring`: it annotates constructors. An annotated constructor identifies a Java initial state from which the conformance checking is required. In the running example, the unique constructor is selected for conformance checking.
  - `@RunStep`: it annotates methods (called *changing methods*) that modify the observed state, i.e., the values of the linked fields (those annotated with `@FieldToFunction/@FieldToLocation`) and the return values of the linked pure methods (those annotated with `@MethodToFunction/@MethodToLocation`). In the running example, the only changing method is *updateLights(Time passedTime)*.
  - `@Param`: it annotates parameters of a changing method. As the annotation `@Monitored`, it has attributes `func` and `args` for specifying the name of a monitored location of the ASM model. When the changing method is executed, the monitored location is set to the value of the actual parameter. In the running example, the formal parameter *passedTime* is linked to the ASM location *passed*.

The linking obtained by the use of the annotations allows to give the following *definitions of conformance* between an instance *obj* of the Java class and the corresponding ASM specification *spec*:

- **State conformance**: *the Java state of obj and the ASM state of spec are conformant* if the values of the fields annotated with `@FieldToFunction/@FieldToLocation` and the values returned by the methods annotated with `@MethodToFunction/@MethodToLocation` are equal to

---

[3] A method is *pure* if it side effect free with respect to the program state: it returns a value but does not assigns values to fields.

the values of the corresponding ASM functions/locations.
- **Step conformance**: *obj is step conformant with spec* if their states are conformant before and after the execution of a Java changing method of *obj* and the execution of a step of *spec*.
- **Run conformance**: *obj is run conformant with spec* if the following conditions hold:
  - the initial state of *obj* is state conformant with one and only one initial state of *spec*;
  - *obj* is step conformant with *spec* upon the execution of each changing method of *obj*.

4.2 Offline conformance checking

The first approach for checking the conformance between the implementation and the ASM formal specification is *model-based testing* [40], in which some tests are derived from the specification according to some coverage criteria and are then executed over the implementation. Since such activity is performed *before* the program deployment, we call it *offline*.

Our model-based testing approach is depicted in the left side of Fig. 4:
1. The tool ATGT [30] derives (using the model checker SPIN or NuSMV) from the ASM formal specification some *abstract test sequences* with the aim of achieving some testing criteria [29]. For example, a test suite satisfies the *basic rule coverage* (BRC) criterion if every rule r is fired in at least one state of a test sequence, and there exists a (possibly different) test sequence in which r does not fire in some state.
2. The abstract test sequences are then *concretized* into *JUnit test cases* for the implementation [8]. The translation takes advantage of the linking described in Sect. 4.1.

We here explain how we have applied the offline conformance checking approach to the running example.

Each coverage criterion produces a set of *test predicates* representing the testing goals that must be covered by the generated tests. We generated 24 test predicates for the specification: 12 have been generated by the BRC criterion, and 12 by the UR criterion that requires that all the updates are covered. One of the test predicates generated by BRC is

phase = GO1STOP2 and passed = ONEHUNDREDTWENTY

that requires to visit the parallel rule in the macro rule r_go1stop2_to_stop1-stop2.

For each test predicate *tp*, we generated an abstract test sequence (i.e., an ASM run) in which *tp* is satisfied in some state of the sequence.

Then, we translated the abstract test sequences in JUnit tests. For example, a generated abstract test sequence is shown in the left part of Fig. 5, and its

```
---- state 0 -----
-- controlled --
phase=STOP1STOP2
stopLight(LIGHTUNIT1)=TRUE
stopLight(LIGHTUNIT2)=TRUE
goLight(LIGHTUNIT1)=FALSE
goLight(LIGHTUNIT2)=FALSE
-- monitored --
passed=FIFTY

---- state 1 -----
-- controlled --
phase=GO2STOP1
stopLight(LIGHTUNIT1)=TRUE
stopLight(LIGHTUNIT2)=FALSE
goLight(LIGHTUNIT1)=FALSE
goLight(LIGHTUNIT2)=TRUE
-- monitored --
passed=ONEHUNDREDTWENTY

---- state 2 -----
-- controlled --
phase=STOP2STOP1
stopLight(LIGHTUNIT1)=TRUE
stopLight(LIGHTUNIT2)=TRUE
goLight(LIGHTUNIT1)=FALSE
goLight(LIGHTUNIT2)=FALSE
-- monitored --
passed=FIFTY

---- state 3 -----
-- controlled --
phase=GO1STOP2
stopLight(LIGHTUNIT1)=FALSE
stopLight(LIGHTUNIT2)=TRUE
goLight(LIGHTUNIT1)=TRUE
goLight(LIGHTUNIT2)=FALSE
-- monitored --
passed=ONEHUNDREDTWENTY
```

```java
public void test() {
  OWTL sut = new OWTL();
  // state 0
  // conformance checking
  assertTrue(sut.redLight1);
  assertTrue(sut.redLight2);
  assertFalse(sut.greenLight1);
  assertFalse(sut.greenLight2);
  // perform step
  sut.updateLights(Time.FIFTY);

  // state 1
  // conformance checking
  assertTrue(sut.redLight1);
  assertFalse(sut.redLight2);
  assertFalse(sut.greenLight1);
  assertTrue(sut.greenLight2);
  // perform step
  sut.updateLights(Time.ONEHUNDREDTWENTY);

  // state 2
  // conformance checking
  assertTrue(sut.redLight1);
  assertTrue(sut.redLight2);
  assertFalse(sut.greenLight1);
  assertFalse(sut.greenLight2);
  // perform step
  sut.updateLights(Time.FIFTY);

  // state 3
  // conformance checking
  assertFalse(sut.redLight1);
  assertTrue(sut.redLight2);
  assertTrue(sut.greenLight1);
  assertFalse(sut.greenLight2);
}
```

Figure 5: Abstract test sequence (left) and JUnit test case (right)

corresponding JUnit test case in the right part.

The concretization works as follows:
- an instance of the class is built using the constructor annotated with @StartMonitoring, and it is associated with the reference variable *sut*;
- the conformance of the initial state is checked (it works as the conformance checking of a generic step described in the following);
- then, for each ASM step, the following instructions are added to the test:
  - fields annotated with @Monitored are updated with the value of the corresponding ASM location;
  - the changing method is called; concrete parameters are retrieved from the values of the locations linked with @Param. In the running example, method *updateLights(Time passedTime)* is called using the value of location *passed* as actual parameter;
  - the conformance checking is executed:
    * for each field *f* annotated with @FieldToFunction/@FieldToLocation, the following assertion is built

      assertEquals(v, sut.f);

      which states that the value of *sut.f* must be equal to *v*, where *v* is the value of the corresponding ASM function/location in the abstract test sequence. For boolean fields, the more compact JUnit assertions *assertTrue(sut.f)* and *assertFalse(sut.f)* are used (depending on whether *v* is *true* or *false*). In the running example, fields *redLight1* and *redLight2* are compared with the values of the locations of the ASM function *stopLight*, and fields *greenLight1* and *greenLight2* are compared with the values of the locations of the ASM function *goLight*;
    * for each method *m* annotated with @MethodToFunction/@MethodToLocation, the following assertion is built:

      assertEquals(v, sut.m());

      which states that the value returned by *sut.m()* must be equal to *v* (where *v* is defined as before). As before, the more compact JUnit assertions *assertTrue(sut.m())* and *assertFalse(sut.m())* are used for boolean methods.

4.3 Online conformance checking

The offline approach described in Sect. 4.2 assures that the implementation is conformant with the specification over the produced tests (i.e., that critical specification behaviors are reproducible also in the implementation). However, the tests usually do not cover all the possible implementation behaviors (neither all the specification behaviors) and, therefore, some implementation faults may be undetected by model-based testing. Thus, for safety-critical systems, one may want to continue checking the conformance also after deployment of the code.

In [7], we proposed a *runtime verification* approach (CoMA) for checking the conformance of a Java code w.r.t. the ASM specification at runtime. The approach is described in the right part of Fig. 4:

1. the *Java program observer* monitors the execution of the Java program and, whenever it detects that a changing method (i.e., a method annotated with @RunStep) is executed, it triggers the execution of simulation of the ASM specification with the ASM simulator;
2. the *conformance analyzer* checks the step conformance between the Java step and the ASM step by comparing the values of the linked fields (those annotated with @FieldToFunction/@FieldToLocation) and methods (those annotated with @MethodToFunction/@MethodToLocation) with those of the corresponding ASM functions/locations. If a violation of conformance is detected, an error is reported.

The runtime framework is also able to check the conformance of nondeterministic systems in which there are multiple states that can be obtained by executing a changing method. For this kind of systems, the ASM simulator must find whether there exists a *next* state that is conformant with the obtained Java state. This can be done by explicitly computing all the next states [8], or by symbolically representing the next states using an SMT solver [10, 13].

5 Case studies

In the past, we applied the two conformance checking techniques to some real-life case studies such as a landing gear system (LGS) [12], the voting system (VS) used by the sensors in the landing gear system [9], a hemodialysis device (HD) [3, 2], and a medical device for measuring the patients' stereoacuity (SD) [1].

Table 1 reports experimental results regarding offline conformance checking of the different case studies; in order to give an idea of the size of the specifications, we report for each model the number of functions (f) and transition rules (r) of each of them. Then, the table reports the number of test predicates (tps), the test suite size, the total test length as the number of states of all the abstract test sequences, and statement and branch coverage of the implementation.

We can see that the number of tests is always much lower than the number of test predicates: this is due to the fact that different tests cover multiple test predicates.

Moreover, we observe that the coverage obtained over the implementation varies with the different case studies. The case studies in which the abstraction gap between the implementation and the specification is not big (as LGS and

Case study	# tps	# tests	Total test length (# states)	Implementation coverage	
				statement	branch
LGS (f: 39, r: 91)	116	22	115	96.9%	72.3%
VS (f: 8, r: 43)	38	10	26	100%	100%
HD (f: 155, r: 648)	980	183	8341	88.5%	58.3%
SD (f: 11, r: 49)	48	9	91	70.2%	41.6%

Table 1: Offline conformance checking results

Technique	Total test length	Implementation coverage		Fault detection		
		statement	branch	F1	F2	F3
Offline checking	115	96.9%	72.3%	NO	YES	YES
CoMA (random inputs)	50	55.6%	31.1%	2%	12%	69%
	500	75.2%	52.4%	2%	81%	100%
	5000	83.2%	60.3%	3%	100%	100%
	50000	83.4%	60.5%	4%	100%	100%

Table 2: Online conformance checking results

VS), have a high coverage, since almost all the implementation details are also reported in the specification. Case studies in which the specification is much more abstract than the implementation (as HD and SD), instead, have a lower coverage: in this case, some implementation details not present in the specification are not covered by the tests (12% (HD) and 30% (SD) of statements are uncovered). Note that this is a well-known limit of model-based testing [40] that, however, mainly focuses on the coverage of the requirements, not of the code. One possible solution is to manually add further tests for covering parts of the code not covered with model-based testing.

Model-based testing is necessary for evaluating the correctness of the implementation, but it is not sufficient, in particular for safety-critical systems. Classical approaches based on theorem proving and formal verification can provide further guarantees, but they are usually difficult to apply and require stronger mathematical skills. We have here shown a runtime verification approach that permits to check the conformance online (i.e., during program execution, after deployment). The approach has the advantage of reusing the theoretical framework of offline conformance checking and, therefore, does not add additional burden to the developer. The online conformance checking technique permits to verify any program execution, in contrast to the offline approach that only checks some selected program executions.

Table 2 reports some experimental results for the runtime verification of the landing gear system [12]. We run the system applying random inputs for 50,

500, 5000, and 50000 steps, and we monitored it with CoMA; we executed 100 experiments and computed the averages of the results over the runs. The table reports the implementation coverage (statement and branch) we obtained using the offline approach and using the random execution; as expected, the coverage obtained with the offline approach is much higher as the generated tests aim at maximizing the model coverage and there is a direct relation between model and implementation coverage; random execution is not able to reach the same coverage results even with 50000 steps.

We also compared the fault detection of the two techniques, by considering three faulty versions of our implementation (the faults were made during the development process): in **F1** a disjunction was used instead of a conjunction, in **F2** a field was updated to the wrong value, and in **F3** a `break` statement was missing in a `switch` statement. We observed that the offline approach can not detect **F1**, as the generated tests execute the faulty code with inputs that make the operands of the disjunction either all true or all false (the fault would only be detected with inputs that make the disjunction true and the conjunction false); CoMA, instead, is able to find it in 4% of the executions (in 50000 steps). The other two faults (that are easier to detect) are identified both by the offline approach and by CoMA with a sufficient number of steps. This experiment shows that runtime verification is particularly suitable to detect subtle faults that do not occur often; this is confirmed by another experiment we did in [9] in which an implementation fault occurs only after five calls of a method.

For internally nondeterministic systems (i.e., the nondeterminism is due to the behaviour of the system, not to its inputs), the offline approach is not applicable. Indeed, in such systems, a sequence generated for covering a given test predicate is only one of the possible sequences that cover the predicate [8]. If the implementation is internally nondeterministic, it could deviate from the expected behaviour and still be correct. Therefore, for internally nondeterministic systems, checking the conformance online is even more important.

6 Related work

The work reported in this paper deals with model-based testing and runtime verification; we here report a brief related work in these two areas.

Model-based testing Surveys about model-based testing (MBT) can be found in [40, 34, 41].

Different formal specifications have been used for testing purposes as, for example, Labelled Transition Systems [39], finite state machines [28], algebraic specification [32], and input-domain notations [23].

A classical approach used in MBT for generating test cases is exploiting the capability of model checkers to return counterexamples for falsified properties [27]; for each *test goal*, a suitable *test predicate* is verified and the returned counterexample (if any) is the abstract test case covering the test goal. The ATGT framework we use for test case generation is based on the model checkers SPIN and NuSMV, for which we provide mappings to their input syntaxes [30, 5]. Other model checkers have been used in combination with other formal notations as, for example, UPPAAL [33] for timed automata, and SAL [38] for Simulink/Stateflow models.

Runtime verification Surveys about runtime verification can be found in [24, 36, 25, 26].

Our runtime conformance checker CoMA (part of the ASMETA framework) is similar to those approaches using operational specifications. In [37], the behavior of a concrete implementation (a Java code) is checked for conformance against its formal specification given as Z model. Another approach that uses ASMs as formal specification for runtime verification is presented in [15]; in that work, *model programs* are used to specify traditional design-by-contract concepts of pre-conditions, post-conditions, and invariants.

Other approaches, instead, perform runtime verification using declarative specifications. *Monitored-oriented programming* (MOP) [21, 22] permits to execute runtime monitoring by means of annotating the code with formal property specifications. The specifications can be written in any formalism for which a logic plug-in has been developed (LTL, ERE, JML, ...). A similar approach is *Lime* [35] that allows to monitor the invocations of the methods of an interface by defining *pre* and *post* conditions, called *call specifications* (CS) and *return specifications* (RS). Specifications can be written as past/future LTL formulas, as regular expressions and as nondeterministic finite automata. Another approach is described in [16], in which traces of programs are examined in order to check if they satisfy some temporal properties expressed in LTL_3, a linear-time temporal logic designed for runtime verification.

7 Conclusions

Formal methods and tools for system design and analysis, such as the ASMs and its ASMETA framework, provide developers with the ability to design and analyze models. Unfortunately, once a model is actually implemented, there is no guarantee that the resulting code conforms to its specification. This is most worsened by the fact that the models and the implementation are artifacts often separately developed in an unrelated way.

In this paper we have proposed a way to bridge the gap between specification and implementation. We have presented an approach to link an implementation with its formal specification by means of code annotations, and have described two techniques for checking whether the implementation conforms to the specification: an offline technique to be used before deployment and based on model-based testing, and an online technique to be used after deployment and based on runtime verification.

We have also reported our experience in applying both techniques to a series of case studies in the context of safety critical systems, where conformance checking is highly demanded, and discussed advantages and shortcomings of one technique w.r.t. the other.

References

[1] P. Arcaini, S. Bonfanti, A. Gargantini, A. Mashkoor, and E. Riccobene. Formal validation and verification of a medical software critical component. In *Formal Methods and Models for Codesign (MEMOCODE), 2015 ACM/IEEE International Conference on*, pages 80–89, Sept 2015.

[2] P. Arcaini, S. Bonfanti, A. Gargantini, A. Mashkoor, and E. Riccobene. Integrating formal methods into medical software development: The ASM approach. *Science of Computer Programming*, pages –, 2017.

[3] P. Arcaini, S. Bonfanti, A. Gargantini, and E. Riccobene. How to assure correctness and safety of medical software: the hemodialysis machine case study. In M. Butler, K.-D. Schewe, A. Mashkoor, and M. Biro, editors, *Abstract State Machines, Alloy, B, TLA, VDM, and Z. 5th International Conference, ABZ 2016, Linz, Austria, May 23-27, 2016, Proceedings*, volume 9675 of *Lecture Notes in Computer Science*, pages 344–359. Springer International Publishing, Cham, 2016.

[4] P. Arcaini, S. Bonfanti, A. Gargantini, and E. Riccobene. Visual notation and patterns for Abstract State Machines. In P. Milazzo, D. Varró, and M. Wimmer, editors, *Software Technologies: Applications and Foundations: STAF 2016 Collocated Workshops: DataMod, GCM, HOFM, MELO, SEMS, VeryComp, Vienna Austria, July 4-8, 2016, Revised Selected Papers*, pages 163–178. Springer International Publishing, Cham, 2016.

[5] P. Arcaini, A. Gargantini, and E. Riccobene. AsmetaSMV: A way to link high-level ASM models to low-level NuSMV specifications. In M. Frappier, U. Glässer, S. Khurshid, R. Laleau, and S. Reeves, editors, *Abstract State Machines, Alloy, B and Z*, volume 5977 of *Lecture Notes in Computer Science*, pages 61–74. Springer Berlin Heidelberg, 2010.

[6] P. Arcaini, A. Gargantini, and E. Riccobene. Automatic review of abstract state machines by meta property verification. In C. Muñoz, editor, *Proceedings of the Second NASA Formal Methods Symposium (NFM 2010)*,

NASA/CP-2010-216215, pages 4–13, Langley Research Center, Hampton VA 23681-2199, USA, April 2010. NASA.

[7] P. Arcaini, A. Gargantini, and E. Riccobene. CoMA: Conformance monitoring of Java programs by Abstract State Machines. In S. Khurshid and K. Sen, editors, *Runtime Verification*, volume 7186 of *Lecture Notes in Computer Science*, pages 223–238. Springer Berlin Heidelberg, 2012.

[8] P. Arcaini, A. Gargantini, and E. Riccobene. Combining model-based testing and runtime monitoring for program testing in the presence of non-determinism. In *2013 IEEE Sixth International Conference on Software Testing, Verification and Validation, Workshops Proceedings, Luxembourg, Luxembourg, March 18-22, 2013*, pages 178–187. IEEE, 2013.

[9] P. Arcaini, A. Gargantini, and E. Riccobene. Offline model-based testing and runtime monitoring of the sensor voting module. In F. Boniol, V. Wiels, Y. Ait Ameur, and K.-D. Schewe, editors, *ABZ 2014: The Landing Gear Case Study*, volume 433 of *Communications in Computer and Information Science*, pages 95–109. Springer International Publishing, 2014.

[10] P. Arcaini, A. Gargantini, and E. Riccobene. Using SMT for dealing with nondeterminism in ASM-based runtime verification. *ECEASST*, 70, 2014.

[11] P. Arcaini, A. Gargantini, and E. Riccobene. SMT-based automatic proof of ASM model refinement. In R. De Nicola and E. Kühn, editors, *Software Engineering and Formal Methods: 14th International Conference, SEFM 2016, Held as Part of STAF 2016, Vienna, Austria, July 4-8, 2016, Proceedings*, Lecture Notes in Computer Science, pages 253–269. Springer International Publishing, Cham, 2016.

[12] P. Arcaini, A. Gargantini, and E. Riccobene. Rigorous development process of a safety-critical system: from ASM models to Java code. *International Journal on Software Tools for Technology Transfer*, 19(2):247–269, 2017.

[13] P. Arcaini, A. Gargantini, and E. Riccobene. SMT for state-based formal methods: the ASM case study. In *6th workshop on Automated Formal Methods (AFM 2017)*, Kalpa Publications in Computing. EasyChair, 2018.

[14] P. Arcaini, A. Gargantini, E. Riccobene, and P. Scandurra. A model-driven process for engineering a toolset for a formal method. *Software: Practice and Experience*, 41:155–166, 2011.

[15] M. Barnett and W. Schulte. Runtime verification of .NET contracts. *J. Syst. Softw.*, 65(3):199–208, Mar. 2003.

[16] A. Bauer, M. Leucker, and C. Schallhart. Runtime verification for LTL and TLTL. *ACM Trans. Softw. Eng. Methodol.*, 20(4):14:1–14:64, Sept. 2011.

[17] S. Bonfanti, M. Carissoni, A. Gargantini, and A. Mashkoor. Asm2C++: A Tool for Code Generation from Abstract State Machines to Arduino. In C. Barrett, M. Davies, and T. Kahsai, editors, *NASA Formal Methods: 9th International Symposium, NFM 2017, Proceedings*, pages 295–301. Springer, 2017.

[18] E. Börger. The abstract state machines method for high-level system design and analysis. In P. Boca, J. P. Bowen, and J. Siddiqi, editors, *Formal Methods: State of the Art and New Directions*, pages 79–116. Springer London, London, 2010.

[19] E. Börger and R. Stärk. *Abstract State Machines: A Method for High-Level System Design and Analysis*. Springer Verlag, 2003.

[20] A. Carioni, A. Gargantini, E. Riccobene, and P. Scandurra. A Scenario-Based Validation Language for ASMs. In *Proceedings of the 1st International Conference on Abstract State Machines, B and Z (ABZ 2008)*, volume 5238 of *Lecture Notes in Computer Science*, pages 71–84. Springer-Verlag, 2008.

[21] F. Chen, M. D'Amorim, and G. Roşu. A formal monitoring-based framework for software development and analysis. In J. Davies, W. Schulte, and M. Barnett, editors, *Formal Methods and Software Engineering: 6th International Conference on Formal Engineering Methods, ICFEM 2004, Seattle, WA, USA, November 8-12, 2004. Proceedings*, pages 357–372. Springer Berlin Heidelberg, Berlin, Heidelberg, 2004.

[22] F. Chen and G. Roşu. Java-MOP: A monitoring oriented programming environment for Java. In N. Halbwachs and L. D. Zuck, editors, *Tools and Algorithms for the Construction and Analysis of Systems: 11th International Conference, TACAS 2005, Held as Part of the Joint European Conferences on Theory and Practice of Software, ETAPS 2005, Edinburgh, UK, April 4-8, 2005. Proceedings*, pages 546–550, Berlin, Heidelberg, 2005. Springer Berlin Heidelberg.

[23] D. M. Cohen, S. R. Dalal, M. L. Fredman, and G. C. Patton. The AETG system: An approach to testing based on combinatorial design. *IEEE Trans. Softw. Eng.*, 23(7):437–444, July 1997.

[24] S. Colin and L. Mariani. Run-time verification. In M. Broy, B. Jonsson, J.-P. Katoen, M. Leucker, and A. Pretschner, editors, *Model-Based Testing of Reactive Systems: Advanced Lectures*, pages 525–555. Springer Berlin Heidelberg, Berlin, Heidelberg, 2005.

[25] N. Delgado, A. Q. Gates, and S. Roach. A taxonomy and catalog of runtime software-fault monitoring tools. *IEEE Transactions on Software Engineering*, 30(12):859–872, 2004.

[26] Y. Falcone, K. Havelund, and G. Reger. A Tutorial on Runtime Verification. In *Engineering Dependable Software Systems*, volume 34 of *NATO Science for Peace and Security Series - D: Information and Communication Security*, pages 141–175. IOS Press, 2013.

[27] G. Fraser, F. Wotawa, and P. E. Ammann. Testing with model checkers: a survey. *Software Testing, Verification and Reliability*, 19(3):215–261, 2009.

[28] A. Gargantini. Conformance testing. In M. Broy, B. Jonsson, J.-P. Katoen, M. Leucker, and A. Pretschner, editors, *Model-Based Testing of Reactive Systems: Advanced Lectures*, pages 87–111. Springer Berlin Heidelberg, Berlin, Heidelberg, 2005.

[29] A. Gargantini and E. Riccobene. ASM-Based Testing: Coverage Criteria and Automatic Test Sequence Generation. *Journal of Universal Computer Science*, 7:262–265, 2001.

[30] A. Gargantini, E. Riccobene, and S. Rinzivillo. Using Spin to Generate Tests from ASM Specifications. In *Abstract State Machines 2003*, volume 2589 of *Lecture Notes in Computer Science*, pages 263–277. Springer Berlin Heidelberg, 2003.

[31] A. Gargantini, E. Riccobene, and P. Scandurra. A Metamodel-based Language and a Simulation Engine for Abstract State Machines. *J. UCS*, 14(12):1949–1983, 2008.

[32] M.-C. Gaudel and P. Le Gall. Testing data types implementations from algebraic specifications. In R. M. Hierons, J. P. Bowen, and M. Harman, editors, *Formal Methods and Testing: An Outcome of the FORTEST Network, Revised Selected Papers*, pages 209–239. Springer Berlin Heidelberg, Berlin, Heidelberg, 2008.

[33] A. Hessel, K. G. Larsen, B. Nielsen, P. Pettersson, and A. Skou. Time-optimal real-time test case generation using Uppaal. In A. Petrenko and A. Ulrich, editors, *Formal Approaches to Software Testing: Third International Workshop on Formal Approaches to Testing of Software, FATES 2003, Montreal, Quebec, Canada, October 6th, 2003. Revised Papers*, pages 114–130. Springer Berlin Heidelberg, Berlin, Heidelberg, 2004.

[34] R. Hierons and J. Derrick. Editorial: special issue on specification-based testing. *Software Testing, Verification and Reliability*, 10(4):201–202, 2000.

[35] K. Kähkönen, J. Lampinen, K. Heljanko, and I. Niemelä. The LIME interface specification language and runtime monitoring tool. In S. Bensalem and D. A. Peled, editors, *Runtime Verification: 9th International Workshop, RV 2009, Grenoble, France, June 26-28, 2009. Selected Papers*, pages 93–100. Springer Berlin Heidelberg, Berlin, Heidelberg, 2009.

[36] M. Leucker and C. Schallhart. A brief account of runtime verification. *The Journal of Logic and Algebraic Programming*, 78(5):293–303, 2009.

[37] H. Liang, J. Dong, J. Sun, and W. Wong. Software monitoring through formal specification animation. *Innovations in Systems and Soft. Eng.*, 5:231–241, 2009.

[38] S. Mohalik, A. A. Gadkari, A. Yeolekar, K. Shashidhar, and S. Ramesh. Automatic test case generation from Simulink/Stateflow models using model checking. *Software Testing, Verification and Reliability*, 24(2):155–180, 2014.

[39] J. Tretmans. Model based testing with Labelled Transition Systems. In R. M. Hierons, J. P. Bowen, and M. Harman, editors, *Formal Methods and Testing: An Outcome of the FORTEST Network, Revised Selected Papers*, pages 1–38. Springer Berlin Heidelberg, Berlin, Heidelberg, 2008.

[40] M. Utting and B. Legeard. *Practical Model-Based Testing: A Tools Approach*. Morgan-Kaufmann, 2006.

[41] M. Utting, A. Pretschner, and B. Legeard. A taxonomy of model-based testing approaches. *Software Testing, Verification and Reliability*, 22(5):297–312, 2012.

Addressing Client Needs for Cloud Computing using Formal Foundations*

Andreea Buga[1], Sorana Tania Nemeș[1], Atif Mashkoor[2]

[1] Johannes Kepler University of Linz, Austria
a.buga|t.nemes@cdcc.faw.jku.at
[2] Software Competence Center Hagenberg GmbH, Austria
atif.mashkoor@scch.at

Abstract. Cloud-enabled large-scale distributed systems orchestrate resources and services from various providers in order to deliver high-quality software solutions to the end users. The space and structure created by such technological advancements are immense sources of information and impose a high complexity and heterogeneity, which might lead to unexpected failures. In this chapter, we present a model that coordinates the multi-cloud interaction through the specification, validation, and verification of a middleware exploiting monitoring and adaptation processes. The monitoring processes handle collecting meaningful data and assessing the state of components, while the adaptation processes restore the system as dictated by the evolution needs and sudden changes in the operating environment conditions. We employ Abstract State Machines to specify the models and we further make use of the ASMETA framework to simulate and validate them. Desired properties of the system are defined and analysed with the aid of the Computation Tree Logic.

1 Introduction

The cloud computing business model offers a wide set of services on a pay-as-you-go payment method, which proves to be efficient for a growing number of businesses. As defined by the National Institute of Standards and Technology (NIST) [34], cloud computing allows "ubiquitous, convenient, on-demand network access to a shared pool of configurable computing resources". In most situations, the client needs to trust the provider that the interaction with the cloud is trustworthy and secure. This can lead further to vendor lock-in issues or lack of client control over privacy policies and quality standards.

---

* The research reported in this chapter has been supported by the Christian-Doppler Society in the frame of the Christian-Doppler Laboratory for Client-Centric Cloud Computing and by the Austrian Ministry for Transport, Innovation and Technology, the Federal Ministry of Science, Research and Economy, and the Province of Upper Austria in the frame of the COMET center SCCH.

The research work of the Christian Doppler Client-Centric Cloud Computing Laboratory (CDCC)[3], headed by Professor Klaus-Dieter Schewe, addresses the lack of client-orientation and the scarcity of formal foundations of cloud services [45, 46]. Its goal is to achieve transparency and reliability of the services, while meeting the client needs. Therefore, various aspects regarding client-cloud interaction have been considered including identity and authorisation management, monitoring of Service Level Agreement (SLA) compliance, detection of malicious security attacks and service content adaptivity to different devices. As a result, several components handling important aspects for the cloud users have been developed as part of the project [28, 30, 42, 48]. All these components have been integrated in a Client-Cloud Interaction Middleware (CCIM), which has been formalized using the Abstract State Machines (ASM) theory [16] and ambient calculus [23]. This middleware will be extended in this chapter with monitoring and adaptation services.

The continuous evolution of computing requirements and the variety of service offerings have led to the need of accessing and composing resources from different providers. A survey focused on cloud buyers and users carried out by RightScale[4] in January 2017 reveals that among the companies relying on clouds, eighty five percent of them use now a multi-cloud strategy [2], running services in an average of four clouds (following both public and private deployment models). The initial research work of CDCC has been expanded in order to cover also the requirements of interoperable Cloud-Enabled Large-Scale Distributed Systems (CELDS). The main goal is to enable interaction of services located on different clouds and to ensure their availability and reliability to the end user. The direction given by the work carried out for the CCIM [18] is expanded with a new abstract machine model organized on three layers responsible for the execution, monitoring and adaptation of services running in CELDS.

The current chapter focuses on the description of the aforementioned client-oriented cloud services and their role in fulfilling meaningful aspects for the users like privacy, security and Quality of Service (QoS). The work further investigates the requirements of CELDS with respect to monitoring and adaptation, which aim to increase the availability and reliability.

The main contribution of this chapter is as follows. First, we propose a fault-tolerant monitoring framework relying on redundancy and partial hierarchy, and discuss a set of metrics to be collected and evaluated. The monitors are considered to be prone to failures and are characterized by a confidence degree value. Second, we define an adaptation layer, which receives information about faulty situations from the monitors. The adaptation solution relies on a case-based repository to build suitable reconfiguration plans in terms of actions and controllers to restore the system to a normal working mode. The workflow of

---

[3] http://cdcc.faw.jku.at/
[4] https://www.rightscale.com/lp/state-of-the-cloud

both the monitoring and adaptation frameworks is captured formally in terms of ASMs, whose behaviour is validated and desired properties are model checked with the aid of the ASMETA toolset [7][5].

The remainder of the chapter is organised as follows. In order to have the overview of the project frame, Section 2 introduces the client-cloud interaction middleware component and its constituent solutions defined for fulfilling the client needs on single clouds. The description completes the specification of the CCIM, whose interaction leads to the formation of CELDS. The layers of CELDS are defined in Section 3. The monitoring and adaptation solutions for CELDS are formally specified in Section 4. The validation and verification procedures for both models are captured in Section 5. Related work is discussed in Section 6 and Section 7 concludes the chapter.

2 Overview of the Client-Cloud Interaction System

Although clouds aim to meet the service requirements of the clients, they offer users a limited control and transparency in the way their data is handled. In response to this drawback, a client-oriented middleware has been proposed [17]. The middleware encompasses a set of specific services described throughout the current section. We present in this section an overview of the offered services of the middleware, which were previously researched and developed at CDCC under the supervision of Professor Klaus-Dieter Schewe.

2.1 Client-Cloud Interaction Middleware

The CCIM defined in [18] proposes a high-level formal model of a novel cloud service component based on ambient ASM, which is able to incorporate the major advantages of ASMs and ambient calculus [23]. The formal models of distributed systems including mobile components are described in two abstraction layers. Therefore, the algorithms of executable components (agents) are specified in terms of ASMs and their communication topology, locality and mobility are described using ambient calculus. The ambient ASM specification gives us a universal way to handle client-cloud interaction independent from particularities of certain cloud services or end-devices, while the instantiation by means of particular ambients results in specifications for particular settings. Thus, the architecture is highly flexible with respect to additional end-devices or cloud services, which would just require the definition of a particular ambient.

This robust architecture model integrates several novel and loosely coupled software solutions (e.g. Client-to-Client Interaction (CTCI) Feature, Identity Management Machine (IdMM), Content Adaptivity, SLA Management, and Security Monitoring Component) into a compound single software component on the client side, see Fig. 1.

---

[5] http://asmeta.sourceforge.net/

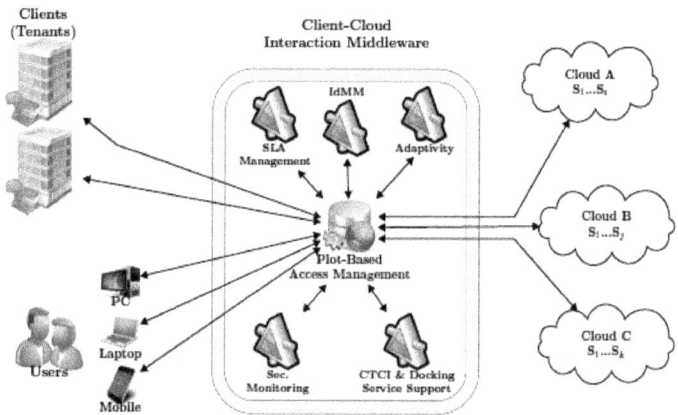

Fig. 1: Distributed deployment model for the proposed CCIM, reprinted from [18]

2.2 Identity Management Solution for the Cloud

Although widely spread, cloud computing still presents some vulnerabilities, which make clients reluctant to trust the providers. The report conducted by [1] named insufficient identity, credential and access management and also account hijacking as main concerns of users with respect to cloud services. While there are several deployment models of an IdMM surveyed by [47,49], cloud providers prefer the approach that keeps on their servers all the data needed for authentication and authorization.

The CCIM offers an IdMM service that can deal with three interaction scenarios [48]. First, the IdMM allows the user to save credentials directly using a cloud service, which shifts the responsibility to the cloud, but might also affect data privacy. Second method permits the user to obfuscate part of the information. A third approach uses federated identity management protocols. This offering of the middleware allows the client to decide on the level of control given to the cloud for storing sensitive information.

2.3 Monitoring of Client-Cloud Interaction using Service Level Agreements

Each cloud provider offers a set of quality measures of its services, which comes in the form of an SLA contract. SLAs stipulate that whenever described properties are not fulfilled, the provider must pay a compensation or a penalty to the client [35]. For this reason, a fair SLA management platform, which ensures a transparent communication between the client and the provider is needed.

The middleware component has been enriched, in this sense, with a framework modelled using a Web Ontology Language (OWL) which allows automated reasoning and monitoring of SLAs. The client is also granted the right to edit, accept or reject an SLA contract [42]. Granting both sides permission to modify the SLA document requires a synchronization policy in order to avoid possible inconsistencies. [42] proposes an approach based on Lamport's synchronization algorithm, which uses logical clocks [31].

2.4 Anomaly-Based Intrusion Detection

As cloud computing accumulates data on the provider side, clouds and their clients become vulnerable to security attacks like malicious insiders, tampering, data breaches, or loss of data. A deep understanding of the problem that enables attacks is necessary to develop countermeasures and secure software.

[30] discusses vulnerabilities in terms of formal language theory and proposes a protocol monitor that mitigates vulnerabilities by validating in- and outgoing messages of a web component, i.e. service provider or consumer. The research proposes a language-based learning approach for anomaly detection on tree-based documents like eXtensible Markup Language (XML) using pushdown automaton. Datatypes XML visibly pushdown automaton (dXVPAs) is inferred to capture the expected language at a particular service interface and is translated for stream validation.

2.5 Adaptivity of Cloud Content

In the context of cloud services, a well-founded system should not tailor specific applications to each type of end-device (desktop computers, laptops, tablets, smartphones, etc.), but guarantee that its clients are able to access the same cloud service and its subsequent output regardless of the used end-device, its operating system and/or distinct hardware characteristics (e.g. processor speed, display size and resolution).

To tackle the problem of providing cloud services to different devices, an adaptivity component is created as part of the middleware application [28]. [8] proposes making use of various internal components to manage the interaction between the clients and the cloud services, so that on-the-fly layout and content adaptation (mostly needed in case of mobile devices) are ensured. The adaptivity component envisions a web application as a communication channel which automatically detects device properties on the client-side and uses them on the server-side to modify the content coming from the cloud.

3 A Management System for Interconnecting Clouds

Interconnected clouds were defined as a model, which permits reallocation of resources and workload transfer through the collaboration of services of different cloud providers in order to achieve the promised service quality measured in terms of performance, availability and agreed SLA [4].

This section introduces the concept of a client-oriented distributed middleware for multi-cloud services and highlights the structure and role of the monitoring and adaptation layers. Multi-cloud systems are already faced with numerous challenges (e.g. complexity, network failures, and bottlenecks) and the addition of other components like monitors and adapters should not interfere with the execution of processes. In order to better understand the intrinsic problems that the framework can face, we focused our attention on an ASM formal model, which can complement the existing work in the area.

3.1 Overview of the Distributed Cloud Interaction Middleware

As illustrated in Fig. 2, the middleware consists of several machines and engines used for supporting the three layers (execution, monitoring and adaptation) of the abstract machine. The execution layer establishes the communication between nodes, handles requests from the users and permits the description of service interfaces [18]. The monitoring layer performs a continuous assessment of the status of the nodes, with a focus on detecting unavailability and crash failures, and communicates information about such issues to the adapters. The adaptation layer analyses a reported issue with previous solved cases and proposes a reconfiguration plan. The efficiency of the adaptation processes is then evaluated by the monitors.

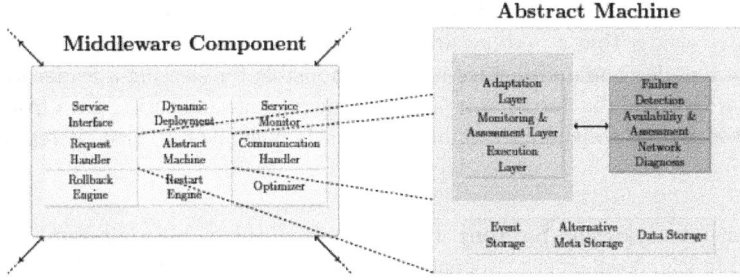

Fig. 2: Layered structure of a middleware component, reprinted from [21]

The collaboration of the monitoring and adaptation layers aims to improve the reliability of the system and to optimize the reconfiguration processes. The three components of the storage refer to the information locally logged in the system. The data part includes low- and high-level metrics collected by the monitors, while adaptation processes are saved in the event storage. Supplementary information (characteristics of the node, uptime of monitors and adapters) are stored in the alternative meta storage. The structure and workflow of monitoring and adaptation processes are extended with a formal specification in Section 4.

3.2 Monitoring Layer

Monitoring refers traditionally to collecting specific data from system components and evaluating them in order to discover execution problems, availability and performance issues. Monitors are components of CELDS and they are also prone to failures. While there is a broad spectrum of issues to monitor, we focus on availability, latency and crash failures. Instead of addressing a wider set of problems, we rather emphasize the accuracy of monitoring processes. Our proposed approach aims to mitigate the risk of monitor failures through enforcing redundancy and applying a confidence function that would correspond to the quality of the diagnoses established by a monitor.

The monitoring layer of the middleware handles the collection of relevant information from the execution. Monitors assess the status of each node of the system. Being part of the CELDS, monitors also face possible failures. In order to avoid the problem of a single point of failure, we opted for assigning a set of monitors to each node of the system and to evaluate possible issues with the aid of a collaborative decision. Also, each monitor is assigned a confidence measure, which reflects its performance. Monitors with low values are stopped by the middleware agent and either restarted or replaced.

We consider that a middleware component handles the assignment of monitors to a node. The middleware is also responsible for electing a leader for each set of monitors. The leader is in charge of acquiring monitoring data from each monitor and establishing a common diagnosis whenever an issue is reported.

Architecture of the System The structure of the monitoring framework depicted in Fig. 3 reflects the dependencies between its components. We opted for a solution composed of different modules and an ASM machine. We assigned the main rule to the middleware, which calls also the execution of the modules. We propose ensuring reliability through redundancy of the monitors. Therefore, a node is evaluated by a set of monitors, which are coordinated by a leader. The leader introduces a hierarchical view to the approach and is responsible for coordinating information obtained from different monitors. Leader module relies on the monitor module, requiring it the necessary information for assessing a collaborative diagnosis. The monitor contains a reference to the node it is assigned to.

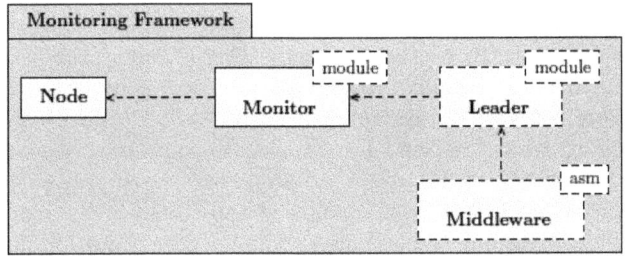

Fig. 3: Structure of the monitoring framework, adapted from [20]

3.3 Adaptation Layer

The adaptation layer reacts to the results of the monitors and identifies changes to CELDS by replacing (sets of) services, relocation of services, deployment and integration of new services, replication, removal of replication, etc. The guideline for the selection of the best alternative is the repair of the encountered problem under presumably optimal performance. Based on the decision which alternative implementation of the CELDS is to be taken, the running system will be (partially) interrupted, rolled back to a consistent state and restarted with the new remaining execution plan.

Regardless of the delivery model associated with the services being provided and consumed in the cloud environment (Software as a Service (SaaS), Platform as a Service (PaaS), Infrastructure as a Service (IaaS), there will be entities and relationships in different layers of the cloud environment that can request or cause adaptations to each other, which in return may lead to conflicting actions. To this purpose, the Adaptation Engine must be general enough to be able to automatically coordinate among the different layers, entities and adaptation options while perpetually reacting to the notifications from the monitoring component. Therefore, once the system recognizes a violation of the expected behaviour, it enters a reconfiguration state and triggers a self-adaptive process in order to reach a stable and correct behaviour.

Architecture of the System As an inner component of the abstract machine included in the middleware, the Adaptation Engine provides a unified solution using a Case-Based Reasoning approach (CBR) [3] enhanced in matters of adaptation actions, their usage and impact on the system. There are two major parts to be considered for the adaptation process: the decision phase defined by solution exploration, identification and maintenance, and the solution management and enactment phase, each with well delimited responsibilities and areas of inference and control.

In the envisioned framework, a case for system adaptation C_r represents a formatted instance of a problem P_r linked to a recorded solving experience S_r. The problem part P_r represents a collection of description features (e.g. response time, price, portability, region, availability, input/output bandwidth) which, as described in [36], are subject to similarity functions and common pattern recognition mechanisms. The adaptation solution is selected based on the collected knowledge of failure scenarios and associated previously validated solutions. The solution part S_r is configured and stored in the repository as a workflow schema detailing the actions and underlying transition dependencies needed to restore the system to a normal execution mode.

An action is an autonomous entity (e.g. a software module) which has the power to act or cause a single update to the system that helps to solve the problem the adaptation system was employed for. Its autonomy and self-awareness imply that its execution is not controlled by the environment or other actions but is, thus, able to deal with unpredictable, dynamically changing, heterogeneous environments while relying fully on existing solutions for CELDS adaptability. Examples of actions for system adaptation include discovery of a suitable matching service to replace the problematic one (by accessing the capabilities of an existing tool) or the dynamical reconfiguration of service calls to the new service. The adaptation is thus a trace of actions that are being executed by employing their capabilities to resources to fulfil a certain adaptation that would bring the system to a normal execution state.

The action's instantiation and execution are handled by linked ActionController loaded based on the defined contract for that particular action. The actions' ordering and dependency on other actions is handled by means of notification/signalling, where every action state change triggers a signal broadcast raised by the parent ActionController. The responsibility of the ActionController lies in the observation of its states, the subsequent action and its states and data, and the detection of undesirable behaviour when executing the action or communicating with other action controllers.

Therefore, the adaptation system consists of a finite set of interacting ActionControllers. Based on the assessed observations and broadcast notifications triggered by actions' execution or failure, the ActionControllers affect the system towards the remediation of the reported problem/failure. The assessment implies either enacting and executing its corresponding action, or ignoring the notification as it is not of interest in the given solution configuration. This model's underlining observer/controller architecture is one realisation of the feedback loop principle, and guides the system behaviour and dynamics for the adaptation to succeed in reaching the intended goals. Fig. 4 depicts the overall structure of the adaptation process once the problem is mapped to previous encountered problems and the attached solution is carried out based on its configuration.

Having ActionControllers to monitor and handle the interaction between the actions of a solution emphasizes new properties of the actions being de-

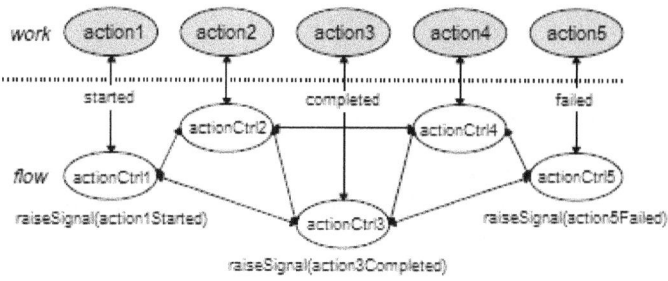

Fig. 4: Overview of workflow schema and execution.

fined in terms of needed input, concrete implementation and resulting output. In addition, it enables the possibility to easily add or substitute any given number of adaptation actions. Once an action is altered or added, the workflow schema needs to reflect the required changes and dependencies with the other configured adaptation actions. These changes are later on translated and visible in the ActionControllers and the interaction between them.

4 Formal Specification of the Monitoring and Adaptation Processes

4.1 Background on the ASM Formal Method

Throughout time, several formal methods have been proposed, each dealing with different properties to model. In order to better understand the suitability of the methods for CELDS we consulted the survey elaborated in [29]. The survey verifies a set of aspects (e.g. supported development phases, tool support, modelling capabilities, and industrial applicability) for each method.

The two most-suitable candidates for modelling a distributed system and its specific characteristics like concurrency and non-determinism are, according to the survey, ASMs and TLA+ [32]. We finally opted for adopting ASMs for a couple of other properties like the assistance throughout the software development process and their suitability for industrial applications. Other important candidates for specifying our proposed monitoring and adaptation services are Petri Nets and the Unified Modelling Language (UML). However, UML still lacks a formal precision and Petri Nets further proved to be more verbose for a set of specific distributed systems applications [15].

ASMs extend the concept of Finite State Machine (FSM) with the possibility to specify synchronous parallel operations and enhance the input and output states with data structures. An ASM machine consists of a tuple $M = (\Sigma, S_0, R, R_0)$, where Σ is the signature (the set of all functions), S_0 is the set of initial states of Σ, R is the set of rule declarations, and R_0 is the main rule of the machine. Rules of an ASM follow the **if** *Condition* **then** *Updates* structure [16], where the *Updates* can consist of a set of parallel assignments of values to locations.

ASMs are able to capture different functions and their structure highlights the separation of concerns principle. Static functions refer specifically to constants, while dynamic functions can be updated during execution. *Controlled* functions are written only by the machine, while *monitored* ones are written by the environment and read by the machine. Both the machine and its environment can update *shared* functions.

The CoreASM [26] execution engine and the ASMETA toolset support the definition of ASM models through their own specific ASM dialects. Each of them permits different functionalities to simulate, check or visualize ASM models. An attempt to unify both dialects led to the development of the Unified Syntax for Abstract State Machine [5], which supports also the translation of models to C++ code designed for Arduino devices [14].

Although the ASM method permits a good formalization it still has a set of drawbacks as follows. Time related aspects, which are critical for real systems cannot be integrated in the model. On a very abstract level, one can constraint a loop block of rules to execute for at most a certain number of times, without actually integrating a time measurement. For validation, the usage of composed functions is limited. In the verification stage, infinite domains have to be removed or replaced by finite sets or enumerations. However, the models are significant and provide important insights on the processes workflow.

4.2 Specification of the Monitoring Solution

As presented in Section 3.2, the formal model of the monitoring layer relies on two modules and an ASM component. The node is defined simply as a domain inside the monitor module and no rules and functions for it are expressed. The following subsections address specific aspects related to each of the components and capture parts of the ASM model[6]. The proposed specifications were built entirely in one step, but another approach would be to incrementally construct the models through refinements as described in [16].

Monitor Module The monitor module represents the core of the solution. After being assigned to a node, the monitor is in the *Active* state from which it sends a heartbeat request to verify the availability of the node and its latency, and moves to the *Wait for response state*. In this state, the monitor verifies two guards. First, it verifies whether a response to its request is received. If so, it checks whether the delay of the response is acceptable. If this condition is satisfied, the monitor moves to the *Collect data* state. If the request has

---

[6] A more detailed specification can be found at http://cdcc.faw.jku.at/staff/abuga/esocc.zip

no response or the delay is too big, the monitor moves to the *Report problem* state. In the *Collect data* state, data about the status of the node is gathered (CPU, memory, storage). The monitor attempts then to *Retrieve data* from a local storage containing previous logs. If the repository is available, the monitor queries it. The monitor moves afterwards to the *Assign diagnosis* state, where it interprets all the available data. In case problems are discovered, the monitor moves to the *Report problem* state, otherwise it transits to the *Log data* state, where meaningful data and operation are locally saved. When an issue is discovered, the monitor modifies a constraint that triggers a request towards the leader of the node to start a collaborative diagnosis. After reporting the issue, the monitor moves to the *Log data* state. The ground model of the module has been detailed previously also in [22].

Leader Module The leader module contains rules for collecting assessments from each of the monitors and building a collective diagnosis. In the current version of the model, we consider that the information submitted by each monitor is equally important for the collective result as shown in Listing 1.

```
rule r_RequestData ($l in Leader) =
  choose $node in Node with (eq(has_leader($node), $l)) do
    forall $m in Monitor with (assigned_node ($m) = $node) do
      if (isDef(diagnosis ($m))) then
        if (diagnosis ($m) = NORMAL) then
          normal_diagnoses ($l) := normal_diagnoses ($l) + 1
        else if (diagnosis ($m) = FAILED) then
          failed_diagnoses ($l) := failed_diagnoses ($l) + 1
        else
          critical_diagnoses ($l) := critical_diagnoses ($l) + 1
        endif
      endif
    endif
```

Listing 1: Request data rule of the leader module

Assessing the status of a node by the leader implies choosing the diagnosis proposed by the majority of the monitors assigned to the node. Listing 2 contains the responsible rule. At the end of a collaborative diagnosis round, each of the monitors updates its confidence degree according to a function defined by [19]. This function takes into account the similarity of the information submitted by a monitor in comparison with the one provided by other monitors assigned to the same node. Therefore, the lower the similarity, the higher the penalty. The rule for recalculating the confidence degree is left abstract for this version of the model. After recalculating the confidence and saving the diagnosis, the counters for failed, critical and normal diagnoses are reset.

```
rule r_AssessNode ($l in Leader) =
  if (max(failed_diagnosis ($l), critical_diagnosis ($l)) = failed_diagnosis ($l) ) then
    if (max(failed_diagnosis ($l), normal_diagnosis ($l)) = failed_diagnosis ($l) ) then
      assessment ($l) := FAILED
  else
    assessment ($l) := NORMAL
  endif
```

```
    else
      if (max(critical_diagnosis ($l), normal_diagnosis ($l)) = critical_diagnosis ($l) ) then
        assessment ($l) := CRITICAL
      else
        assessment ($l) := NORMAL
      endif
    endif
```

Listing 2: Assess node rule of the leader module

Middleware ASM The middleware component orchestrates the processes of the monitor and leader modules and manages the workflow of the system. It initializes the functions and contains the main rule, which calls for the execution of rules belonging to the monitor and leader modules. The middleware assigns a set of monitors to each node and elects a leader (these processes occur only once as in each of the corresponding rules there is a guard verifying whether monitors were already assigned to a node and whether a leader has already been elected). The middleware ensures that when a monitor wants to report a problem (trigger_gossip), the corresponding leader moves to the *Evaluate* state and starts to collect diagnoses from all the monitors assigned to the node reported as having problems. The middleware also dismisses monitors whose confidence degree is below the accepted minimum value.

```
rule r_MiddlewareProgram =
  if (middleware_state(self) = EXECUTING) then
    par
      r_AssignMonitorsToNode []
      r_ElectLeader []
      forall $m in Monitor with (trigger_gossip($m)) do
        let ($n = assigned_node($m)) in
          let ($l = has_leader($n)) in
            if (leader_state($l) = IDLE_LEADER) then
              leader_state($l) := EVALUATE
            endif
          endlet
        endlet
      forall $mon in Monitor with (confidence_degree($mon) < min_confidence_degree) do
        r_DismissMonitor [$mon]
    endpar
  endif
```

Listing 3: Main rule of the middleware agent

4.3 Specification of the Adaptation Solution

Based on the overall specification of the adaptation framework described in Section 3.3, we define the specific states and transitions of the adaptation processes, with emphasis on one of the main modules, the actionController module.

ActionController Module The actionController module represents the core of the action enactment process of the system and bares the responsibilities that come with the action's observation and triggering, compliant to the executed workflow schema. The actionController can pass through several states by various rules and guards.

```
rule r_AcknowledgeNotificationReceived($c in Controller,$broadcaster in Controller) =
  if (controller_state($c) = NOTIFICATION_RECEIVED) then
    seq
      controller_state($c) := ASSESS_NOTIFICATION
      par
        acknowledged_controllers($broadcaster) := acknowledged_controllers($broadcaster) + 1
        r_HandleNotification[$c]
      endpar
    endseq
  endif
```

Listing 4: Acknowledge notification ASM rule.

At initialization, the actionController is in the *Passive/Waiting notification* state. Once a notification arises, the actionController acknowledges the received notification in disregard of the actual sender, after which it moves to the *Assess notification* state. The rule responsible for acknowledging a notification is captured in Listing 4.

In order to assess the received notification, the actionController must first validate the received notification. If the notification is compliant with its contract, the actionController broadcasts the notification/signal that the underlying action is bound to start its execution, as captured in Listing 5.

```
rule r_BroadcastNotification($c in Controller, $n in Notification) =
  forall ($neighbor in Controller) then
    if (not(id($c) = id($neighbor)))
      seq
        acknowledged_controllers($c) := 1
        par
          controller_state($c) := WAITING_FOR_ACKNOWLEDGEMENT
          AcknowledgeNotificationReceived[$neighbor, $c]
        endpar
      endseq
    endif
  endforall
```

Listing 5: Broadcast notification ASM rule.

Every actionController instantiated as part of the same adaptation session must receive and acknowledge the broadcast notification. Once the notification is acknowledged by all neighbouring actionControllers, the actionController should proceed to the execution of its action. The rule responsible for triggering the associated adaptation action is captured partially in Listing 6.

```
rule r_TriggerAction($c in Controller) =
  seq
    while (controller_state($c) = RUNNING_ACTION)
      wait
    if (action_completed($c))
      par
```

```
      r_BroadcastNotification[$c, ACTION_COMPLETED]
      r_AwaitAcknowledgement[$c]
      if (acknowledged_controllers($n) = numberOfControllers)
        par
          r_ClearNotificationEcho[$c]
          controller_state($c) := WAITING_NOTIFICATION
        endpar
      else
        par
          controller_state($c) := CONTROLLER_ACKNOW_FAILED
          AssessDataAndStatus
        endpar
      endif
    endpar
  else
    par
      BroadcastNotification[$c, ACTION_FAILED]
      ...
```

Listing 6: Trigger action ASM rule

Regardless of the output of the executed action, one notification will be broadcast signalling the success or failure of this particular system update. There is no linked track of the actionControllers' order to execution. Therefore, if at least one action fails or one actionController does not acknowledge any of the sent notifications, the adaptation is abruptly terminated. The component data and status are afterwards assessed and logged accordingly. If the associated action's successful execution and acknowledgement by all the other controllers are fulfilled, the actionController reaches again the initial state. This initial state is reached again either when the associated action's execution and acknowledgement by all the other controllers are fulfilled or when the received notification is not bound to influence the actionController in question. This is a clear indication of the continuous character of the adaptation process which takes place in the background of service execution.

5 Validation and Verification of the Specifications

This section presents the practicability of the ASM method in reasoning about the system requirements by applying different validation and verification activities. In Section 5.1, the usefulness of scenario-based validation is explained and examples of such simulated scenarios are provided. Section 5.2 explains how model checking techniques are applied using the AsmetaSMV tool and presents the verification of some classical temporal properties specified on the models of the adaptation and monitoring solutions.

5.1 Validation

Validating the models enables us to check whether the system behaves as expected and the models correctly capture the intended requirements. We performed random and interactive simulation with the aid of AsmetaS simulator [27]. The AsmetaV validator allowed us to also build and execute scenarios of expected behaviours. A scenario is usually associated with a specific execution path and can be used for testing the state of the system after a set of transitions.

Inconsistency errors were detected during simulation with the aid of the AsmetaS tool. For example, more than one system failure can be reported in a short time frame. And although a schema is locked while its associated solution is executed, a parallel execution of simultaneous adaptations may try to update system parts or components with different values at the same time. Triggering simultaneously multiple adaptations within the system is then supported by transaction specific operations where every solution is annotated with extensive knowledge on the area of inference in the system of each subsequent action, which would later on be considered in the adaptation decision phase.

Model simulation leads sometimes to verbose traces, from which it is hard to discover possible design issues. For such situations, we opted for the use of validation scenarios defined with the aid of the Avalla language introduced in [24]. The scenarios allow the description of particular execution paths and through their validation we can check whether boolean constraints are fulfilled. We present in Listing 7 a simple scenario for determining whether the leader correctly evaluates the diagnosis of a node observed by three monitors. The first monitor indicates a high latency, and thus triggers the leader to evaluate the status of the node. At the time of the request, the second and third monitor did not carry out a full monitoring cycle and they cannot send any evaluation. For this situation, the leader takes into consideration only the available information from the first monitor and evaluates correctly that the node has failed. This situation, however, highlights the fact that insufficient data might lead to inconsistent evaluations.

```
set assigned_monitors(node_1) := [monitor_1, monitor_2, monitor_3];
set has_leader(node_1) := leader_1;
set leader_state(leader_1) := IDLE_LEADER;
set has_leader(node_1) := leader_1;
step
set heartbeat_response_arrived(heartbeat_1) := false;
set heartbeat_response_arrived(heartbeat_2) := true;
set heartbeat_response_arrived(heartbeat_3) := true;
set heartbeat_latency(heartbeat_2) := 5;
set heartbeat_latency(heartbeat_3) := 7;
step
set heartbeat_response_arrived(heartbeat_1) := true;
set heartbeat_latency(heartbeat_1) := 21;
set monitor_measurements(monitor_2) := [("Latency", 5), ("CPU Usage", 10), ("Storage Usage",
    15), ("Memory Usage", 10), ("Bandwidth", 50)];
set monitor_measurements(monitor_3) := [("Latency", 7), ("CPU Usage", 40), ("Storage Usage",
    15), ("Memory Usage", 10), ("Bandwidth", 30)];
step
set is_repository_available(monitor_2) := true;
set is_repository_available(monitor_3) := false;
step
step
step
check assessment(leader_1) = FAILED;
```

Listing 7: AsmetaV simple scenario for leader diagnosis validation

5.2 Verification

Correctness and reliability of the monitoring and adaptation solutions were guaranteed by verifying classical temporal properties (e.g. reachability, safety, correctness). Through safety properties, we ensure that something bad will not occur. Reachability properties verify whether a certain state can be reached from the initial state of the system. These together with correctness and component specific properties are expressed with Computation Tree Logic (CTL) formulas and are verified with the aid of the AsmetaSMV Eclipse plugin [6], which also relies on the NuSMV model checker [25]. NuSMV imposes a set of constraints on the model as it works only with finite sets and domains. Hence, we had to simplify the initial models in order to verify them. Several properties targeting different modules of the monitoring and adaptation solutions are listed below. This list of properties extends the one previously defined in [20].

Monitor safety property. First property, showcased in Listing 8, indicates that a monitor which waits for a reply for its heartbeat request and which receives a response whose delay did not exceed the maximum limit, moves immediately to the *Collect data* state. This also ensures that a monitor cannot falsely report a problem.

```
CTLSPEC (forall $m in Monitor with ag((monitor_state($m) = WAIT_FOR_RESPONSE and
heartbeat_response_arrived($m) and not(heartbeat_timeout($m)))
implies ax(monitor_state($m) = COLLECT_DATA)))
```

Listing 8: Monitor reachability property

Leader reachability property. Addressing the collaboration between a monitor and a leader, we verify that for all the situations in which a monitor wants to report a problem, the leader eventually moves to the *Evaluate* state. We used a static function for a leader in this case as we could not apply a composed function to point towards the leader state of a leader assigned to a monitor inside the CTL formula.

```
CTLSPEC (forall $m in Monitor with ag((trigger_gossip($m) = true)
implies ef(leader_state(leader_1) = EVALUATE )))
```

Listing 9: Leader reachability property

ActionController reachability property. Considering the collaboration between an ActionController and its subsequent action, we verify that for all the situations in which an action wants to start its execution, the ActionController eventually moves from the state *WaitingForAcknowledgement* to *ActionRunning*.

```
CTLSPEC (forall $a in Action with ag((trigger execute($a) = true)
implies ef (actionController state(actionController 1) = ACTION_RUNNING )))
```

Listing 10: ActionController reachability property

ActionController liveness property. Any ActionController which is assigned to an action, being thus in the *ActionRunning* state, eventually reaches the state where it is ready to be removed, after the adaptation process is successfully completed. We ensure in this case that an adaptation cycle is eventually completed.

CTLSPEC (**forall** $ac **in** ActionController **with** ag((actionController_state($ac) = ACTION_RUNNING) implies ef(actionController_state($ac)=READY_FOR_REMOVAL)))

Listing 11: ActionController fairness property

Monitor safety properties. The following property verifies that any identified issue (e.g. a high usage value for CPU and memory) is reported to the leader, which triggers afterwards a collaborative diagnosis. If the monitor does not identify an issue, it must not reach the *ReportProblem* state. By checking both these properties, the monitors are prevented to report false positives.

CTLSPEC (**forall** $m **in** Monitor **with** ag((is_problem_discovered($m)) implies ax(monitor_state($m) = REPORT_PROBLEM))) **CTLSPEC** (**forall** $m **in** Monitor **with** ag(monitor_state($m) = ASSIGN_DIAGNOSIS and (not(is_problem_discovered($m))) implies ex(monitor_state($m) = LOG_DATA)))

Listing 12: Monitor safety properties

Leader fairness property. This property of a leader deals with reaching a correct assessment. It verifies that the leader agent starts and resets its counters for a diagnosis to zero at the beginning and end of each voting cycle. Therefore, before inquiring the monitors, the leader is not biased towards a specific diagnosis, but it rather starts from the premise there is an equal chance that the observed node is in a normal, critical or failed situation.

CTLSPEC (**forall** $l **in** Leader **with** ag((leader_state($l) = IDLE_LEADER) implies ax(failed_diagnoses($l) = 0 and critical_diagnoses($l) = 0 and normal_diagnoses($l) = 0)))

Listing 13: Leader fairness property

Leader property. The leader collects information from all the monitors assigned to the node it is responsible for. As each monitor submits only one analysis value, the number of diagnoses the leader gathers in the assessment is equal to the number of monitors assigned to the node. The following CTL property verifies this assumption and ensures that the leader collects the exact number of diagnoses from the monitors.

CTLSPEC (**forall** $l **in** Leader **with** ag((leader_state($l) = ASSESS) implies ax(failed_diagnoses($l) + critical_diagnoses($l) + normal_diagnoses($l) = 3)))

Listing 14: Leader property

Listing 15 displays the positive results for the evaluation of some of the monitor properties (Listing 8, Listing 9, Listing 12, Listing 13, Listing 14) tested for a system consisting of a node with three assigned monitors.

6 Related Work

Previous work in the area of CELDS services focuses on providing robust infrastructures, which enable easy resource and service transition among different cloud providers [39, 41, 43]. A higher-level approach has been previously addressed through the development of the TOSCA language, which allows a standardized description of cloud services [13]. Our project extends the existent work and addresses the need of formalizing cloud-specific processes. While previous work focuses on service execution, our approach highlights the importance of correct processes.

While the adoption of formal methods in real systems is highly arguable due to the steep learning curve and large time project setting, they can indicate serious problems of the system that might reside from either the analysis or design phase. A relevant work to adopt formal specifications was carried out by Amazon Web Services, which integrated the TLA+ formal language to verify its infrastructure and discover design flaws that cause bugs [38]. The ASM method we adopted accompanied us just for the analysis and design phase of creating the monitoring and adaptation layers for the distributed CCIM, and is limited to verifying the correct behaviour of the monitors and adaptors given the expected requirements.

```
> NuSMV −dynamic −coi −quiet C:\Work\Specs\ASMeta_Specs\code\Verification\
    SingleModelVerification.smv
−− specification ((AG ((!heartbeat_timeout(monitor_2) & (monitor_state(monitor_2) =
WAIT_FOR_RESPONSE & heartbeat_response_arrived(monitor_2))) −>
AX monitor_state(monitor_2) = COLLECT_DATA) & AG ((!heartbeat_timeout(monitor_1) &
(heartbeat_response_arrived(monitor_1) & monitor_state(monitor_1) = WAIT_FOR_RESPONSE))
−> AX monitor_state(monitor_1) = COLLECT_DATA) & AG ((!heartbeat_timeout(monitor_3)
& (monitor_state(monitor_3) = WAIT_FOR_RESPONSE & heartbeat_response_arrived(monitor_3
))) −> AX monitor_state(monitor_3) = COLLECT_DATA)) is true
−− specification ((AG (trigger_gossip(monitor_2) −> EF leader_state(leader_1) = EVALUATE)
& AG (trigger_gossip(monitor_3) −> EF leader_state(leader_1) = EVALUATE))
& AG (trigger_gossip(monitor_1) −> EF leader_state(leader_1) = EVALUATE)) is true
−− specification ((AG (is_problem_discovered(monitor_1) −> AX monitor_state(monitor_1) =
REPORT_PROBLEM) & AG (is_problem_discovered(monitor_2) −> AX
monitor_state(monitor_2) = REPORT_PROBLEM)) & AG (is_problem_discovered(monitor_3)
−> AX monitor_state(monitor_3) = REPORT_PROBLEM)) is true
−− specification ((AG ((monitor_state(monitor_1) = ASSIGN_DIAGNOSIS
& !is_problem_discovered(monitor_1)) −> EX monitor_state(monitor_1) = LOG_DATA)
& AG ((monitor_state(monitor_3) = ASSIGN_DIAGNOSIS & !is_problem_discovered(monitor_3))
−> EX monitor_state(monitor_3) = LOG_DATA)) & AG ((monitor_state(monitor_2) =
ASSIGN_DIAGNOSIS & !is_problem_discovered(monitor_2)) −> EX monitor_state(monitor_2)
= LOG_DATA)) is true
−− specification AG (leader_state(leader_1) = IDLE_LEADER −>
AX ((critical_diagnoses(leader_1) = 0 & failed_diagnoses(leader_1) = 0)
& normal_diagnoses(leader_1) = 0)) is true
−− specification AG (leader_state(leader_1) = ASSESS −> AX (failed_diagnoses(leader_1)
+ critical_diagnoses(leader_1)) + normal_diagnoses(leader_1) = 3) is true
```

Listing 15: Verification of AsmetaSMV properties

Modelling cloud systems has been proposed by the MODACloud project in order to obtain self-adaptive multi-cloud applications [39]. The project relies on CloudML[7] language for modelling the runtime processes and specifying the data, QoS models and monitoring operation rules [10]. Resulting specifications are called models@runtime and allow bidirectional adaptation of models and execution based on updates performed at any of the both sides.

mOSAIC uses a Service Oriented Architecture (SOA) approach for cloud-based applications described by [40]. It contains a brokering agent at the middleware level, dealing with maintaining promised SLAs and Key Performance Indicators (KPIs) at the infrastructure and application level. Reservoir cloud federation relies on Lattice framework for monitoring services [44], which follows a publish-subscribe model.

The ASM formal method has been previously used for specifying the behaviour of adaptive systems. For instance, Arcaini et al. propose in [9], an ASM model for analyzing MAPE-K loops of self-adapting systems that follow a decentralized architecture. Flexibility and robustness to silent node failures of the specification is validated and verified with the aid of the ASMETA toolset. MAPE-K loops are important also for understanding the monitoring processes and their role for ensuring self-adapting systems. Ma et al. introduced the notion of Abstract State Services based on ASMs and described it for a flight booking over a cloud service case study [33].

Formal modelling was also used for specifying grids. ASM contributed to the description of job management and service execution in [11] and, as a further extension, in [12]. Specification of grids in terms of ASMs has been proposed also by [37], with a focus in underlining differences between grids and normal distributed systems.

7 Conclusion

The advances of distributed systems aim to respond to ever growing requirements of clients. However, existing cloud and multi-cloud solutions propose services focused on the providers, rather than on the clients. For this reason, adoption of cloud services has been hindered. Shifting the focus towards a client-oriented platform for multi-clouds is, therefore, the goal of our research work. The premise was a middleware enhanced with security, privacy, adaptivity, and QoS measurements.

In this chapter, we described a distributed version of the middleware and detailed the monitoring and adaptation layers, which complement and ensure a reliable execution. The highlighted workflow of the processes was consolidated with ASM specification models which capture the intended requirements of the system. Through thorough analysis of the model, we can identify design flaws, that otherwise, would propagate further to the implementation phase of the software development. Thus, the models were subject to validation by different scenarios and verification of specific meta-properties that reflect their quality.

---

[7] http://cloudml.org/

Properties of the solution are expressed in terms of CTL formulas and verified on the NuSMV version of the models.

In the next steps of our work, we aim to enrich the model with the communication between the monitors and the adaptation component. The monitors should be able to submit data related to an issue to the adapters, which in return should request an evaluation of the system after the enactment of an adaptation plan. We also propose building a weighted diagnosis, using the confidence degree of a monitor as its weight. In this way, monitors with a lower confidence degree will have a smaller contribution to the final evaluation. The extensions will be added to the current versions of the ASM models.

References

1. The treacherous 12 - cloud computing top threats in 2016. Technical report, Cloud Security Alliance, February 2016. https://downloads.cloudsecurityalliance.org/assets/research/top-threats/Treacherous-12_Cloud-Computing_Top-Threats.pdf.
2. State of the cloud report. Technical report, RightScale, 2017. http://assets.rightscale.com/uploads/pdfs/RightScale-2017-State-of-the-Cloud-Report.pdf.
3. Aamodt Agnar and Plaza Enric. Case-based Reasoning: Foundational issues, methodological variations, and system approaches. *AI Commun.*, 7(1):39–59, March 1994.
4. Tomonori Aoyama and Hiroshi Sakai. Inter-cloud computing. *Business & Information Systems Engineering*, 3(3):173, 2011.
5. Paolo Arcaini, Silvia Bonfanti, Marcel Dausend, Angelo Gargantini, Atif Mashkoor, Alexander Raschke, Elvinia Riccobene, Patrizia Scandurra, and Michael Stegmaier. Unified syntax for Abstract State Machines. In *Abstract State Machines, Alloy, B, TLA, VDM, and Z - 5th International Conference, ABZ 2016, Linz, Austria, May 23-27, 2016, Proceedings*, pages 231–236, 2016.
6. Paolo Arcaini, Angelo Gargantini, and Elvinia Riccobene. AsmetaSMV: A way to link high-level ASM models to low-level NuSMV specifications. In *Abstract State Machines, Alloy, B and Z, Second International Conference, ABZ 2010, Orford, QC, Canada, February 22-25, 2010. Proceedings*, pages 61–74, 2010.
7. Paolo Arcaini, Angelo Gargantini, Elvinia Riccobene, and Patrizia Scandurra. A model-driven process for engineering a toolset for a formal method. *Softw. Pract. Exper.*, 41(2):155–166, February 2011.
8. Paolo Arcaini, Roxana-Maria Holom, and Elvinia Riccobene. ASM-based formal design of an adaptivity component for a cloud system. *Formal Asp. Comput.*, 28(4):567–595, 2016.
9. Paolo Arcaini, Elvinia Riccobene, and Patrizia Scandurra. Formal design and verification of self-adaptive systems with decentralized control. *ACM Trans. Auton. Adapt. Syst.*, 11(4):25:1–25:35, January 2017.
10. Alexander Bergmayr, Alessandro Rossini, Nicolas Ferry, Geir Horn, Leire Orue-Echevarria, Arnor Solberg, and Manuel Wimmer. The evolution of CloudML and its applications. In *Proceedings of the 3rd International Workshop on Model-Driven Engineering on and for the Cloud 18th International Conference on Model*

Driven Engineering Languages and Systems (MoDELS 2015), Ottawa, Canada, September 29, 2015., pages 13–18, 2015.
11. Alessandro Bianchi, Luciano Manelli, and Sebastiano Pizzutilo. A distributed Abstract State Machine for grid systems: A preliminary study. In Peter Ivnyi and Barry H.V. Topping, editors, *Proceedings of the Second International Conference on Parallel, Distributed, Grid and Cloud Computing for Engineering*. Civil-Comp Press, April 2011.
12. Alessandro Bianchi, Luciano Manelli, and Sebastiano Pizzutilo. An ASM-based model for grid job management. *Informatica (Slovenia)*, 37(3):295–306, 2013.
13. Tobias Binz, Uwe Breitenbücher, Oliver Kopp, and Frank Leymann. TOSCA: portable automated deployment and management of cloud applications. In *Advanced Web Services*, pages 527–549. 2014.
14. Silvia Bonfanti, Marco Carissoni, Angelo Gargantini, and Atif Mashkoor. Asm2C++: A tool for code generation from Abstract State Machines to arduino. In *NASA Formal Methods - 9th International Symposium, NFM 2017, Moffett Field, CA, USA, May 16-18, 2017, Proceedings*, pages 295–301, 2017.
15. Egon Börger. Modeling distributed algorithms by Abstract State Machines compared to Petri nets. In *Abstract State Machines, Alloy, B, TLA, VDM, and Z - 5th International Conference, ABZ 2016, Linz, Austria, May 23-27, 2016, Proceedings*, pages 3–34, 2016.
16. Egon Börger and Robert F. Stark. *Abstract State Machines: A Method for High-Level System Design and Analysis*. Springer-Verlag New York, Inc., Secaucus, NJ, USA, 2003.
17. Károly Bósa. An ambient ASM model of client-to-client interaction via cloud computing and an anonymously accessible docking service. In *Software Technologies - 8th International Joint Conference, ICSOFT 2013, Reykjavik, Iceland, July 29-31, 2013, Revised Selected Papers*, volume 457 of *Communications in Computer and Information Science*, pages 235–255. Springer, 2014.
18. Károly Bósa, Roxana-Maria Holom, and Mircea Boris Vleju. A formal model of client-cloud interaction. In *Correct Software in Web Applications and Web Services*, pages 83–144. 2015.
19. Andreea Buga and Sorana Tania Nemes. A formal approach for failure detection in large-scale distributed systems using abstract state machines. In *Database and Expert Systems Applications - 28th International Conference, DEXA 2017, Lyon, France, August 28-31, 2017, Proceedings, Part I*, pages 505–513, 2017.
20. Andreea Buga and Sorana Tania Nemes. Formalizing monitoring processes for large-scale distributed systems using abstract state machines. In *Proceedings of the 1st International Workshop on Formal Approaches for Advanced Computing Systems*. Springer, September 2017. To appear.
21. Andreea Buga and Sorana Tania Nemes. Towards modeling monitoring of smart traffic services in a large-scale distributed system. In *Proceedings of the 7th International Conference on Cloud Computing and Services Science - Volume 1: CLOSER,*, pages 483–490. INSTICC, ScitePress, 2017.
22. Andreea Buga, Sorana Tania Nemes, and Klaus-Dieter Schewe. Conceptual modelling of autonomous multi-cloud interaction with reflective semantics. In *Conceptual Modeling - 36th International Conference, ER 2017, Valencia, Spain, November 6-9, 2017, Proceedings*, pages 120–133, 2017.
23. Luca Cardelli and Andrew D. Gordon. Mobile ambients. *Theoretical Computer Science*, 240(1):177–213, 2000.

24. Alessandro Carioni, Angelo Gargantini, Elvinia Riccobene, and Patrizia Scandurra. A scenario-based validation language for ASMs. In *Proceedings of the 1st International Conference on Abstract State Machines, B and Z*, ABZ '08, pages 71–84, Berlin, Heidelberg, 2008. Springer-Verlag.
25. Alessandro Cimatti, Edmund Clarke, Fausto Giunchiglia, and Marco Roveri. Nusmv: a new symbolic model checker. *International Journal on Software Tools for Technology Transfer*, 2(4):410–425, Mar 2000.
26. Roozbeh Farahbod, Vincenzo Gervasi, and Uwe Glässer. Coreasm: An extensible ASM execution engine. *Fundam. Inform.*, 77(1-2):71–103, 2007.
27. Angelo Gargantini, Elvinia Riccobene, and Patrizia Scandurra. A metamodel-based language and a simulation engine for abstract state machines. *J. UCS*, 14(12):1949–1983, 2008.
28. Roxana Holom. *Formal Modeling of a Client-Cloud Interaction Middleware with respect to Adaptivity*. PhD thesis, Johannes Kepler Universität Linz, September 2016.
29. Felix Kossak and Atif Mashkoor. *How to Select the Suitable Formal Method foran Industrial Application: A Survey*, pages 213–228. Springer International Publishing, Cham, 2016.
30. Harald Lampesberger. *Language-Based Anomaly Detection in Client-Cloud Interaction*. PhD thesis, Johannes Kepler Universität Linz, June 2016.
31. Leslie Lamport. Time, clocks, and the ordering of events in a distributed system. *Communications of the ACM*, 21(7):558–565, 1978.
32. Leslie Lamport. TLA+: whence, wherefore, and whither. In *First NASA Formal Methods Symposium - NFM 2009, Moffett Field, California, USA, April 6-8, 2009.*, page 3, 2009.
33. Hui Ma, Klaus-Dieter Schewe, and Qing Wang. An abstract model for service provision, search and composition. In *4th IEEE Asia-Pacific Services Computing Conference, IEEE APSCC 2009, Singapore, December 7-11 2009, Proceedings*, pages 95–102, 2009.
34. Peter M. Mell and Timothy Grance. Sp 800-145. the NIST definition of cloud computing. Technical report, Gaithersburg, MD, United States, 2011.
35. Nikola Milanovic. *Non-Functional Properties in Service Oriented Architecture: Requirements, Models and Methods: Requirements, Models and Methods*. Premier reference source. Information Science Reference, 2011.
36. Sorana Tania Nemes and Andreea Buga. Towards a case-based reasoning approach to dynamic adaptation for large-scale distributed systems. In David W. Aha and Jean Lieber, editors, *Case-Based Reasoning Research and Development - 25th International Conference, ICCBR 2017, Trondheim, Norway, June 26-28, 2017, Proceedings*, LNCS, pages 257–271. Springer, 2017.
37. Zsolt Németh and Vaidy Sunderam. Characterizing grids: Attributes, definitions, and formalisms. *Journal of Grid Computing*, 1(1):9–23, 2003.
38. Chris Newcombe, Tim Rath, Fan Zhang, Bogdan Munteanu, Marc Brooker, and Michael Deardeuff. How Amazon web services uses formal methods. *Commun. ACM*, 58(4):66–73, March 2015.
39. Elisabetta Di Nitto, Peter Matthews, Dana Petcu, and Arnor Solberg, editors. *Model-Driven Development and Operation of Multi-Cloud Applications. The MODAClouds Approach*. Springer Briefs in Applied Sciences and Technology. Springer International Publishing, 2017.

40. Dana Petcu, Ciprian Crăciun, Marian Neagul, Silviu Panica, Beniamino Di Martino, Salvatore Venticinque, Massimiliano Rak, and Rocco Aversa. Architecturing a sky computing platform. In *Proceedings of the 2010 International Conference on Towards a Service-based Internet*, ServiceWave'10, pages 1–13, Berlin, Heidelberg, 2011. Springer-Verlag.
41. Dana Petcu, Beniamino Di Martino, Salvatore Venticinque, Massimiliano Rak, Tamás Máhr, Gorka Esnal Lopez, Fabrice Brito, Roberto Cossu, Miha Stopar, Svatopluk Šperka, and Vlado Stankovski. Experiences in building a mOSAIC of clouds. *Journal of Cloud Computing: Advances, Systems and Applications*, 2(1):12, May 2013.
42. Mariam Rady. *A Client-Centric Model For Managing Availability and Performance Conditions of Service Level Agreements in Cloud Computing*. PhD thesis, Johannes Kepler Universität Linz, August 2015.
43. Benny Rochwerger, David Breitgand, Amir Epstein, David Hadas, Irit Loy, Kenneth Nagin, J. Tordsson, C. Ragusa, Massimo Villari, Stuart Clayman, Eliezer Levy, A. Maraschini, Philippe Massonet, H. Muñoz, and Giovanni Toffetti. Reservoir - when one cloud is not enough. *IEEE Computer*, 44(3):44–51, 2011.
44. Benny Rochwerger, David Breitgand, Eliezer Levy, Alex Galis, Kenneth Nagin, Ignacio Martín Llorente, Rubén S. Montero, Yaron Wolfsthal, Erik Elmroth, Juan A. Cáceres, Muli Ben-Yehuda, Wolfgang Emmerich, and Fermín Galán. The Reservoir model and architecture for open federated cloud computing. *IBM Journal of Research and Development*, 53(4):4, 2009.
45. Klaus-Dieter Schewe, Károly Bósa, Harald Lampesberger, Ji Ma, Mariam Rady, and Mircea Boris Vleju. Challenges in cloud computing. *Scalable Computing: Practice and Experience*, 12(4):385–390, 2011.
46. Klaus-Dieter Schewe, Károly Bósa, Harald Lampesberger, Ji Ma, and Mircea Boris Vleju. The Christian Doppler laboratory for client-centric cloud computing. In *2nd Workshop on Software Services (WoSS 2011)*, Timisoara, Romania, June 2011.
47. Edwin Sturrus. Identity and access management in a cloud computing environment. Master's thesis, Econometric Institute, Erasmus School of Economics, Erasmus University Rotterdam, August 2011. https://thesis.eur.nl/pub/10422/MA-5
48. Mircea Boris Vleju. *Client-Centric Identity and Access Management in Cloud Computing*. PhD thesis, Johannes Kepler Universität Linz, July 2015.
49. Bernd Zwattendorfer, Thomas Zefferer, and Klaus Stranacher. An overview of cloud identity management-models. In *WEBIST 2014 - Proceedings of the 10th International Conference on Web Information Systems and Technologies, Volume 1, Barcelona, Spain, 3-5 April, 2014*, pages 82–92, 2014.

From Concepts in Non-Monotonic Reasoning to High-Level Implementations Using Abstract State Machines and Functional Programming

Christoph Beierle[1], Steven Kutsch[1], Gabriele Kern-Isberner[2]

[1]Dept. of Computer Science, University of Hagen, 58084 Hagen, Germany
[2]Dept. of Computer Science, TU Dortmund, 44221 Dortmund, Germany

Abstract. Default rules of the form "if A then usually B" are a powerful means in knowledge representation to establish plausible relationships between formulas A and B. Formal-logically, they can be expressed by qualitative conditionals $(B|A)$ to which conditional-logical standards should be applied in principle. So, when reasoning nonmonotonically based on conditional knowledge bases consisting of a set of conditionals, a rich structure going beyond classical logic is required, e.g. ranking functions that assign a degree of implausibility to each possible world. System P and system Z are popular approaches for that, using the set of all ranking functions and the unique Pareto-minimal ranking function accepting a knowledge base, respectively. However, though the logical quality of both systems is very high, implementations of them that can be used for practical reasoning are rare. In this paper, we demonstrate that two well-known declarative programming paradigms provide excellent tools for implementing such non-monotonic formalisms: abstract state machines and functional programming. For both approaches, we show how core functionalities that are common to system P and system Z can be implemented. Moreover, for functional programming, we present a full high-level Haskell implementation of both system P and system Z which is used as a backend in the conditional reasoning tool InfOCF. In both cases, our aim is to keep the operational program code as close as possible to the formal definitions. This closeness supports the validation of the implementation.

1 Introduction

Knowledge representation is one of the major topics in artificial intelligence. While in knowledge representation, non-standard logics supporting non-monotonic reasoning have been studied for a long time, especially in recent years a specific research focus has been on the implementation of non-monotonic formalisms. In this paper, we address some relevant core concepts in non-monotonic reasoning. The first one, system P [1], is often considered to be the gold standard for non-monotonic inference relations from default rules. The second one, system Z [33], can be viewed as a refinement of system P that allows for more, but still plausible, inferences. For these concepts, given as abstract mathematical definitions, we develop high-level operational, fully executable realizations employing two different formalisms, abstract state machines and functional programming, respectively.

Default rules of the form "If A then *usually / normally / preferably B*" play an important role in the area of knowledge representation and reasoning. They establish plausible relationships between two formulas A and B and are often called *conditionals*. A set of such rules can be used to represent the (commonsense and generic) knowledge of a reasoning agent.

Example 1. The following four sentences describe plausible relations in the domain of birds:

- "Birds usually fly."
- "Penguins are usually birds."
- "Penguins usually don't fly."
- "Birds usually have wings."

A rational agent whose knowledge base is given by such a set of sentences should be able to reason and to draw inferences based on these sentences. While such knowledge bases may contain all relevant rules for an agent, they usually do not contain enough information to represent all plausible beliefs that a reasoning agent, operating based on this knowledge, should have. For instance, while believing that birds usually fly seems to be an immediate consequence from the sentences given in Example 1, the situation is not so clear regarding e.g. the question whether penguins having wings usually do not fly. Thus, for a reasoning agent it is essential to extend a knowledge base to what is called a complete *epistemic state*, containing all beliefs necessary to answer arbitrary questions [21]. There are many ways to inductively complete a knowledge base and to represent the resulting epistemic state of an agent, e.g. using probability distributions [32], possibility theory [13], or ordinal conditional functions (OCF), also called ranking functions [35, 36]. Based on the explicit information specified in the knowledge base, these approaches assign a probability, possibility, or implausibility value, respectively, to each possible world in order to be able to compare different possible worlds accordingly. Thanks to the induced ordering of the possible worlds, sentences as given in Example 1 can then be evaluated and used for inference.

An inference relation based on rules as given in Example 1 is nonmonotonic since the rules are not strict, but also allow for exceptions. System P is an axiomatic system providing a set of desirable properties for nonmonotonic inference relations [1]. It induces an inference relation by taking all models of a knowledge base according to the given semantics into account. System Z [33] is an approach to relax this condition by defining an inference relation based on a single preferred OCF-model of a knowledge base.

Default rules and OCFs have been applied in multiple areas. For instance, in [23] qualitative conditionals and OCFs are used to model psychological phenomena in language processing discovered in recent experiments, and [28] describes a combination of OCF-based belief revision and reinforcement learning applied to object recognition in computer vision. The experimental results mentioned in [31] show that in practice knowledge bases of thousands of conditionals can be processed.

Among the most prominent implementations of non-monotonic reasoning are answer set programming systems that are based on logic programming, see e.g. [17, 27, 16]. On the other hand, little use has been made of other declarative or high-level approaches for implementing non-monotonic reasoning. The aim of this paper is to show that both abstract state machines [19, 10] as well as functional programming are excellent tools for implementing non-monotonic formalisms, especially when the objective is to minimize the gap between the formal definitions on the one hand and the operational program code on the other hand. For the two well-known non-monotonic entailment relations given by system P and system Z, we give a fully operational ASM specification for the core functionalities of system P and system Z entailment, i.e., for checking the consistency of a conditional knowledge base by computing the so-called tolerance partitioning. For both system P and system Z, we present the complete code of a high-level implementation in Haskell [30] that is used as a backend in an interactive system for evaluating and comparing various non-monotonic inference relations that have been proposed for conditional knowledge bases. While there are other implementations of system Z such as Z-log [31] that focus on exploring the computational complexity of system Z, the main objective of our HaskPZ implementation is to be close to the formal definitions of the underlying concepts and algorithms.

The starting point of the concept of abstract state machines was to extend and to improve Turing's thesis about the notion of an algorithm. Gurevich's "sequential ASM thesis" states that every sequential algorithm, on any level of abstraction, can be viewed as a sequential ASM [19]. ASMs have been employed for formal specification, modelling, and analysis in a large variety of different areas, including many programming and design languages, hardware and software systems, and various application domains, cf. [8, 10]. Klaus-Dieter Schewe has made numerous contributions to computer science, and he has also played a major role in the use, application, and further development of ASMs, including his work on extensions of the sequential ASM thesis to non-deterministic, parallel, and concurrent abstract state machines, see e.g. [9, 14, 15, 34]. We dedicate this article to Klaus-Dieter, with our sincere congratulations and best wishes on the occasion of his 60th birthday.

The rest of this paper is organized in the following way. After briefly recalling the required background of conditional logic in Section 2, we present our ASM approach for realizing system P and system Z in Section 3, describe the design, implementation, and use of HaskPZ in Section 4, and in Section 5, we conclude and point our further work.

2 Conditionals, Ranking Functions, System P, and System Z

Let $\mathcal{L}$ be a propositional language, generated by a finite set Σ of atoms $a, b, c, \ldots$. We denote formulas of $\mathcal{L}$ with uppercase letters $A, B, C, \ldots$. In formulas, we omit the *and*-connective, writing AB instead of $A \wedge B$. We indicate negation of a formula with overlining, i.e. $\overline{A}$ means $\neg A$. The set of possible worlds Ω contains all propositional interpretations over $\mathcal{L}$, these interpretations can easily be identified with the complete conjunctions over Σ mentioning each atom exactly once, either in positive or negative form. For $\omega \in \Omega$, $\omega \models A$ means that the propositional formula $A \in \mathcal{L}$ holds in the possible world ω.

2.1 Conditionals and Ranking Functions

To formalize the idea of plausible, probable, or possible connections between propositions, we introduce a new binary operator | to form conditionals.

Definition 1 (Conditionals). *Let $A, B \in \mathcal{L}$. Then $(B|A)$ is the conditional formalizing the conditional rule "if A then (usually) B". A is called the* antecedent *and B is called the* consequent. *The language of all conditionals over a propositional language $\mathcal{L}$ is denoted by $(\mathcal{L} \mid \mathcal{L})$. If $(B|A)$ is a conditional, then we call $(\overline{B}|A)$ its negation.*

We will commonly use sets of conditionals as knowledge bases for our calculations.

Definition 2 (Conditional Knowledge Base). *Let $\mathcal{L}$ be a propositional language. A set*

$$\mathcal{R} = \{(B_1|A_1), \ldots, (B_n|A_n)\}$$

where every $A_i, B_i \in \mathcal{L}$ for $i \in \{1, \ldots, n\}$, is called a (conditional) knowledge base.

Example 2 ($\mathcal{R}_{birds}$). We formalize the four sentences from Example 1 as conditionals. Let $\Sigma = \{b(\text{birds}), p(\text{penguins}), f(\text{flying}), w(\text{wings})\}$. The knowledge base $\mathcal{R}_{birds} = \{r_1, r_2, r_3, r_4\}$ consists of the four conditionals:

$$\begin{aligned} r_1 &= (f|b) & \text{"birds usually fly"} \\ r_2 &= (b|p) & \text{"penguins are usually birds"} \\ r_3 &= (\overline{f}|p) & \text{"penguins usually don't fly"} \\ r_4 &= (w|b) & \text{"birds usually have wings"} \end{aligned}$$

It is crucial to understand that a conditional $(B|A)$ is different from the material implication $A \Rightarrow B \equiv \neg A \vee B$. According to DeFinetti [11], conditionals are three-valued objects and can be represented as *generalized indicator functions*:

$$(B|A)(\omega) = \begin{cases} 1 & \text{if } \omega \models AB & \text{(verification)} \\ 0 & \text{if } \omega \models A\overline{B} & \text{(falsification)} \\ u & \text{if } \omega \models \overline{A} & \text{(not applicable)} \end{cases} \quad (1)$$

In order to give appropriate semantics to conditionals, they are usually considered within richer structures such as *epistemic states* [21]. Beside certain (logical) knowledge, epistemic states also allow for the representation of preferences, beliefs, or assumptions of an intelligent agent. Basically, an epistemic state allows one to use meta-information to compare formulas or worlds with respect to plausibility, possibility, necessity, probability, etc.

Spohn's *ordinal conditional functions*, OCFs [35], also called *ranking functions* are capable of representing epistemic states.

Definition 3 (ordinal conditional functions, OCFs). *An ordinal conditional function is a function $\kappa : \Omega \to \mathbb{N}$ such that there is at least one possible world ω with $\kappa(\omega) = 0$.*

Ranking functions assign a degree of implausibility to every possible world. The higher $\kappa(\omega)$, the less plausible ω is considered by κ. Note that for each κ, at least one world must be most plausible, i.e. must have rank 0. An OCF κ can be extended to arbitrary formulas $A \in \mathcal{L}$ by

$$\kappa(A) = \begin{cases} \min\{\kappa(\omega) \mid \omega \models A\} & \text{if } A \text{ is satisfiable} \\ \infty & \text{otherwise} \end{cases} \quad (2)$$

and to conditionals $(B|A) \in (\mathcal{L} \mid \mathcal{L})$ by:

$$\kappa((B|A)) = \begin{cases} \kappa(AB) - \kappa(A) & \text{if } \kappa(A) \neq \infty \\ \infty & \text{otherwise} \end{cases} \quad (3)$$

Note that $\kappa((B|A)) \geqslant 0$ since any ω satisfying AB also satisfies A and therefore $\kappa(AB) \geqslant \kappa(A)$.

Since ranking functions provide non-logical information on the plausibility of worlds, they are suitable for representing an epistemic state of a reasoning agent. So, we can define the acceptance of a conditional by an agent in epistemic state κ.

Definition 4 (Acceptance of Conditionals). *Let κ be a ranking function. The conditional $(B|A)$ is accepted by κ, denoted by $\kappa \models (B|A)$, iff*

$$\kappa(AB) < \kappa(A\overline{B}). \quad (4)$$

Thus, a conditional is accepted iff its verification is considered strictly more plausible than its falsification.

We say that κ accepts a knowledge base $\mathcal{R}$, denoted by $\kappa \models \mathcal{R}$, iff $\kappa \models (B|A)$ for every $(B|A) \in \mathcal{R}$. A knowledge base is *consistent*, iff a ranking function exists that accepts it [33].

ω	$\kappa(\omega)$	ω	$\kappa(\omega)$	ω	$\kappa(\omega)$	ω	$\kappa(\omega)$
$bpfw$	2	$b\overline{p}fw$	0	$\overline{b}pfw$	2	$\overline{b}\overline{p}fw$	0
$bpf\overline{w}$	2	$b\overline{p}f\overline{w}$	1	$\overline{b}pf\overline{w}$	2	$\overline{b}\overline{p}f\overline{w}$	0
$bp\overline{f}w$	1	$b\overline{p}\overline{f}w$	1	$\overline{b}p\overline{f}w$	2	$\overline{b}\overline{p}\overline{f}w$	0
$bp\overline{f}\overline{w}$	1	$b\overline{p}\overline{f}\overline{w}$	1	$\overline{b}p\overline{f}\overline{w}$	2	$\overline{b}\overline{p}\overline{f}\overline{w}$	0

Table 1. A ranking function accepting the knowledge base $\mathcal{R}_{birds}$ from Example 2.

Example 3. Consider the knowledge base $\mathcal{R}_{birds}$ from Example 2. Table 1 shows a ranking function κ that accepts every conditional in $\mathcal{R}_{birds}$. For instance, $\kappa \models (\overline{f}|p)$ because $\kappa(p\overline{f}) = \min_{\omega \models p\overline{f}} \kappa(\omega) = 1 < 2 = \min_{\omega \models pf} \kappa(\omega) = \kappa(pf)$.

Every ranking function induces a non-monotonic inference relation between formulas. This relation is based on the acceptance of conditionals in Definition 4.

Definition 5 (Ranking Function Inference). *Let $A, B \in \mathcal{L}$ and κ a ranking function. Then B is a non-monotonic inference of A by κ, denoted by $A \mathrel{|\!\sim}_\kappa B$, iff the conditional $(B|A)$ is accepted by κ, i.e., iff $\kappa(AB) < \kappa(A\overline{B})$.*

Note that ranking function inference is a special instance of preferential entailment [29] because each ranking function induces a preferential relation on worlds in a straightforward way.

2.2 System P and p-entailment

A common benchmark for non-monotonic inference relations is the axiom system P [1]. The postulates of system P given in [26] provide what most researchers regard as a core any non-monotonic system should satisfy; in [22], Hawthorne and Makinson call these axioms the "industry standard" for qualitative nonmonotonic inference:

$$
\begin{array}{rl}
\textit{Reflexivity or Inclusion} &: \quad A \mathrel{|\!\sim} A \\[4pt]
\textit{Cut} &: \quad \dfrac{A \wedge B \mathrel{|\!\sim} C,\ A \mathrel{|\!\sim} B}{A \mathrel{|\!\sim} C} \\[8pt]
\textit{Cautious Monotony} &: \quad \dfrac{A \mathrel{|\!\sim} B,\ A \mathrel{|\!\sim} C}{A \wedge B \mathrel{|\!\sim} C} \\[8pt]
\textit{Right Weakening} &: \quad \dfrac{A \mathrel{|\!\sim} B,\ B \models C}{A \mathrel{|\!\sim} C} \\[8pt]
\textit{Left Logical Equivalence} &: \quad \dfrac{\models A \equiv B,\ A \mathrel{|\!\sim} C}{B \mathrel{|\!\sim} C} \\[8pt]
\textit{Or} &: \quad \dfrac{A \mathrel{|\!\sim} C,\ B \mathrel{|\!\sim} C}{A \vee B \mathrel{|\!\sim} C}
\end{array}
$$

An important result about system P is that it induces an inference relation, called system P inference, that coincides with the inference relation, called *p-entailment*, that takes every ranking function accepting a given knowledge base into account (for details, see [12]).

Definition 6 (p-entailment). *[18] Let $A, B \in \mathcal{L}$ and $\mathcal{R}$ a knowledge base. Then B is p-entailed from A in the context of $\mathcal{R}$, denoted by $A \mathrel|\!\!\sim_p^\mathcal{R} B$, iff $A \mathrel|\!\!\sim_\kappa B$ for every κ accepting $\mathcal{R}$.*

Since a knowledge base is only consistent if a ranking function accepting it exists, this form of inference can be implemented by testing the consistency of the knowledge base augmented by the negated query conditional.

Proposition 1. *[18] Let $\mathcal{R}$ be a consistent knowledge base. Then*

$$A \mathrel|\!\!\sim_p^\mathcal{R} B \quad \text{iff} \quad \mathcal{R} \cup \{(\overline{B}|A)\} \text{ is inconsistent.} \tag{5}$$

For checking the consistency of $\mathcal{R}$, a special partition of $\mathcal{R}$ based on the notion of *tolerance* can be used. Intuitively, a conditional r is tolerated by a set of conditionals $\mathcal{R}$, iff there is a world w that satisfies r and does not falsify any $r' \in \mathcal{R}$ (as defined by (1)).

Definition 7 (Tolerance). *[18] A conditional $(D|C)$ is tolerated by a knowledge base $\mathcal{R}$, iff there is $w \in \Omega$ satisfying the formula*

$$CD \land \bigwedge_{(B|A) \in \mathcal{R}} (A \Rightarrow B).$$

Example 4. The knowledge base $\mathcal{R}_{birds}$ from Example 2 tolerates the conditionals $(f|b)$ and $(w|b)$ because the world $w = b\overline{p}fw$ verifies both conditionals and falsifies no other conditional from $\mathcal{R}_{birds}$, therefore $w \models bfw \land (p \Rightarrow \overline{f}) \land (p \Rightarrow b)$. None of the other conditionals is tolerated by $\mathcal{R}_{birds}$.

The notion of tolerance induces a partition of the conditionals in $\mathcal{R}$ that yields a consistency test for $\mathcal{R}$.

Definition 8 (Ordered Partition). *[18] Let $\mathcal{R}$ be a set of conditionals. $\mathcal{R}_p = (\mathcal{R}_0, \dots, \mathcal{R}_k)$ is an ordered partition, iff $\{\mathcal{R}_0, \dots, \mathcal{R}_k\}$ is a partition of $\mathcal{R}$ and for every $0 \leqslant i \leqslant k$, every $r \in \mathcal{R}_i$ is tolerated by the union $\bigcup_{j=i}^{k} \mathcal{R}_j$.*

Proposition 2. *[33] $\mathcal{R}$ is consistent iff there is an ordered partition for $\mathcal{R}$.*

Thus, since p-entailment can be reduced to a consistency check, p-entailment can also be reduced to the question whether an ordered partition of $\mathcal{R}$ exists. Moreover, the construction of an ordered partition is also crucial for system Z.

2.3 System Z

The condition for p-entailment is rather strict as it takes all ranking models of a knowledge base into account, possibly disallowing inferences that may still be considered plausible, although they do not hold in all ranking models of $\mathcal{R}$, but e.g. in a subset of preferred ranking models of $\mathcal{R}$. The idea of system Z [33] is to define a plausible inference relation taking only a uniquely defined "best" or most preferred model into account. For any consistent knowledge base $\mathcal{R}$, System Z defines a unique ranking function accepting $\mathcal{R}$. While in general, there are several different ordered partitions of $\mathcal{R}$, there is a uniquely defined inclusion-maximal ordered partition for every consistent $\mathcal{R}$, that is, every conditional is in the lowest possible subset. Based on this inclusion-maximal ordered partition $\mathcal{R}_p = (\mathcal{R}_0, \ldots, \mathcal{R}_k)$, the function $Z : \mathcal{R} \to \{0, \ldots, k\}$ is defined by:

$$Z(r) = i \quad \text{iff} \quad r \in \mathcal{R}_i \qquad (6)$$

With this function the System Z ranking function $\kappa_\mathcal{R}^Z$ is defined as [33]

$$\kappa_\mathcal{R}^Z(\omega) = \begin{cases} 0 & \text{iff } \omega \text{ does not falsify any } (B|A) \in \mathcal{R} \\ \max_{(B|A)\in\mathcal{R}} \{Z((B|A))|\omega \models A\overline{B}\} + 1 & \text{otherwise.} \end{cases}$$

$$(7)$$

It is straightforward to show that $\kappa_\mathcal{R}^Z \models \mathcal{R}$.

Because the ranking function $\kappa_\mathcal{R}^Z$ defined by system Z is based on the inclusion-maximal partition satisfying the tolerance relation, it can be shown that it is the, with respect to assigned ranks, Pareto-minimal ranking function accepting $\mathcal{R}$ [33].

Example 5. The ranking function κ in Table 1 is the ranking function $\kappa_\mathcal{R}^Z$ using the inclusion-maximal ordered partition $\mathcal{R}_{birds} = (\{(f|b), (w|b)\}, \{(b|p), (\overline{f}|p)\})$.

While p-entailment takes all ranking functions accepting $\mathcal{R}$ into account, z-entailment is the inference relation induced by this "best" model $\kappa_\mathcal{R}^Z$.

Definition 9 (System Z inference; z-entailment). *Let $A, B \in \mathcal{L}$ and $\mathcal{R}$ a knowledge base. Then B is z-entailed from A in the context of $\mathcal{R}$, denoted by $A \mid\!\sim_z^\mathcal{R} B$, iff $A \mid\!\sim_{\kappa_\mathcal{R}^Z} B$.*

The following example illustrates that z-entailment enables plausible inferences which are not possible with p-entailment.

Example 6. A question we might want to answer based on the knowledge in $\mathcal{R}_{birds}$ is whether winged penguins are still unable to fly, that is whether from wp we can plausibly infer $\overline{f}$ in the context of $\mathcal{R}$, denoted by $wp \mid\!\sim^\mathcal{R} \overline{f}$.

If we add the conditional $(f|wp)$, representing the negation of the query conditional $(\overline{f}|wp)$ to $\mathcal{R}_{birds}$, the ordered partition

$(\{(f|b),(w|b)\},\{(b|p),(\overline{f}|p)\},\{(f|wp)\})$ respects the tolerance condition. Therefore $\mathcal{R}_{birds} \cup \{(f|wp)\}$ is consistent and $wp \hspace{0.1em}\mid\hspace{-0.5em}\sim_p^{\mathcal{R}} \overline{f}$.

In contrast, using the ranking function $\kappa_\mathcal{R}^Z$ listed in Table 1 we see that $\kappa(p\overline{f}w) = 1 < 2 = \kappa(pfw)$ and therefore $wp \hspace{0.1em}\mid\hspace{-0.5em}\sim_p^{\mathcal{R}} \overline{f}$.

3 Using Abstract State Machines for Implementing System P and System Z

In previous work, we have already used abstract state machines for modelling and analyzing knowledge representation and reasoning methods. In [5], the ASM method is used to develop a high-level specification of CONDOR, a system that models the reasoning component of an intelligent agent whose knowledge is represented as a set of conditionals expressing default rules. A refinement of this ASM employing ranking functions for the representation of the agent's epistemic state is given in [6], which in [7] is further refined to a fully operational specification, called CONDOR@AsmL, using the AsmL language [20]. In [7], a verification proof is given that CONDOR@AsmL correctly implements belief revision and belief update based on c-revisions [24, 25]. In this section, we will reuse part of the ASM presented in [7] and show that it correctly implements the core functionality for both system P inference and system Z inference, i.e., checking the consistency of a knowledge base $\mathcal{R}$ and computing its inclusion-maximal ordered partition.

3.1 Logical Formulas and Knowledge Bases

Given a set Σ of propositional variables, Formula is the set of all propositional sentences over Σ, and World is the set of all possible worlds that can be distinguished using Σ, i.e., the set of all complete conjunctions over Σ. Since in the ASM developed in [7], conditionals are called rules, we will use these two terms interchangeably in this section. The central set of conditionals (Rule) and the power set of the set of conditionals (RuleSet) are implemented by:

```
structure Rule
    Conclusion as Formula
    Premise as Formula

type RuleSet = Set of Rule
```

Without giving their complete AsmL code here, we assume that the two auxiliary functions

```
verify(w as World, in_R as RuleSet) as RuleSet
falsify(w as World, in_R as RuleSet) as RuleSet
```

```
buildPartition(in_R as RuleSet, in_partition as Map of Integer to RuleSet)
                                         as Map of Integer to RuleSet
// recursively build proper partition
 var rules as RuleSet = in_R
 var partition as Map of Integer to RuleSet = in_partition
 let tolerating_worlds = {w | w in Omega where falsify(w,rules) = {} }
 let tolerated_rules = {r | w in tolerating_worlds, r in verify(w,rules)}
 step
  if tolerated_rules = {}
  then partition := {->}    // proper partition does not exist
  else                       // extend current partition
    let next_index = Size(partition)   // next index, starts with 0
    step partition := partition + {next_index -> tolerated_rules}
         rules := rules - tolerated_rules
    step if rules <> {}   // partition remaining rules recursively
         then partition := buildPartition(rules, partition)
 step return partition
```

Fig. 1. AsmL code for determining an ordered partition of a set of conditionals [7]

satisfy, for any world w in World and any set of rules $\mathcal{R}$ in RuleSet the following two conditions:

$$\text{verify}(w, \mathcal{R}) = \{r \in \mathcal{R} \mid w \text{ verifies } r\} \quad (8)$$
$$\text{falsify}(w, \mathcal{R}) = \{r \in \mathcal{R} \mid w \text{ falsifies } r\} \quad (9)$$

Note that (8) and (9) are ensured easily by implementing some fundamental primitives of propositional logic. Furthermore, the global AsmL variable Omega contains all complete conjunctions over the propositional variables Σ, and the global Asml variable Partition will hold an ordered partition of a set of rules checked for consistency, as will be explained in more detail in the next subsections.

3.2 Computing the Inclusion-Maximal Ordered Partition

Figure 1 shows the AsmL code for computing the inclusion-maximal ordered partition. The binary function buildPartition(in_R,p) returns an ordered partition (implemented as an injective function mapping natural numbers to sets of rules) of in_R if it exists, and the empty map, in AsmL denoted by {->}, otherwise. It takes a set of rules in_R (still to be partitioned) and a partition p (of those rules that have already been assigned to a particular partition set R_m). Initially, in_R contains all given rules and p is the empty function {->}.

Proposition 3. *(i) For any given set of rules $\mathcal{R}$, buildPartition($\mathcal{R}$,{->}) computes an ordered partition of $\mathcal{R}$ if it exists, and the empty map {->} if no ordered partition of $\mathcal{R}$ exists. (ii) Furthermore, if $\mathcal{R}$ is consistent, the computed partition is an inclusion-maximal ordered partition of $\mathcal{R}$.*

Proof. For showing (i), we will use the invariant INV($\mathcal{S}$, p) for recursive calls of buildPartition($\mathcal{S}$, p), where $\mathcal{S}$ is a set of rules, and p is a partition of a set of rules:

INV($\mathcal{S}$, p):
- $\mathcal{R} = \mathcal{S} \cup \bigcup_{i \to R_i \in p} R_i$
- p is an ordered partition of $\bigcup_{i \to R_i \in p} R_i$
- every rule $r \in \mathcal{S}$ is tolerated by $\bigcup_{i \to R_i \in p} R_i$

Let buildPartition(in_R, in_p) either denote the initial call buildPartition($\mathcal{R}$,{->}) or any subsequent recursive call caused directly or indirectly by this initial call. We will prove that INV(in_R, in_p) holds for any of these calls and show that the conclusion of (i) in the proposition follows.

For the initial call, INV(in_R, in_p) holds trivially since in_p is empty. So let buildPartition(in_R, in_p) be any subsequent call and assume that INV(in_R, in_p) holds for this call. The first two let-constructs in Fig. 1 ensure that tolerating_worlds contains all worlds that do not falsify any of the rules in in_R, and that tolerated_rules contains all rules from in_R verified by some of these worlds. Using Definitions 7 and 8, we thus have:

$$\text{tolerated\_rules} = \{r \mid r \in \text{in\_R},\ r \text{ is tolerated by in\_R}\}$$

If tolerated_rules is empty, no ordered partition of in_R exists according to Definition 8, and together with INV(in_R, in_p) we conclude that in_p can not be extended to an ordered partition of $\mathcal{R}$. Thus, there is no ordered partition of $\mathcal{R}$ and the initial call buildPartition($\mathcal{R}$,{->}) terminates, returning the empty map.

If tolerated_rules is not empty, in_p is an ordered partition

$$\{0 \to R_0,\ 1 \to R_1,\ \ldots,\ m{-}1 \to R_{m-1}\}$$

where $m = $ Size(in_p). According to Fig. 1, partition and rules are set to:

partition = $\{0 \to R_0,\ 1 \to R_1,\ \ldots,\ m{-}1 \to R_{m-1},\ m \to $ tolerated_rules$\}$
rules = in_R $\setminus$ tolerated_rules

From INV(in_R, in_p) and Definitions 7 and 8, we conclude that for these values INV(rules, partition) holds. If rules is empty, the computation terminates and INV(rules, partition) implies the conclusion of (i) in the proposition; if rules is not empty, the invariant holds for the recursive call buildPartition(rules, partition).

It remains to be shown that buildPartition($\mathcal{R}$,{->}) terminates. This follows from the fact that in each recursive invocation of buildPartition, the set of rules in the first argument is decreased by at least one rule since tolerated_rules is checked to be non-empty.

To show (ii), consider the first two let constructs in Fig. 1. They ensure that at any recursive call, tolerated_rules contains *all* conditionals from the set of conditionals still to be partitioned that are tolerated by the set of conditionals still to be partitioned. Since the complete set denoted by `tolerated_rules` is used to build the next partition set, it follows that the computed ordered partition is inclusion-maximal. □

3.3 Consistency Check

The AsmL code for inferring the consistency of a set of rules is now straightforward; it is given in Fig. 2.

```
isConsistent(in_R as RuleSet) as Boolean
  step Partition := buildPartition(in_R, {->})
  step return Partition <> {->}  // consistent iff Partition is non-empty
```

Fig. 2. AsmL code for inferring the consistency of a set of conditionals

Corollary 1. *For any given set of rules* **in_R**, *isConsistent(in_R) returns true if* **in_R** *is consistent, and false otherwise.*

Proof. This is a direct consequence of Propositions 2 and 3 since `isConsistent(in_R)` calls `buildPartition(in_R,{->})`, writes the result to `Partition` and checks whether this is the empty map or not. □

The AsmL code presented here covers the core functionalities of system P and system Z. In the next section, we will demonstrate how also functional programming can be used to achieve a fully operational implementation of conditional reasoning that is close to the mathematically defined concepts.

4 Implementation of System P and System Z in Haskell

In this section, we will develop an implementation, called `HaskPZ`, of system P and system Z in Haskell [30]. The implementation of `HaskPZ` can be split into three parts. In the first part we will pay attention to the underlying datatypes that represent various formal parts of a logical language (Section 4.1). The second part shows how the inclusion-maximal ordered partition is computed (Section 4.2) and describes the implementation of the consistency check algorithm as well as its usage for implementing p-entailment (Section 4.3). Section 4.4 describes how $\kappa_\mathcal{R}^Z$ is calculated, represented, and used to realize z-entailment. Finally, in Section 4.5, we describe different ways how `HaskPZ` can be used for reasoning tasks.

4.1 Logical Formulas and Knowledge Bases

The basic type class is `Atom`. Together with the type `Interpretation`, representing possible worlds, it is the basis of a logical system.

```
type Interpretation a = a -> Bool

class (Ord a) => Atom a where
  evalA  :: a -> Interpretation a -> Bool
  printA :: a -> String
```

From this foundation, different types of formulas can be implemented. The type class `Formula` encapsulates these types of formulas and gives them a common interface.

```
class (Atom a) => Formula a f | f -> a where
  evalF    :: f -> Interpretation a -> Bool
  printF   :: f -> String
  getAtoms :: f -> [a]
```

In our implementation we need literals, conjunctions of literals, and formulas in disjunctive normal form (DNF), i.e. disjunctions of conjunctions of literals. Any standardized representation of an arbitrary formula would work here. The decision for DNFs is mainly founded by compatibility to other reasoning systems. For ease of modeling we represent conjunctions as lists of literals and DNFs as lists of conjunctions. All these types of formulas have suitable `Formula` instances.

```
data Literal a = Pos a
               | Neg a
               deriving (Eq, Ord)

type Conjunction a = [Literal a]

type DNF a = [Conjunction a]
```

Conditionals can not yet be expressed in this framework. We define them outside of this framework as pairs of formulas in disjunctive normal form and provide a three-valued datatype for evaluating them.

```
type Conditional a = (DNF a, DNF a)

data ConditionalIndicatorValue = Verified | Falsified | NotApplicable
                                 deriving (Show, Eq)

evalConditional :: Atom a => Conditional a
                          -> Interpretation a
                          -> ConditionalIndicatorValue
evalConditional c w = case (evalF (fst c) w, evalF (snd c) w) of
                        (True,True)   -> Verified      -- AB
                        (False,True)  -> Falsified     -- A!B
                        (_,False)     -> NotApplicable -- !A
```

Note that the function `evalConditional` directly implements the *generalized indicator function* as given in (1).

This foundation of types is general enough to build many kinds of classical logical systems. In this paper, we only implement a propositional language by defining propositions as a type with a suitable `Atom` instance.

```
data Proposition = A | B | C | D | E | F | G | H | I | J | K | L | M
                 | N | O | P | Q | R | S | T | U | V | W | X | Y | Z
                 | Prop String
                 deriving (Show, Eq, Ord)

instance Atom Proposition where
  evalA p i = i p
  printA (Prop s) = s
  printA p = map toLower $ show p
```

By defining the type Proposition with data constructors A, ..., Z, we can easily define formulas using A, B, ..., Z as generic propositional variables. The generic formula $a \vee b$ for example, can be written as a list of positive literals over the two propositions A and B, i.e. [Pos A, Pos B]. The additional constructor Prop allows us to work with arbitrary strings as atomic propositions. For example *"this is an animal"* $\wedge$ *"this is a bird"* translates to [[Pos (Prop "this is an animal")],[Pos (Prop "this is a bird")]].

In the Atom instance of the type Proposition, an interpretation i is just a function of type (Proposition -> Bool).

We represent a knowledge base as a record type, containing all the necessary information.

```
class (Atom a) => KnowledgeBase k a | k -> a where
  name         :: k -> String
  signature    :: k -> [a]
  conditionals :: k -> [Conditional a]
  printKB      :: k -> String

data PropositionalKnowledgeBase =
    PKB { pKBname      :: String
        , pKBsignature :: [Proposition]
        , pKBconditionals :: [Conditional Proposition] }
```

Example 7. The knowledge base $\mathcal{R}_{birds}$ from Example 2 is represented as:

```
c1 = ([[Pos F]],[[Pos B]])
c2 = ([[Pos B]],[[Pos P]])
c3 = ([[Neg F]],[[Pos P]])
c4 = ([[Pos W]],[[Pos B]])

kb_birds = PKB { pKBname = "birds"
               , pKBsignature = [B,P,F,W]
               , pKBconditionals = [c1, c2, c3,c4] }
```

Based on an actual knowledge base with a fixed and finite signature, we can generate the set Ω of all possible worlds as a finite list of functions, making use of the bijection between complete conjunctions in Ω and functions of the type $\omega : \Sigma \to Bool$. These functions make use of the *closed world assumption* and assign False to every atom not in the signature. The function bigOmega generates this list of functions, by generating all possible combinations of True and False of length $|\Sigma|$ and constructing a closure for every combination using the function omega. This closure realizes a lookup, returning the Boolean value of the argument in this interpretation and False if the argument is not part of the signature.

```
omega :: (Eq a) => ([Proposition],[Bool]) -> Interpretation a
omega w = (\a -> case elemIndex a (fst w) of
                    Just x  -> (snd w) !! x
                    Nothing -> False) -- closed world assumption

bigOmega :: PropositionalKnowledgeBase -> [Interpretation Proposition]
bigOmega kb = map (\x -> omega (props,x)) $
                    combinations propcount [False, True]
  where props = signature kb
        propcount = length props

combinations :: (Num a, Ord a) => a -> [b] -> [[b]]
combinations n _ | n <= 0 = [[]]
combinations 1 xs = map (:[]) xs
combinations n xs = (:) <$> xs <*> combinations (n-1) xs
```

4.2 Computing the Inclusion-Maximal Ordered Partition

To implement the consistency check according to Proposition 2 we need to know whether a conditional is tolerated by a knowledge base. We implement this Boolean function using two nested list comprehensions which immediately follow from Definition 7.

```
tolerated :: Conditional Proposition -> PropositionalKnowledgeBase -> Bool
tolerated c kb =
    not $ null [w | w <- bigOmega kb, evalConditional c w == Verified,
                    null [c' | c' <- conditionals kb,
                               evalConditional c' w == Falsified] ]
```

We use the function bigOmega to generate the list of possible worlds as described above. If $c = (B|A)$, then the generated list is empty exactly when there is no $\omega \in \Omega$ for which $\omega \models AB$ and $\omega \not\models \overline{CD}$ for any $c' = (D|C) \in \mathcal{R}$.

Using the function tolerated we will implement a recursive Haskell function orderedPartition computing an inclusion-maximal ordered partition of a knowledge base analogously to the ASM function buildPartition presented in Figure 1. Since we need to account for inconsistent knowledge bases, for which no ordered partition exists, we use the Maybe type and return Nothing for inconsistent knowledge bases.

```
orderedPartition :: PropositionalKnowledgeBase
                   -> Maybe [[Conditional Proposition]]
orderedPartition kb = fmap reverse (pkb [] $ conditionals kb)
  where pkb parts [] = Just parts
        pkb parts l = let tcs = toleratedcs l
                       in if (tcs == [])
                          then Nothing        -- KB inconsistent
                          else pkb (tcs:parts) $ l \\ tcs
          toleratedcs l =
            [c | c <- l
               , tolerated c (defaultPKB { pKBsignature = signature kb
                                         , pKBconditionals = l })]
```

The locally defined function toleratedcs uses a list comprehension to construct the sublist only containing conditionals tolerated by the original list. It is necessary to use the function defaultPKB to construct a knowledge base matching the type of tolerated. The function pkb handles the bookkeeping such as the construction of the list of sublists and the actual recursion. It then returns the constructed ordered partition, or Nothing if there are no conditionals left that are tolerated by the knowledge base. The resulting list of sublists needs to be reversed so that the list starts with the set of conditionals being tolerated by the complete knowledge base.

4.3 Consistency Check and System P

Since the function orderedPartition realizes a consistency test for knowledge bases, we use it to implement p-entailment from Definition 6 by testing the consistency of the knowledge base after adding the negated query conditional.

```
p_entails :: PropositionalKnowledgeBase -> DNF Proposition
             -> DNF Proposition -> Bool
p_entails kb ant con = isNothing $ orderedPartition kb'
  where kb' =
    defaultPKB
        { pKBsignature = (signature kb),
          pKBconditionals = ((negateDNF con),ant) : (conditionals kb) }
```

Thus, for a knowledge base $\mathcal{R}$ and propositional formulas A and B (transformed in DNF) the function call p_entails R A B returns True iff the p-entailment $A \mathrel{\vert\!\sim}_p^{\mathcal{R}} B$ holds.

4.4 System Z

The computation of the unique minimal model of the knowledge base $\mathcal{R}$ is at the core of system Z. The higher order function calcZ uses the function orderedPartition to generate the function $Z : \mathcal{R} \to \mathbb{N}$ as defined in (6).

```
calcZ :: PropositionalKnowledgeBase -> (Conditional Proposition -> Int)
calcZ kb = case orderedPartition kb of
              Just parts -> (\c -> fromJust $ findIndex (elem c) parts)
              Nothing -> error "KB inconsistent"
```

The returned function uses findIndex to determine the index of the partition containing its argument. Using the function fromJust is save in this case, since the only function using the returned function is guaranteed to only pass it conditionals contained in the original knowledge base and therefore also contained in one of the sublists in the result of orderedPartition. The use of error in the case of an inconsistent knowledge base is justified by the use case of HaskPZ detailed in Section 4.5. HaskPZ can be used in an interactive session, where the error is simply printed as a message, or as a backend to another program, that expects an error code in the case of an inconsistent knowledge base.

The function kappa_z generates the ranking function $\kappa_{\mathcal{R}}^Z : \Omega \to \mathbb{N}$ according to (7).

```
kappa_z :: PropositionalKnowledgeBase
           -> (Interpretation Proposition -> Int)
kappa_z kb = (\w -> if null $ falsifiedCs w
                    then 0
                    else maximum [z c | c <- falsifiedCs w] + 1)
    where
```

```
         z = calcZ kb
         falsifiedCs w = [c | c <- conditionals kb
                            , evalConditional c w == Falsified]
```

The returned function models the formal definition of a ranking function closely, since ranking functions are defined as functions between interpretations and positive integers, cf. Definition 3. Using this function we can implement z-entailment $A \mathrel{|\!\sim}_Z^{\mathcal{R}} B$, i.e. checking whether A entails B in the context of the knowledge base $\mathcal{R}$ using the unique ranking function $\kappa_{\mathcal{R}}^Z$. We realize this relationship by implementing Definition 4.

```
z_entails :: PropositionalKnowledgeBase
             -> DNF Proposition -> DNF Proposition -> Bool
z_entails kb ant cons = min_kappa verifWorlds < min_kappa falsWorlds
    where worlds = bigOmega kb
          kappa = kappa_z kb
          verifWorlds = [w | w <- worlds
                            , evalF ant w
                            , evalF cons w]
          falsWorlds  = [w | w <- worlds
                            , evalF ant w
                            , evalF (negateDNF cons) w]
          min_kappa l = minimum $ map kappa l
```

We use list comprehensions to determine the worlds that verify or falsify the conjunction of the antecedence and the consequence. From those we select the minimal κ-value. If the minimal rank of the verifying worlds is smaller than the minimal rank of the falsifying worlds, the inference `ant` $\mathrel{|\!\sim}_{\kappa_{\mathcal{R}}^Z}$ `cons` holds.

In the implementation of `HaskPZ` we make heavy use of features like list comprehensions and higher order functions to stay close to the formal definitions. List comprehensions are used to construct lists of objects that have the properties required by the definitions. We model interpretations as functions from signatures to boolean values and ranking functions as functions from interpretations to non-negative integers. All of this helps to see the close connections between runnable code and formal definition, and it makes arguing about its correctness easy.

4.5 Using `HaskPZ`

There are two ways of using `HaskPZ`. It can be used for interactively working with knowledge bases and queries in a `ghci` session by importing the relevant modules, or as a backend that writes results to files in machine readable form. This section details the work flow in both cases.

HaskPZ in ghci The interactive *Read Eval Print Loop* `ghci` of the Glasgow Haskell Compiler[1] can be used to evaluate Haskell expressions. This allows us to interactively calculate ranking functions and answer queries. We start with a file named `birds.hs` containing the knowledge base $\mathcal{R}_{birds}$ from Example 7 together with the imports of modules containing the needed functionality.

---
[1] www.haskell.org

```
import Data.Logic.SystemP
import Data.Logic.SystemZ
import Text.PrettyPrint.TruthTable

c1 = ([[Pos F]],[[Pos B]])
c2 = ([[Pos B]],[[Pos P]])
c3 = ([[Neg F]],[[Pos P]])
c4 = ([[Pos W]],[[Pos B]])

kb = PKB { pKBname = "birds"
         , pKBsignature = [B,P,F,W]
         , pKBconditionals = [c1, c2, c3, c4] }
```

For convenient use of the functions p_entails and z_entails we can define operators already containing the knowledge base kb.

```
(|~p) = p_entails kb
(|~z) = z_entails kb
```

Installing HaskPZ with Haskells package manager cabal[2] ensures that the modules defining the functions for calculating ranking functions and answering queries are known to ghci after loading the file containing the knowledge base. We are now able to interactively work with the knowledge base. We can calculate the ranking function $\kappa_\mathcal{R}^Z$ using the function printTruthTable.

```
> printTruthTable kb
b p f w|( f  |  b )|( b  |  p )|( !f  |  p )|( w  |  b )|kappa_z
-------|---------|---------|----------|---------|-------
0 0 0 0|    u    |    u    |    u     |    u    |   0
0 0 0 1|    u    |    u    |    u     |    u    |   0
0 0 1 0|    u    |    u    |    u     |    u    |   0
0 0 1 1|    u    |    u    |    u     |    u    |   0
0 1 0 0|    u    |    -    |    +     |    u    |   2
0 1 0 1|    u    |    -    |    +     |    u    |   2
0 1 1 0|    u    |    -    |    -     |    u    |   2
0 1 1 1|    u    |    -    |    -     |    u    |   2
1 0 0 0|    -    |    u    |    u     |    -    |   1
1 0 0 1|    -    |    u    |    u     |    +    |   1
1 0 1 0|    +    |    u    |    u     |    -    |   1
1 0 1 1|    +    |    u    |    u     |    +    |   0
1 1 0 0|    -    |    +    |    +     |    -    |   1
1 1 0 1|    -    |    +    |    +     |    +    |   1
1 1 1 0|    +    |    +    |    -     |    -    |   2
1 1 1 1|    +    |    +    |    -     |    +    |   2
-------|---------|---------|----------|---------|-------
  Z(r) |    0    |    1    |    1     |    0    |
```

The result is produced using the boxes-library[3]. It lists every possible world in the first column, followed by a column showing the ConditionalIndicatorValue (+ = verified, - = falsified, u = not applicable) of every conditional in the knowledge base. The bottom line of the table above lists the value of the function Z for every conditional, and the last column shows the calculated ranking function $\kappa_\mathcal{R}^Z$.

Using the two operators for inference we can answer the queries from Example 6 in the context of the knowledge base using the different semantics.

```
> [[Pos P, Pos W]] |~p [[Neg F]]
```

---

[2] www.haskell.org/cabal
[3] hackage.haskell.org/package/boxes

```
False
> [[Pos P, Pos W]] |~z [[Neg F]]
True
```

HaskPZ as a backend Currently, `HaskPZ` is used as a backend in a conditional reasoning tool called `InfOCF`[4], that produces files like `birds.hs`. These files also contain a `main` function that either writes some result to a file that can be read by `InfOCF` or, in the case of inference, terminates with a return code to indicate the inference result.

If we add the line

```
main = exportOCF kb
```

to the end of the file `birds.hs` and run the program using `runhaskell`, which compiles the program and executes the action `exportOCF kb` we just added to `birds.hs`, the following output is written to the file `birds_systemz.ocf`.

```
p,b,f,w
0,0,0,0;0
0,0,0,1;0
0,0,1,0;0
0,0,1,1;0
0,1,0,0;1
0,1,0,1;1
0,1,1,0;1
0,1,1,1;0
1,0,0,0;2
1,0,0,1;2
1,0,1,0;2
1,0,1,1;2
1,1,0,0;1
1,1,0,1;1
1,1,1,0;2
1,1,1,1;2
```

This file is then read by `InfOCF` and interpreted as a ranking function that can be compared with other ranking functions produced by different backends, like a SICStus Prolog[4] component that calculates various sets of c-representations [24, 25, 2, 3] by solving a constraint satisfaction problem derived from a conditional knowledge base.

To get the return code indicating the result of a query, we use either `p_entails_rc` or `z_entails_rc` as our definition of main.

```
p_entails_rc :: PropositionalKnowledgeBase -> DNF Proposition
                -> DNF Proposition -> IO ()
p_entails_rc kb ant con = if p_entails kb ant con
                          then exitSuccess
                          else exitFailure

z_entails_rc :: PropositionalKnowledgeBase -> DNF Proposition
                -> DNF Proposition -> IO ()
z_entails_rc kb ant con = if z_entails kb ant con
                          then exitSuccess
                          else exitFailure
```

The return code is read by `InfOCF`, where it can be further processed depending on the calling function.

---

[4] www.sicstus.sics.se

5 Conclusions

While logic programming has been used extensively in knowledge representation and reasoning, little use of functional programming has been made in those areas. Likewise, the high-level approach of abstract state machines has only been used rarely for this task, in particular for nonmonotonic and uncertain reasoning. In this paper, we filled this gap by demonstrating that both abstract state machines and functional programming enable elegant, high-level implementations of non-monotonic reasoning. We choose two of the most popular approaches for nonmonotonic reasoning, system P and system Z, and described basic implementations for both, using abstract state machines on the one hand, and Haskell on the other. Since both reasoning systems are based on (non-Boolean) conditionals and the rich epistemic structure of ranking functions, non-classical challenges had to be tackled. Both abstract state machines and Haskell make use of high-level functional and declarative programming techniques to keep the executable code close to the formal definitions. These features make it easy to follow the code based on the formalisms and help convince the programmer and the user of the correctness of the implementation. For the declarative Haskell language, we presented full implementations of the non-monotonic inference formalisms system P and system Z.

The foundational type definitions in our Haskell implementation make it easy to formulate knowledge bases by hand and through automated code generation. This makes `HaskPZ` usable as an experimentation environment and as a backend to systems like `InfOCF` [4] that also provide other nonmonotonic inference relations, e.g. based on c-representations [24, 2, 3].

Our current work includes extending the foundational types to more general frameworks for representing further logics and additional inference relations in abstract state machines and Haskell.

References

1. E. W. Adams. *The Logic of Conditionals: An Application of Probability to Deductive Logic*. Synthese Library. Springer Science+Business Media, Dordrecht, NL, 1975.
2. C. Beierle, C. Eichhorn, and G. Kern-Isberner. Skeptical inference based on c-representations and its characterization as a constraint satisfaction problem. In M. Gyssens and G. R. Simari, editors, *Foundations of Information and Knowledge Systems - 9th International Symposium, FoIKS 2016, Linz, Austria, March 7–11, 2016. Proceedings*, volume 9616 of *LNCS*, pages 65–82. Springer, 2016.
3. C. Beierle, C. Eichhorn, G. Kern-Isberner, and S. Kutsch. Skeptical, weakly skeptical, and credulous inference based on preferred ranking functions. In G. A. Kaminka, M. Fox, P. Bouquet, E. Hüllermeier, V. Dignum, F. Dignum, and F. van Harmelen, editors, *Proceedings 22nd European Conference on Artificial Intelligence, ECAI-2016*, volume 285, pages 1149–1157. IOS Press, 2016.
4. C. Beierle, C. Eichhorn, and S. Kutsch. A practical comparison of qualitative inferences with preferred ranking models. *KI*, 31(1):41–52, 2017.

5. C. Beierle and G. Kern-Isberner. Modelling conditional knowledge discovery and belief revision by Abstract State Machines. In E. Börger, A. Gargantini, and E. Riccobene, editors, *Abstract State Machines 2003 – Advances in Theory and Applications (ASM'2003)*, volume 2589 of *LNCS*, pages 186–203. Springer, 2003.
6. C. Beierle and G. Kern-Isberner. An ASM refinement and implementation of the Condor system using ordinal conditional functions. In A. Prinz, editor, *Proceedings 14th International Workshop on Abstract State Machines (ASM'2007)*. Agder University College, Grimstad, Norway, 2007.
7. C. Beierle and G. Kern-Isberner. A verified AsmL implementation of belief revision. In E. Börger, M. Butler, J. P. Bowen, and P. Boca, editors, *Abstract State Machines, B and Z, First International Conference, ABZ 2008, London, UK, September 16-18, 2008. Proceedings*, volume 5238 of *LNCS*, pages 98–111. Springer, 2008.
8. E. Börger and K. Schewe. Specifying transaction control to serialize concurrent program executions. In Y. A. Ameur and K. Schewe, editors, *Abstract State Machines, Alloy, B, TLA, VDM, and Z - 4th International Conference, ABZ 2014, Toulouse, France, June 2-6, 2014. Proceedings*, volume 8477 of *LNCS*, pages 142–157. Springer, 2014.
9. E. Börger and K. Schewe. Concurrent abstract state machines. *Acta Inf.*, 53(5):469–492, 2016.
10. E. Börger and R. Stärk. *Abstract State Machines: A Method for High-Level System Design and Analysis*. Springer-Verlag, 2003.
11. B. DeFinetti. *Theory of Probability*, volume 1,2. John Wiley & Sons, 1974.
12. D. Dubois and H. Prade. Conditional Objects as Nonmonotonic Consequence Relations. In *Principles of Knowledge Representation and Reasoning: Proceedings of the Fourth International Conference (KR'94)*, pages 170–177, San Francisco, CA, USA, 1996. Morgan Kaufmann Publishers.
13. D. Dubois and H. Prade. Possibility theory and its applications: Where do we stand? In J. Kacprzyk and W. Pedrycz, editors, *Springer Handbook of Computational Intelligence*, pages 31–60. Springer, Berlin, 2015.
14. F. Ferrarotti, K. Schewe, L. Tec, and Q. Wang. A logic for non-deterministic parallel abstract state machines. In M. Gyssens and G. R. Simari, editors, *Foundations of Information and Knowledge Systems - 9th International Symposium, FoIKS 2016, Linz, Austria, March 7-11, 2016. Proceedings*, volume 9616 of *LNCS*, pages 334–354. Springer, 2016.
15. F. Ferrarotti, K. Schewe, L. Tec, and Q. Wang. A new thesis concerning synchronised parallel computing - simplified parallel ASM thesis. *Theor. Comput. Sci.*, 649:25–53, 2016.
16. M. Gebser, B. Kaufmann, and T. Schaub. Conflict-driven answer set solving: From theory to practice. *Artif. Intell.*, 187:52–89, 2012.
17. M. Gelfond. Answer sets. In F. van Harmelen, V. Lifschitz, and B. W. Porter, editors, *Handbook of Knowledge Representation*, volume 3 of *Foundations of Artificial Intelligence*, pages 285–316. Elsevier, 2008.
18. M. Goldszmidt and J. Pearl. Qualitative probabilities for default reasoning, belief revision, and causal modeling. *Artificial Intelligence*, 84:57–112, 1996.
19. Y. Gurevich. Sequential abstract-state machines capture sequential algorithms. *ACM Trans. Comput. Log.*, 1(1):77–111, 2000.
20. Y. Gurevich, B. Rossman, and W. Schulte. Semantic essence of AsmL. *Theoretical Computer Science*, 343(3):370–412, 2005.

21. J. Halpern. *Reasoning About Uncertainty*. MIT Press, 2005.
22. J. Hawthorne and D. Makinson. The quantitative/qualitative watershed for rules of uncertain inference. *Studia Logica*, 86(2):247–297, 2007.
23. M.-B. Isberner and G. Kern-Isberner. A formal model of plausibility monitoring in language comprehension. In *Proceedings of the 29th International FLAIRS Conference (FLAIRS'16)*, pages 662 – 667, 2016.
24. G. Kern-Isberner. *Conditionals in nonmonotonic reasoning and belief revision*, volume 2087 of *LNAI*. Springer, 2001.
25. G. Kern-Isberner. A thorough axiomatization of a principle of conditional preservation in belief revision. *Ann. of Math. and Artif. Intell.*, 40(1-2):127–164, 2004.
26. S. Kraus, D. Lehmann, and M. Magidor. Nonmonotonic reasoning, preferential models and cumulative logics. *Artificial Intelligence*, 44:167–207, 1990.
27. N. Leone, G. Pfeifer, W. Faber, F. Calimeri, T. Dell'Armi, T. Eiter, G. Gottlob, G. Ianni, G. Ielpa, C. Koch, S. Perri, and A. Polleres. The DLV system. In S. Flesca, S. Greco, N. Leone, and G. Ianni, editors, *Logics in Artificial Intelligence, European Conference, JELIA 2002, Cosenza, Italy, September, 23-26, Proceedings*, volume 2424, pages 537–540. Springer, 2002.
28. T. Leopold, G. Kern-Isberner, and G. Peters. Belief revision with reinforcement learning for interactive object recognition. In *Proceedings 18th European Conference on Artificial Intelligence, ECAI'08*, 2008.
29. D. Makinson. General theory of cumulative inference. In M. Reinfrank et al., editors, *Non-monotonic Reasoning*, pages 1–18. Springer Lecture Notes on Artificial Intelligence 346, Berlin, 1989.
30. S. Marlow (editor). Haskell 2010 Language Report. www.haskell.org/onlinereport/haskell2010/, 2010. Accessed: 2017-06-23.
31. M. Minock and H. Kraus. Z-log: Applying system-z. In *Proceedings of the 8th European Conference on Logics in Artificial Intelligence (JELIA'02)*, pages 545 – 548, 2002.
32. J. Pearl. *Probabilistic Reasoning in Intelligent Systems*. Morgan Kaufmann Publishers Inc., San Francisco, CA, USA, 1988.
33. J. Pearl. System Z: A natural ordering of defaults with tractable applications to nonmonotonic reasoning. In R. Parikh, editor, *Proceedings of the 3rd conference on Theoretical aspects of reasoning about knowledge (TARK1990)*, pages 121–135, San Francisco, CA, USA, 1990. Morgan Kaufmann Publishers Inc.
34. K. Schewe and Q. Wang. A simplified parallel ASM thesis. In J. Derrick, J. S. Fitzgerald, S. Gnesi, S. Khurshid, M. Leuschel, S. Reeves, and E. Riccobene, editors, *Abstract State Machines, Alloy, B, VDM, and Z - Third International Conference, ABZ 2012, Pisa, Italy, June 18-21, 2012. Proceedings*, volume 7316 of *LNCS*, pages 341–344. Springer, 2012.
35. W. Spohn. Ordinal conditional functions: a dynamic theory of epistemic states. In W. Harper and B. Skyrms, editors, *Causation in Decision, Belief Change, and Statistics, II*, pages 105–134. Kluwer Academic Publishers, 1988.
36. W. Spohn. *The Laws of Belief: Ranking Theory and Its Philosophical Applications*. Oxford University Press, Oxford, UK, 2012.

Part IV
Miscellaneous

Recent Developments in Armstrong codes

Attila Sali*

Alfréd Rényi Institute of Mathematics
Hungarian Academy of Sciences
Budapest, Hungary
`sali.attila@renyi.mta.hu`

Dedicated to the 60th birthday of Klaus-Dieter Schewe

Abstract. Armstrong codes were introduced as Armstrong instances of certain collections of functional dependencies with restriction on domain sizes. They are interesting in coding theory point of view, as well. The main restriction is that a q-ary code $\mathcal{C}$ is Armstrong, if the minimum distance is taken in "every possible direction", that is if the minimum distance is d, then for every possible d-subset X of coordinate positions there exists two codewords $\mathbf{x}, \mathbf{y} \in \mathcal{C}$ such that $\mathbf{x}$ and $\mathbf{y}$ differ in exactly the positions of X. This puts a strong restriction on the length of the code. The maximum length of a q-ary Armstrong code of given minimum distance is investigated. Extremal constructions use design-type techniques, as well as polynomials over finite fields. Upper bounds apply advanced coding-theory methods.

1 Introduction

Interaction between combinatorics and database theory may be dated back to the beginning of 1980's. As an early trace of this interaction, one might consider that in query optimization acyclic joins were used that are related to notions of acyclicity in hypergraphs; see [5, 22, 9, 3, 26]. Although some of the references cited above are not of combinatorics flavour, hypergraphs are nevertheless one of the main topics in combinatorics.

Other combinatorial studies involved problems of existence and minimal size of Armstrong instances of certain dependency systems. These included methods of extremal combinatorics, graph theory, polynomials over finite fields, finite geometries and design theory.

Probabilistic combinatorics was applied in studying keys of random databases [14, 15].

Recently, extremal combinatorial problems have surfaced again in connection with key systems in incomplete databases, or in the presence of NULL values [28, 31, 32], namely in studying certain and possible keys.

---

* Research was partially supported by Hungarian National Research, Development and Innovation Office – NKFIH, K116769

The present paper is dedicated to te 60th birthday of Klaus-Dieter Schewe and aims to show how database theory can motivate research in the theory of error correcting codes. In fact, this direction originates in papers of Klaus-Dieter [29, 36, 37].

In what follows we use the terminology of the book [1]. We assume familiarity with basic concepts, such as relational schema, attributes, relation, functional dependency. Relational databases may satisfy several kinds of dependencies, most fundamental of them are functional dependencies, in particular key dependencies. If $\mathbb{R}$ denotes the set of attributes, then $K \subseteq \mathbb{R}$ is a *key*, if k is a minimal set of attributes such that the functional dependency $K \to \mathbb{R}$ holds, that is values taken by a record or tuple in attributes of K determine all other values taken by the tuple.

The following problem is a classical problem in schema design. Given a collection Σ of functional dependencies, what other dependencies hold in a database instance that satisfies Σ? A way of solving this problem is the construction of an *Armstrong instance* for Σ, that is a database that satisfies a functional dependency $X \to Y$ if and only if $\Sigma \models X \to Y$. Here $\models$ denotes logical implication. The existence of Armstrong instance for a set of functional dependencies was proved by Armstrong [2] and Demetrovics [11].

Further investigations [12, 13, 16, 17, 39, 19, 20, 18, 23] concentrated on the minimum size of an Armstrong instance, since it is a good measure of the complexity of the collection of dependencies or system of minimal keys in question. A useful Armstrong relation for a constraint set should be of minimal size. Indeed, the smaller the relation, the easier it is to understand for humans, redundant tuples do not add any new information about the constraint set.

Basic assumption in relational database theory that the *domain* of each attribute is unbounded, countably infinite set. However, in the study of *Higher Order Data model* [29, 34, 36, 37] the question of bounded domains arises naturally. In fact, if a minimal key system of *nested attributes* contains only *counter attributes*, then the possible number of tuples in an Armstrong instance is bounded from above. Nested attributes are defined by succesively applying several kinds of constructors, starting from simple attributes. Counter attributes are special subattributes where the constructors are applied to the null attribute.

Another reason to consider bounded domains comes from real life databases. In many cases the domain of an attribute is a well defined finite set, for example in car rental, the class of cars can take values from the set {subcompact, compact, mid-size, full-size, SUV, sports-car, van}. Same kind of finiteness may occur in the case of job assignments, schedules, etc.

Thalheim [40] investigated the maximum number of keys in the case of bounded domains and showed that having restrictions on the sizes of domains makes a significant difference.

It is natural to ask what can be said about Armstrong instances if attribute A_i has a domain of size q. An easy fact is that keys form a *Sperner system*, that is if K, K' are distinct keys (with respect to some collection Σ of functional dependencies), then $K \not\subseteq K'$. The main question of this section was introduced in [37] and investigated in papers [30, 38, 7, 8]. Let $\mathcal{K}_n^k$ denote the collection of all k-subsets of an n-element attribute set **R**. It was proven in [17] that there exists a unique collection of functional dependencies on **R** such that the collection of keys is $\mathcal{K}_n^k$.

Definition 1.1. *Let $q > 1$ and $k > 1$ be given natural numbers. Let $f(q, k)$ be the maximum such n that there exists an Armstrong instance for $\mathcal{K}_n^k$ being the system of minimal keys such that each attribute's domain is of size at most q. Define $f(q, k)$ to be the maximum n such that $\mathcal{K}_n^k$ has an Armstrong instance, under the additional condition that the cardinality of the domain of each attribute is at most q.*

It is clear that for a meaningful Armstrong instance we need at least two distinct symbols, so $q > 1$ is necessary. On the other hand the minimal Armstrong instance for $\mathcal{K}_n^1$ uses only two symbols for arbitrary n [16], hence $f(q, k)$ could be well defined only for $k > 1$. It is an easy exercise for the reader to prove that for $q, k > 1$ $f(q, k)$ is indeed bounded.

Definition 1.2. *Let $\mathcal{K}$ be a Sperner system of keys.*

$$\mathcal{K}^{-1} = \{A \subsetneq \mathbf{R} : \not\exists K \in \mathcal{K} \text{ such that } K \subseteq A \\ \text{and } A \text{ is maximal subject to this condition}\} \quad (1)$$

is the collection of antikeys corresponding to $\mathcal{K}$.

The following basic fact is known [16].

Proposition 1.1. **A** *is an Armstrong instance for $\mathcal{K}$ iff the following two properties hold:*

(K) *There are no two rows of **A** that agree in all positions for any $K \in \mathcal{K}$ and*

(A) *For every $A \in \mathcal{K}^{-1}$ there exist two rows of **A** that agree in all positions of A.*

It is helpful to view an Armstrong instance for $\mathcal{K}_n^k$ as key system using at most q symbols as a q-ary code $\mathcal{C}$ of length n, where codewords are the tuples, or rows of the instance. Using $(\mathcal{K}_n^k)^{-1} = \mathcal{K}_n^{k-1}$ we obtain

(md) $\mathcal{C}$ has minimum Hamming distance at least $n - k + 1$ by **(K)**, that is any two codewords differ in at least $n - k + 1$ positions.

(di) For any set of $k - 1$ coordinates there exist two codewords that agree exactly there by **(A)**.

A $k - 1$-set of coordinates could be considered as a 'direction', so in $\mathcal{C}$ the minimum distance is *attained in all directions*. Such a code $\mathcal{C}$ is called an *Armstrong(q, k, n) code*, denoted by $\mathrm{Arm}(q, k, n)$. For example, the rows of the $(k + 1) \times (k + 1)$ identity matrix form an Armstrong$(2, k, k + 1)$ code, so Armstrong codes **do** exist.

Remark 1.1. Let $q > 1$ and $k > 1$ be given natural numbers. Then $f(q,k)$ is the maximum such n that an Armstrong(q,k,n) code exists.

We survey recent results about $f(q,k)$. The paper is organized as follows. Section 2 contains general upper bounds. Section 3 is dedicated to general lower bounds. The special cases $k = 3, 4$ are treated in Section 4 and some new results are presented. Finally, Section 5 explores some generalizations and lists open problems.

2 Upper bounds

There are two main upper bound theorems. One is for all pairs q, k, the other is for large k.

Theorem 2.1 (G.O.H. Katona, K.-D. Schewe, Sali [30]).

$$f(q,k) \leq q(k-1)\left(1 + \frac{q-1}{\sqrt{\frac{2(qk-q-k+2)^{k-1}}{(k-1)!}} - q}\right) \qquad (2)$$

holds.
If $5 \leq k$ and $2 \leq q$ then the upper bound in (2) can be improved to

$$f(q,k) \leq q(k-1) \qquad (3)$$

with the following exceptions: $(k,q) = (5,2), (5,3), (5,4), (5,5), (6,2)$.

In order to prove the upper bounds two estimates on n were given that are functions of the number of codewords: $a_{q,k}(m)$ being a decreasing, while $b_{q,k}(m)$ being an increasing function of m. $a_{q,k}(m)$ uses actually a version of Turán's famous theorem on the maximum number of edges of an m-vertex graph without a $q + 1$-vertex complete subgraph, we need the minimum number of edges in the complement of such a graph.

Lemma 2.1. *The number of equal pairs in a sequence of length m that contains elements of $\{1, 2, \ldots, q\}$ is at least*

$$\frac{m}{2}\left(\frac{m}{q} - 1\right) \qquad (4)$$

Let m denote the length of an $Arm(q,k,n)$ code. It can be considered as an $m \times n$ matrix. The number of pairs of equal entries in each column is at least (4). Thus number of pairs of equal entries in the $Arm(q,k,n)$ is:

$$n\frac{m}{2}\left(\frac{m}{q} - 1\right) \qquad (5)$$

If one pair of rows had k equal entries it would contradict the assumption about the minimum distance. Thus:

$$n\frac{m}{2}\left(\frac{m}{q}-1\right) \leq (k-1)\binom{m}{2} \tag{6}$$

This inequality can be written in the following form:

$$n \leq \frac{(k-1)\binom{m}{2}}{\frac{m}{2}\left(\frac{m}{q}-1\right)} = q(k-1)\left(1+\frac{q-1}{m-q}\right) \tag{7}$$

Let $a_{q,k}(m)$ denote the right-hand side of the inequality, thus:

$$a_{q,k}(m) = q(k-1)\left(1+\frac{q-1}{m-q}\right) \tag{8}$$

On the other hand, $b_{q,k}(m)$ is derived from

$$\binom{n}{k-1} \leq \binom{m}{2}. \tag{9}$$

that holds for any Armstrong(q, k, n)-code of size m, since for any two $k-1$-subsets of coordinates there must exist a pair of codewords that agree there by (di), and these pairs of codewords must be different for different $k-1$-subsets. Therefore, if α is the solution of the equation

$$a_{q,k}(m) = b_{q,k}(m) \tag{10}$$

in m then $a_{q,k}(\alpha) = b_{q,k}(\alpha)$ is a universal (independent of m) upper bound for n. The paper [30] also contains an exact result and and upper bound that has recently [8] been proved sharp.

Proposition 2.1. $f(q,2) = \binom{q+1}{2}$ and $f(q,3) \leq 3q-1$.

The bound in (3) looks nice and one may be tempted to guess that it is sharp, or close to being sharp. However, for k large enough with respect to q, one can give better upper bound. An Arm(q, k, n) can be embedded into the $(q-1)n$ dimensional space and then can be considered as a spherical code. That way existing bounds for spherical codes can be applied for Arm(q, k, n)'s.

Theorem 2.2 (Székely, Sali [38]). For $k > k_0(q)$ we have

$$f(q,k) < (q-\log q)k. \tag{11}$$

Let $l = k-1$ and m be the size of an Arm(q, k, n) code $\mathcal{C}$. This can be embedded into the $(q-1)n$ dimensional space by defining a bijective mapping $s : \{0, 1, 2, \ldots, q-1\} \to \mathbb{R}^{q-1}$ of the symbols to the vertices of a regular simplex centered at the origin, and then obtaining a new code from Arm(q, k, n) by juxtaposition of coordinates of vectors that are images of symbols of codewords under s. Let $\mathcal{D}$ be the spherical code obtained by normalizing these vectors from $\mathbb{R}^{(q-1)n}$ so they are unit vectors. Since $\mathcal{D}$ is obtained from an $\mathrm{Arm}(q,k,n)$ which by definition has minimum distance $n - l$, $\mathcal{D}$ has minimum angle ϕ with

$$\cos \phi = \frac{lq - n}{(q-1)n} \qquad \sin \frac{\phi}{2} = \sqrt{\frac{q(n-l)}{2(q-1)n}} \qquad (12)$$

(12),(9) and the upper bound of Rankin [33] give:

$$\sqrt{2\binom{n}{l}} < m \leq \sqrt{\frac{\pi}{2}(q-1)^3 n^3 \frac{lq-n}{(q-1)n}} \left(\sqrt{\frac{q(n-k+1)}{(q-1)n}}\right)^{-(q-1)n} (1 + o(1)) \qquad (13)$$

After writing $l = cn$ and approximating $\binom{n}{cn}$ we get:

$$\sqrt{2}\left(\frac{1}{c^c(1-c)^{1-c}}\right)^{\frac{n}{2}} < \sqrt{\frac{\pi}{2}(q-1)n\sqrt{(cq-1)n}} \left(\sqrt{\frac{q-1}{q(1-c)}}\right)^{(q-1)n} \qquad (14)$$

(14) holds for large n only if:

$$\frac{1}{c^c(1-c)^{1-c}} < \left(\frac{q-1}{q(1-c)}\right)^{q-1} \qquad (15)$$

If an $\mathrm{Arm}(q,k,n)$ exists then a $\mathrm{Arm}(q,k,n')$ also exists if $k < n' < n$. If (15) does not hold for some c then $n = \frac{k-1}{c}$ is an upper bound for $f(q,k)$ if q is sufficiently large. It can be shown that (15) does not hold if $c = \frac{1}{q - \log q}$.

In [30], Theorem 3.3, it was shown that if an $\mathrm{Arm}(2,k,n)$ exists, and $k \geq 7$, then $n \leq 2(k-1)$. Later, Blokhuis, Brouwer and Sali [7] improved this to

Theorem 2.3. *If an Armstrong code $\mathrm{Arm}(2,k,n)$ exists, then we have asymptotically $n \leq 1.224k$.*

The proof is based on the following observation. Let $d = n - k + 1$.

Proposition 2.2. *Let $A(n,d)$ and $A(n,d,w)$ denote the maximum size of a binary code of word length n, minimum distance d (and constant weight w). Suppose an $\mathrm{Arm}(2,k,n)$ exists. Then $2\binom{n}{d} \leq A(n,d)A(n,d,d)$.*

Proof. If $\mathcal{C}$ is an $\mathrm{Arm}(2,k,n)$ and we look at all spheres of radius d around code words, then we see each difference at least twice. Here a sphere of radius d around a codeword is the set of codewords differing from it in exactly d positions.

Let $H_2(x) = -x\log_2(x) - (1-x)x\log_2(1-x)$ be the binary entropy function, then we have $\frac{1}{n}\log_2\binom{n}{\alpha n} \approx H_2(\alpha)$. Let $d = \delta n$. Let $\kappa_0 = \kappa_0(\delta)$ be such that a code of length n with constant weight d and minimum distance d has size at most $2^{\kappa_0 n}$. Let $\kappa_1 = \kappa_1(\delta)$ be such that an arbitrary code with length n and minimum distance d has size at most $2^{\kappa_1 n}$.

Proposition 2.2 says that if an $\mathrm{Arm}(2,k,n)$ exists, then $2\binom{n}{d} \leq 2^{(\kappa_0 + \kappa_1)n}$. Hence $H_2(\delta) \leq \kappa_0(\delta) + \kappa_1(\delta)$. Various known bounds on $\kappa_0(\delta)$ and $\kappa_1(\delta)$ now give upper bounds for n/k for Armstrong codes that result in Theorem 2.3.

3 Lower bounds

First lower bounds were given in [30] using greedy construction. That is pick a pair of codewords for a given $k-1$ subset of positions such that they agree exactly at that $k-1$ positions. Then rule out the balls of radii $n-k$ around the two codewords. Note that a ball of radius $n-k$ around a codeword is the set of codewords that differ from it in at most $n-k$ positions. If enough codewords are left, then we can pick a pair for the next $k-1$ subset of positions, etc. In order to complete the plan above we need the following lemma.

Lemma 3.1. *Let Q be the set of q^n q-ary codewords of length n, furthermore let K be a $k-1$ subset of coordinate positions. Assume that $n \geq k$. Then Q can be partitioned into q^{n-1} classes of size q each, that any two codewords of the same class agree exactly on the positions of K.*

Using Lemma 3.1 the following was proven.

Theorem 3.1 (G.O.H. Katona, K.-D. Schewe, Sali [30]).

1. *Given $q > 4$, there is k_0 such that for every $k > k_0$ and for every $n < \frac{1}{2} k \log q$ we have $n \leq f(q, k)$.*
2. *There exists k_0 and $c > 1$ constants, that for $k > k_0$, and $\lfloor ck \rfloor \leq f(2, k)$.*

The significance of the second statement of Theorem 3.1 is that the constant c is strictly greater than 1. Previously only the rows of $n \times n$ identity were known binary Armstrong codes. In [7] this constant was calculated

Proposition 3.1. *An $Arm(2, k, n)$ exists when $n \leq 1.12k$.*

For general n and k these are the only constructive lower bounds. However, using Lovász' Local Lemma [21], we can give a better lower bound.

Theorem 3.2 (Székely, Sali [38]). *For $k > k_0(q)$ we have*

$$\frac{\sqrt{q}}{e} k < f(q, k) \tag{16}$$

There is still a large gap between the coefficient of k in (16) and (11).

Questions can be asked in other settings, as well. One may fix k and let q go to infinity, or we may think about both k and q tend to infinity. This last problem has not been investigated yet. The first one is the subject of the next section.

4 Small k

The $k = 2$ case was completely solved in [30]] see Proposition 2.1. However, $k = 3$ led to interesting connection with design theoretic methods.

4.1 $k = 3$

Let us assume that an $\mathrm{Arm}(q,3,n)$ exists. If we think about it as an $n \times m$ matrix whose rows are the codewords, then we have that for every pair of columns there are two rows that agree there and that any two rows agree at most at two columns. Each column contains at most q distinct symbols, so we may view columns as *partitions* of the set of rows in at most q parts. A partition covers a pair (of rows) r, s if r and s are in the same partite set. Thus $\mathrm{Arm}(q, 3, n)$ exists iff n partitions of $\{1, 2, \ldots, m\}$ exist such that

i) Any pair of $\{1, 2, \ldots, m\}$ is covered by at most two partitions,
ii) For any two partitions there exists at least one pair that they both cover.

A similar concept [25] to this was introduced in connection with finding minimum sized Armstrong instances of uniform key systems. It was started by a conjecture of Demetrovics, Füredi and Katona [12].

Definition 4.1. *A collection of partitions of $\{1, 2, \ldots, m\}$ is called orthogonal double cover (of K_m) if it satisfies the following two properties:*

1. *for any two partitions there is exactly one pair of elements, which is covered by both,*
2. *each pair of elements is covered by exactly two different partitions.*

It is easy to see that if an orthogonal double cover exists than the number of partitions (n) is equal to m. The following conjecture was formulated in [12]. (It was posed in other terms, since the notion of orthogonal double cover was introduced later, in [25].)

Conjecture 4.1 (Demetrovics, Katona, Füredi [12]). There exists an orthogonal double cover of the n-element set by n partitions provided $n \geq 7$.

Another conjecture was stated in the same paper. Its solution has a surprising connection to Armstrong codes.

Theorem 4.1 (Ganter and Gronau [24]). *If $n = 3r + 1$, then there exists an orthogonal double cover of the n-element set by n partitions that have one 1-element class and r of the 3-element classes.*

Bennett and Wu [6] verified that Conjecture 4.1 is true.

Theorem 4.1 establishes a lower bound for $f(q, 3)$. The Armstrong instance provided there has $r + 1$ symbols in every column, and has $3r + 1$ columns. That is, $q = r + 1$ and $n = 3q - 2$, thus $f(q, 3) \geq 3q - 2$. It is easy to see that $f(2, 3) = 4$. This lead Sali [35] to conjecture that $f(q, 3) = 3q - 2$. However, Y.M. Chee, H. Zhang and X. Zhang [8] noted that when $m \geq 2$, an orthogonal double cover of K_{6m+2} by $6m + 2$ copies of $2mK_3 \cup K_2$ was constructed by Gronau et al. [27]. This gives an $\mathrm{Arm}(2m + 1, 3, 6m + 2)$ Armstrong code and hence $f(2m + 1, 3) \geq 6m + 2$. Thus, $f(q, 3) = 3q - 1$ for all odd $q \geq 5$. The main result of paper [8] is that $f(q, 3) = 3q - 1$ holds for even $q \geq 6$, if

$q \neq 14, 16, 20$. Their method is that instead of orthogonal double covers they construct extODC's, that is sets of partitions satisfying properties i) and ii) above. Their construction applies *resolvable triple systems* and *group divisible designs*.

4.2 $k = 4$

In this case we do not have exact bounds. In fact, no construction has been known before other than the trivial case of the identity matrix. In this section we give a linear lower bound based on polynomials over finite fields. Also a slightly better upper bound, than the specialization of (2) to $k = 4$ is given. Furthermore we review several methods that are available for estimating the size of an Armstrong code.

Lower bound. An idea of Füredi [12] is modified for the lower bound construction. The original problem was to minimize the number of codewords. Now we want to minimize the number of symbols used in the codewords. Let p be a prime number and let $\mathcal{F}_r^p$ denote the collection of polynomials over the finite field $\mathbb{Z}_p$ that are of degree r. Let $\mathcal{H}_4^p$ denote the set of polynomials that are the squares of polynomials of $\mathcal{F}_2^p$ that have main coefficient equal to 1.

$$\mathcal{H}_4^p = \{(x^2 + ax + b)^2 : a, b \in \{0, 1, \ldots, p-1\}\}$$

Thus $|\mathcal{H}_4^p| = p^2$. A code $\mathcal{C}$ is constructed as follows. $\mathcal{C} = \{\mathbf{c}^1, \mathbf{c}^2 \ldots, \mathbf{c}^{p^2}\}$ so that if $\mathcal{H}_4^p = \{h_1, h_2, \ldots, h_{p^2}\}$ then $\mathbf{c}_j^i = h_i(j) \pmod{p}$.

Proposition 4.1. *Code $\mathcal{C}$ defined above is an $\text{Arm}(\frac{p+1}{2}, 4, p)$ code.*

Proof. By definition the code length is p. Since symbols are values of squares of polynomials, they are quadratic residues in $\mathbb{Z}_p$, so there are $\frac{p+1}{2}$ of them. If two codewords $\mathbf{c}^{i_1}$ and $\mathbf{c}^{i_2}$ agree on coordinate positions $j_1, j_2, \ldots j_b$, then the polynomial $f(x) = h_{i_1}(x) - h_{i_2}(x)$ has distinct roots $j_1, j_2, \ldots j_b$. However, $f(x)$ is of degree at most 3, since $h_{i_1}(x), h_{i_2}(x) \in \mathcal{F}_4^p$ and have leading coefficient 1. So $b \leq 3$, that is the minimum distance of code $\mathcal{C}$ is at least $n - 3$. In order to finish the proof we need to show that for any three distinct coordinate positions j_1, j_2, j_3 there exist two codewords that agree exactly there.

Claim. For every $f(x) \in \mathcal{F}_3^p$ such that $f(x)$ has three distinct roots and leading coefficient 1, there exists $g(x), h(x) \in \mathcal{H}_4^p$ satisfying $4f(x) = g(x) - h(x)$.

Indeed, let $f(x) = (x - t_1)(x - t_2)(x - t_3)$, we need $g = (x^2 + ax + b)^2$ and $h = (x^2 + cx + d)^2$ such that

$$(x^2 + ax + b)^2 - (x^2 + cx + d)^2 = 4(x - t_1)(x - t_2)(x - t_3).$$

This is equivalent with

$$\begin{aligned}((x^2 + ax + b) + (x^2 + cx + d))&\left((x^2 + ax + b) - (x^2 + cx + d)\right) \\ &= 4(x - t_1)(x - t_2)(x - t_3),\end{aligned} \tag{17}$$

that is
$$(2x^2 + (a+c)x + b + d)((a-c)x + b - d) = 4(x - t_1)(x - t_2)(x - t_3) \quad (18)$$

The left-hand side of (18) is a product of a polynomial of degree 2 and a polynomial of degree 1. The right-hand side of (18) can also be viewed this way and we may put:
$$(2x^2 + (a+c)x + b + d) = 2(x - t_1)(x - t_2) = \sqrt{g(x)} + \sqrt{h(x)}$$
and
$$((a-c)x + b - d) = 2(x - t_3) = \sqrt{g(x)} - \sqrt{h(x)}$$
by solving these for $\sqrt{g(x)}$ and $\sqrt{h(x)}$ we get:
$$\sqrt{g(x)} = x - t_3 + (x - t_1)(x - t_2) \quad (19)$$
and
$$\sqrt{h(x)} = -x + t_3 + (x - t_1)(x - t_2). \quad (20)$$

To prove that $\mathcal{C}$ satisfies Armstrong property (**di**) let $\{t_1, t_2, t_3\}$ be a 3-subset of coordinate positions. Apply Claim 4.2 for $f(x) = (x - t_1)(x - t_2)(x - t_3)$ to find $g(x)$ and $h(x)$ in $\mathcal{H}_4^p$ such that $4f(x) = g(x) - h(x)$. Now, the codewords defined by $g(x)$ and $h(x)$ agree exactly at the coordinate positions $\{t_1, t_2, t_3\}$.

Proposition 4.1 provides Armstrong codes showing that if $q = p$ is an odd prime then $f(q, 4) \geq 2q - 1$. Applying Chebishev's Theorem we obtain lower bound for any large enough q.

Theorem 4.2 (Chebyshev). *For every positive integer q there exists a p prime such that $q \leq p \leq 2q$.*

A generalized form of Chebyshev's theorem can be used to find primes in a smaller range around a number to find primes for large enough numbers.

Theorem 4.3. *For every real $\epsilon > 0$ and sufficiently large q there is p a prime such that $q \leq p \leq (1 + \epsilon)q$.*

The value of $f(q, 4)$ is monotonically increasing in q since an Arm(q, k, n) is also an Arm$(q + t, k, n)$. Thus $f(q, 4) \geq 2p - 1$ where p is the greatest prime number that is less or equal to q. So,
$$\forall \epsilon > 0 \;\; \exists q_0 \;\; \text{if} \;\; q > q_0 \;\; \text{then} \;\; f(q, 4) \geq \frac{2q - 1}{1 + \epsilon}. \quad (21)$$

Upper bound. Inequality (2) specializes to
$$f(q, 4) \leq 3q \left(1 + \frac{q - 1}{\sqrt{\frac{2(3q-2)^3}{6}} - q} \right) \quad (22)$$

The right hand side of (22) is larger than $3q + \sqrt{q}$. We can improve this bound slightly. The basic inequality (9) can be written in the form for $k = 4$

$$\sqrt{2\binom{n}{3}} \leq m. \tag{23}$$

We use Delsarte's Linear Programming Bound [10]. In context of association schemes we follow the book of E. Bannai and T. Ito [4].

Definition 4.2. *A symmetric association scheme $A = \{X, R\}$ is a finite set X and $R = \{R_0, R_1, \ldots, R_d\} \subseteq \mathcal{P}(X \times X)$ such that:*

1. $R_0 = \{(x, x) : x \in X\}$
2. *If $(x, y) \in R_i$, then $(y, x) \in R_i$ for all $i \leq d$*
3. *R partitions $X \times X$*
4. *For every $h, i, j \leq d$ the relations R_h, R_i, R_j satisfy that for each $(x, y) \in R_h$, the number of elements $z \in X$ such that $(x, z) \in R_i$ and $(z, y) \in R_j$ is always the same: $p_{i,j}^h$*

Association schemes can be considered as a graph with vertex set X, and edges $(x, y) \in X \times X$. The relations R_i represent a coloring of the complete graph on X into $d+1$ colors. The fourth condition in the definition of association schemes means that for fixed values of $0 \leq h, i, j \leq d$ every edge colored R_h is in the same number of triangles that are coloured with colours R_h, R_i, R_j.

Definition 4.3. *The Hamming scheme on $\mathbb{F}_q^n$ is the association scheme given by: $R = \{R_0, R_1, \ldots, R_n\}$ such that $(x, y) \in R_i$ if and only if $d_H(x, y) = i$*

The Hamming scheme is an association scheme with

$$p_{i,j}^h = \sum_{\delta=0}^{\lfloor i+j-h/2 \rfloor} (q-2)^{i+j-h-2\delta}(q-1)^\delta \binom{h}{j-\delta}\binom{j-\delta}{h-i+\delta}\binom{n-h}{\delta}$$

Definition 4.4. *The matrices $A_0, A_1, \ldots, A_d$ are the associate matrices of an association scheme $(X, \{R_0, R_1, \ldots, R_d\})$, where A_i is the indicator matrix of R_i, that is entry (x, y) of A_i is 1 if $(x, y) \in R_i$, and 0 otherwise.*

The associate matrices with matrix multiplication form an algebra called the Bose-Mesner algebra that has a basis $\{E_0, E_1, \ldots, E_d\}$ of pairwise orthogonal idempotent matrices

1. $E_i E_j$ is the zero matrix for all $i \neq j$
2. $E_i^2 = E_i$.

Definition 4.5. *The first eigenmatrix P, and the second eigenmatrix Q of a Bose-Mesner algebra are matrices that satisfy:*

1. $A_i = \sum_{j=0}^d P_{ji} E_j$

2. $E_i = \frac{1}{|X|} \sum_{j=0}^{d} Q_{ij} A_j$

Q can be represented in terms of Krawtchouk polynomials.

Definition 4.6. *The Krawtchouk polynomial $K_t(x)$ is defined as:*

$$K_t(x) = \sum_{i=0}^{t} \binom{x}{i}\binom{n-x}{t-i}(-1)^i(q-1)^{t-i}$$

Proposition 4.2. $Q_{i,t} = K_t(i)$.

Definition 4.7. *The distribution vector a of an association scheme (X, R) and an $Y \subseteq X$ is a vector of length $d+1$ such that*

$$a_i = \frac{|(Y \times Y) \cap R_i|}{|Y|}. \qquad (24)$$

a_i can be thought of as the average degree of a vertex in Y in the graph of the association scheme.

It follows from the definition that $a_i \geq 0$ for every i, and

$$\sum_{i=0}^{d} a_i = |Y| \qquad (25)$$

In the case of the Hamming scheme $X = \mathbb{F}_q^n$ and if a code has minimum distance r it follows that $a_1 = a_2 = \cdots = a_{r-1} = 0$

Theorem 4.4 (Delsarte [10]). *If a is a distribution vector of a subset Y of an association scheme with second eigenmatrix Q, then $aQ \geq 0$.*

Delsarate's linear program is the following:

1. $a_0 = 1$
2. $a_i = 0$ for $1 \leq i \leq r$
3. $a_i \geq 0$ for $r \leq i \leq n$
4. $aQ \geq 0$
   $\max \sum_{i=1}^{d} a_i$

An optimal solution for Delsarte's linear program gives an upper bound $m \leq \sum a_i$ for the size of a code with given minimum distance r. In the case of an Armstrong code we have $r = n - k + 1$, putting of $k = 4$ we get that

$$Q_{0,t} + a_{n-3}Q_{n-3,t} + a_{n-2}Q_{n-2,t} + a_{n-1}Q_{n-1,t} + a_n Q_{n,t} \geq 0 \qquad (26)$$

has to stand for every $1 \leq t \leq n-3$. More specifically

$$\binom{n}{t}(q-1)^t +$$

$$a_{n-3}\left[(-1)^{t-3}\binom{n-3}{t-3}(q-1)^3 + (-1)^{t-2}3\binom{n-3}{t-2}(q-1)^2\right.$$
$$\left. +(-1)^{t-1}3\binom{n-3}{t-1}(q-1) + (-1)^t\binom{n-3}{t}\right]$$

$$+a_{n-2}\left[(-1)^{t-2}\binom{n-2}{t-2}(q-1)^2\right. \tag{27}$$
$$\left. +(-1)^{t-1}2\binom{n-2}{t-1}(q-1) + (-1)^t\binom{n-2}{t}\right]$$

$$+a_{n-1}\left[(-1)^{t-1}\binom{n-1}{t-1}(q-1) + (-1)^t\binom{n-1}{t}\right]$$

$$+a_n\left[(-1)^t\binom{n}{t}\right] \geq 0$$

We only consider (26) for $t = 1$, so the maximum we get will be an upper bound on the maximum of the full linear program.

The constraints on $a_{n-3}, a_{n-2}, a_{n-1}, a_n$ in case $t = 1$ are:

$$n(q-1) + a_{n-3}[3(q-1) - (n-3)] + a_{n-2}[2(q-1) - (n-2)] + a_{n-1}[(q-1) - (n-1)] + a_n(-n) \geq 0 \tag{28}$$

Rewriting (28) of form

$$n(q-1) \geq \sum \alpha_i a_i \tag{29}$$

the coefficients α_i of a_i's are:

1. $\alpha_n = n$
2. $\alpha_{n-1} = n - q$
3. $\alpha_{n-2} = n - 2q$
4. $\alpha_{n-3} = n - 3q$

And the objective function is:

$$\max (a_{n-3} + a_{n-2} + a_{n-1} + a_n + a_0) \tag{30}$$

Given any feasible solution $a = (a_{n-3}, a_{n-2}, a_{n-1}, a_n)$ another feasible solution $a^* = (a^*_{n-3}, a^*_{n-2}, a^*_{n-1}, a^*_n)$ can be obtained where $a^*_{n-3} = a_{n-3} + a_{n-2} + a_{n-1} + a_n$, $a^*_{n-2} = 0$, $a^*_{n-1} = 0$, $a^*_n = 0$. a^* is indeed feasible since a_{n-3} has the smallest coefficient in (29). The values of the objective function of a and a^* are the same. Thus, without loss of generality we may restrict our attention to the special case of $a_n = a_{n-1} = a_{n-2} = 0$. (29) simplifies to

$$(n - 3q)a_{n-3} \leq n(q-1) \tag{31}$$

We may assume that $n \geq 3q$, otherwise we already have $f(q,4) \leq 3q$. Using $a_{n-3} \geq m$ and (23) we get

$$\sqrt{2\binom{n}{3}} \leq m \leq \frac{n(q-1)}{n-3q} \qquad (32)$$

This means that

$$\frac{(n-3)^3}{3} \leq \frac{n^2q^2}{(n-3q)^2} \qquad (33)$$

Let $n = 3q + c(q) = 3q + c$ and assume also that $c \geq \sqrt{q} + 3$. (33) is rewritten as

$$c^2 \leq \frac{3n^2q^2}{(3q+\sqrt{q})^2}. \qquad (34)$$

After cross multiplication and expansion we get

$$c^2(27q^3 + 27q^2\sqrt{q} + 9q^2 + q\sqrt{q}) \leq 3(3q+c)^2q^2. \qquad (35)$$

(35) implies that

$$c^2(9q + 9\sqrt{q} + 9) \leq 9q^2 + 6qc + c^2 \qquad (36)$$

(36) in turn clearly implies $c < \sqrt{q}$, thus we get that

$$n < 3q + \sqrt{q} + 3 \qquad (37)$$

holds for arbitrary q.

If q is large enough we can prove better bound. The Armstrong code can be partitioned into q sets of codewords by looking at their symbol on a given position. Let m_i denote the number of codewords that begin with the symbol $i \leq q$. We have

$$m = m_1 + m_2 + \cdots + m_q \qquad (38)$$

Because of the Armstrong property to every 3 different positions that are not the (fixed) first column there is a pair of codewords that agree on these three columns, furthermore the two codewords of this pair are in different partitions, otherwise they would meet in 4 positions, contradicting the Armstrong property. This means that

$$\binom{n-1}{3} \leq \sum_{1 \leq i < j \leq q} m_i m_j = \frac{1}{2}\left(\left(\sum m_i\right)^2 - \sum m_i^2\right) \qquad (39)$$

We know that $\sum m_i = m$ and this means that

$$\frac{1}{q}\sum_{i=1}^{n} m_i^2 \geq \left(\frac{m}{q}\right)^2 \qquad (40)$$

Therefore

$$\frac{1}{2}\left(\left(\sum m_i\right)^2 - \sum m_i^2\right) \leq \frac{1}{2}\left(m^2\left(1 - \frac{1}{q}\right)\right) \qquad (41)$$

And thus we get
$$\binom{n-1}{3} \leq \frac{1}{2}\left(m^2\left(1-\frac{1}{q}\right)\right) \tag{42}$$

Now inequality (42) can be written in place of (9) into equation (10), since (42) is also a monotone increasing bound for n as a function of m. Replacing $\binom{n-1}{3}$ with $\frac{(n-3)^3}{6}$ in (42) we get $n \leq b'_{q,4}(m) = \sqrt[3]{3m^2(1-\frac{1}{q})} + 3$. Similarly to Section 2 for the solution α of

$$b'_{q,4}(m) = a_{q,4}(m) \tag{43}$$

$b'_{q,4}(\alpha) = a_{q,4}(\alpha)$ is a universal upper bound for n. That is,

$$\sqrt[3]{3m^2(1-\frac{1}{q})} + 3 = 3q\left(1 + \frac{q-1}{m-q}\right). \tag{44}$$

After some simple algebraic manipulations (44) holds if

$$m^3 - 3qm^2 + (3q - 9q^4(q-1)^2)m - q^3 = 0. \tag{45}$$

(45) is a cubic equation in m, it is easy to see that the LHS is negative for $m = 0$ and that the derivative of LHS has one negative and one positive root, since the derivative is also negative at $m = 0$. Thus (45) has only one positive root. Writing this root $\alpha = q^\beta$ we get that the order of magnitude of the terms in (45) are $q^{3\beta}, q^{2\beta+1}, q^{\beta+1}, q^3, q^{6+\beta}$. So for large enough q equality can hold only if $\alpha = \Theta(q^3)$. However, if $\alpha = \Theta(q^3)$, then $a_{q,4}(\alpha) = 3q\left(1 + \Theta(\frac{1}{q^2})\right) \leq 3q + 1$, for large enough q.

5 Generalizations, conclusions

We have seen that the determination of $f(q, k)$ raises interesting coding theory and design theory problems, furthermore algebraic combinatorics methods are also used. The current best general lower bound is probabilistic, while the upper bound uses spherical codes

$$\frac{\sqrt{q}}{e}k < f(q,k) < (q - \log q)k \tag{46}$$

holds for $k > k_0(q)$. It is an intriguing question what is the true coefficient of k in $f(q, k)$ if k is large enough. For fixed (small) k we have seen that $f(q, 2) = \binom{q+1}{2}$ and $f(q, 3) = 3q - 1$. This latter was proved recently by disproving a conjecture of the current author. New result in the present paper is that for $\forall \epsilon \exists q_0 = q_0(\epsilon)$ such that for $q > q_0$

$$\frac{2q-1}{1+\epsilon} < f(q,4) < 3q + 1 \tag{47}$$

holds.

Chee, Zhang and Zhang introduced a generalization of Armstrong codes [8], they correspond to Armstrong instances of *branching dependencies* introduced by Demetrovics et.al. [19].

Definition 5.1. *Let $X \subset \mathbf{R}$ be a set of attributes and $y \in \mathbf{R}$ be an attribute. For positive integers $s \leq t$ we say that y (s,t)-depends on X iff for any $t+1$ rows $r_1, r_2, \ldots r_{t+1}$ of the database table*

$$|\{d_i|_{\{x\}} : 1 \leq i \leq t+1\}| \leq s \text{ for each } x \in X \tag{48}$$

implies that

$$|\{d_i|_{\{y\}} : 1 \leq i \leq t+1\}| \leq t \tag{49}$$

holds.

Given $1 \leq s \leq t$ an (s,t)-*dependent* key K is a subset of the attribute set $\mathbf{R}$ such that (s,t)-dependencies $K \xrightarrow{(s,t)} y$ holds for all $y \in \mathbf{R}$.

Definition 5.2. *A q-ary code $\mathcal{C}$ of length n is called an $(q, k, n)_{s,t}$-Armstrong code if*

(i) for any $t+1$ codewords of $\mathcal{C}$ there exists at most $k-1$ coordinate positions such that the codewords have at most s distinct symbols in each position, respectively, and

(ii) for any $k-1$ coordinate positions there exist $t+1$ codewords such that they have at most s distinct symbols in each coordinate position respectively, but there exists a coordinate position where all these $t+1$ codewords have pairwise distinct symbols.

The Armstrong code of Definition 5.2 can be considered as an Armstrong instance with n attributes. The first property says that each k-subset of attributes is an (s,t)-dependent key, while the second property states these are minimal keys. Thus a $(q, k, n)_{s,t}$-Armstrong code is Armstrong instance for $\mathcal{K}_n^k$ as (s,t)-dependent minimal key system. Clearly, $q > s, t$ and $k > 1$ is needed for a meaningful $(q, k, n)_{s,t}$-Armstrong code. Note that a $(q, k, n)_{1,1}$-Armstrong code is an Arm(q, k, n).

Let $q > t \geq s \geq 1$ and $k > 1$, then $f_{s,t}(q,k)$ denotes the largest n such that a $(q, k, n)_{s,t}$-Armstrong code exists. The analogues of inequalities (6) and (9) are as follows. Let ϕ be the least number of submultisets $S \subset M$ of size $t+1$ with at most s distinct elements, where M ranges over all multisets of size m over $\{q, 2, \ldots, q\}$.

Proposition 5.1 ([8]). *Let $\mathcal{C}$ be a $(q, k, n)_{s,t}$-Armstrong code of size m. Then*

$$\binom{n}{k-1} \leq \binom{m}{t+1} \text{ and } n \cdot \phi \leq (k-1)\binom{m}{t+1}. \tag{50}$$

The proof of Proposition 5.1 is straightforward extension of that of inequalities (6) and (9). As in [30] the two inequalities of (50) give two upper bounds on $f_{s,t}(q,k)$, one increasing another one decreasing, so an universal upper bound exists where the two upper bounds intersect. However, it is impossible to get explicit bounds in most of the cases. So [8] gives results in some special cases. The analogue of the first equality of Proposition 2.1 holds with a proof analogous to the original.

Proposition 5.2 ([8]).

$$f_{1,t}(q,2) = \binom{qt+1}{t+1}. \tag{51}$$

An interesting special case exists when another exact bound is given.

Theorem 5.1 ([8]).
$$f_{2,2}(q,4) = 2q - 1 \tag{52}$$
for all integers $q \geq 3$.

To prove the upper bound a good lower bound of number ϕ of (50) is given. The lower bound construction uses the classical near 1-factorizaion of the complete graph K_{2q-1}.

In general the probabilistic method of [38] can be applied using Lovász Local Lemma to prove the lower bound

Proposition 5.3 ([8]). *Let $t \geq 1$ and q,k satisfy $q > 4e^2 \sqrt[k]{2e(t+1)^2}$. Then a $(q,k,n)_{1,t}$-Armstrong code exists for $n \leq \sqrt[2k]{\frac{1}{2e(t+1)^2}} \frac{\sqrt{q}}{e} k$, that is*

$$f_{1,t}(q,k) \geq \sqrt[2k]{\frac{1}{2e(t+1)^2}} \frac{\sqrt{q}}{e} k. \tag{53}$$

There are many questions left open for generalized Armstrong codes. The most important is to give an upper bound for $f_{s,t}(q,k)$ in general. Also, no construction is known in the case $s > 1$ other than the $s = t = 2$, $k = 4$ case. It is plausible that a greedy construction may work, however the calculations were already pretty complex in the "ordinary" Armstrong case of $s = t = 1$. As Armstrong codes have both design and code theoretic properties one believes that finding good lower bounds requires extensive application of design theory techniques.

References

1. ABITEBOUL, S., HULL, R., AND VIANU, V. *Foundations of Databases.* Addison-Wesley, 1995.
2. ARMSTRONG, W. W. Dependency structures of database relationships. *Information Processing* (1974), 580–583.
3. AUSIELLO, G., D'ATRI, A., AND MOSCARINI, M. Chordality properties on graphs and minimal conceptual connections in semantic data models. *Journal of Computer and System Sciences 33*, 2 (1986), 179 – 202.
4. BANNAI, E., AND ITO, T. *Algebraic combinatorics.* Benjamin/Cummings Menlo Park, 1984.
5. BEERI, C., FAGIN, R., MAIER, D., AND YANNAKAKIS, M. On the desirability of acyclic database schemes. *J. ACM 30* (July 1983), 479–513.
6. BENNETT, F. E., AND WU, L. On minimum matrix representation of closure operations. *Discrete Applied Mathematics 26*, 1 (1990), 25 – 40.

7. BLOKHUIS, A., BROUWER, A. E., AND SALI, A. Note on the size of binary armstrong codes. *Designs, codes and cryptography* (2014), 1–4.
8. CHEE, Y. M., ZHANG, H., AND ZHANG, X. Complexity of dependences in bounded domains, armstrong codes, and generalizations. *IEEE Transactions on Information Theory 61*, 2 (2015), 812–819.
9. D'ATRI, A., AND MOSCARINI, M. On the recognition and design of acyclic databases. In *Proceedings of the 3rd ACM SIGACT-SIGMOD symposium on Principles of database systems* (New York, NY, USA, 1984), PODS '84, ACM, pp. 1–8.
10. DELSARTE, P. *An algebraic approach to the association schemes of coding theory.* Philips Res. Rep. Suppl. 10, 1973.
11. DEMETROVICS, J. On the equivalence of candidate keys with Sperner systems. *Acta Cybernetica 4* (1979), 247–252.
12. DEMETROVICS, J., FÜREDI, Z., AND KATONA, G. O. H. Minimum matrix reperesentation of closure operetions. *Discrete Applied Mathematics 11* (1985), 115–128.
13. DEMETROVICS, J., AND GYEPESI, G. A note on minimum matrix reperesentation of closure operetions. *Combinatorica 3* (1983), 177–180.
14. DEMETROVICS, J., KATONA, G., MIKLÓS, D., SELEZNJEV, O., AND THALHEIM, B. The average length of keys and functional dependencies in (random) databases. In *Database Theory - ICDT'95, 5th International Conference, Prague, Czech Republic, January 11-13, 1995* (1995), G. Gottlob and M. Vardi, Eds., vol. 893 of *LNCS*, pp. 266–279.
15. DEMETROVICS, J., KATONA, G., MIKLÓS, D., SELEZNJEV, O., AND THALHEIM, B. Asymptotic properties of keys and functional dependencies in random databases. *Theoretical Computer Science 190*, 2 (1998), 151–166.
16. DEMETROVICS, J., AND KATONA, G. O. H. Extremal combinatorial problems in relational data base. In *Fundamentals of Computing Theory (FCT 1981)*, no. 117 in LNCS. Springer-Verlag, Berlin, 1981, pp. 110–119.
17. DEMETROVICS, J., AND KATONA, G. O. H. Extremal combinatorial problems of database models. In *MFDBS 87, 1st Symposium on Mathematical Fundamentals of Database Systems, Dresden, GDR, January 19-23, 1987* (1987), J. Biskup, J. Demetrovics, J. Paradaens, and B. Thalheim, Eds., no. 305 in LNCS, pp. 99–127.
18. DEMETROVICS, J., AND KATONA, G. O. H. A survey of some combinatorial results concerning functional dependencies in databases. *Annals of Mathematics and Artificial Intelligence 7* (1993), 63–82.
19. DEMETROVICS, J., KATONA, G. O. H., AND SALI, A. The characterization of branching dependencies. *Discrete Applied Mathematics 40* (1992), 139–153.
20. DEMETROVICS, J., KATONA, G. O. H., AND SALI, A. Design type problems motivated by database theory. *Journal of Statistical Planning and Inference 72* (1998), 149–164.
21. ERDŐS, P., AND LOVÁSZ, L. Problems and results on 3-chromatic hypergraphs and some related questions. In *Finite and Infinite Sets* (1974), vol. 10 of *Colloq. Math. Soc. János Bolyai*, pp. 609–627.
22. FAGIN, R. Degrees of acyclicity for hypergraphs and relational database schemes. *J. ACM 30* (July 1983), 514–550.
23. FÜREDI, Z. Perfect error-correcting databases. *Discrete Applied Mathematics 28* (1990), 171–176.

24. GANTER, B., AND GRONAU, H.-D. O. F. On two conjectures of Demetrovics, Füredi and Katona concerning partitions. *Discrete Mathematics 88* (1987), 149–155.
25. GANTER, B., GRONAU, H.-D. O. F., AND MULLIN, R. C. On orthogonal double covers of K_n. *Ars Combinatoria 37* (1994), 209–221.
26. GRAHNE, G., AND RÄIHÄ, K.-J. Characterizations for acyclic database schemes. In *Advances in Computing Research* (1986), F. Preparata and P. Kanellakis, Eds., JAI Press Inc., Greenwich, Conn., p. 1941.
27. GRONAU, H.-D. O. F., MULLIN, R. C., AND SCHELLENBERG, P. J. On orthogonal double covers of k_n and a conjecture of chung and west. *J. Combin. Des. 3*, 3 (1995), 213–231.
28. HARTMANN, S., LECK, U., AND LINK, S. On codd families of keys over incomplete relations. *Comput. J. 54*, 7 (2011), 1166–1180.
29. HARTMANN, S., LINK, S., AND SCHEWE, K.-D. Weak functional dependencies in higher-order datamodels. In *Foundations of Information and Knowledge Systems* (2004), D. Seipel and J. M. Turull Torres, Eds., vol. 2942 of *Springer LNCS*, Springer Verlag.
30. KATONA, G. O. H., SALI, A., AND SCHEWE, K.-D. Codes that attain minimum distance in all possible directions. *Central Eruopean J. of Math. 6* (2008), 1–11.
31. KÖHLER, H., LECK, U., LINK, S., AND PRADE, H. Logical foundations of possibilistic keys. In *Logics in Artificial Intelligence - 14th European Conference, JELIA 2014, Funchal, Madeira, Portugal, September 24-26, 2014. Proceedings.* (2014), pp. 181–195.
32. KÖHLER, H., LINK, S., AND ZHOU, X. Possible and certain sql keys. *PVLDB 8*, 11 (2015), 1118–1129.
33. RANKIN, R. A. The closest packing of spherical caps in n dimensions. *Proceedings of the glasgow mathematical association 2* (1955), 145–146.
34. SALI, A. Minimal keys in higher-order datamodels. In *Foundations of Information and Knowledge Systems* (2004), D. Seipel and J. M. Turull Torres, Eds., vol. 2942 of *Springer LNCS*, Springer Verlag.
35. SALI, A. Coding theory motivated by relational databases. In *Semantics in Data and Knowledge Bases*. Springer, 2011, pp. 96–113.
36. SALI, A., AND SCHEWE, K.-D. Counter-free keys and functional dependencies in higher-order datamodels. *Fundamenta Informaticae 70* (2006), 277–301.
37. SALI, A., AND SCHEWE, K.-D. Keys and Armstrong databases in trees with restructuring. *Acta Cybernetica 18* (2008), 529–556.
38. SALI, A., AND SZÉKELY, L. On the existence of armstrong instances with bounded domains. In *Foundations of Information and Knowledge Systems. FoIKS 2008* (2008), S. Hartmann and G. Kern-Isberner, Eds., vol. 4932 of *LNCS*, Springer, pp. 151–157.
39. SELEZNJEW, O., AND THALHEIM, B. On the numbers of shortest keys in relational databases on non-uniform domains. *Acta Cybernetica 8*, 3 (1988), 267–271.
40. THALHEIM, B. The number of keys in relational and nested relational databases. *Discrete Applied Mathematics 40* (1992), 265 – 282.

Misunderstandings and Their Overwhelming Success - An Essay in Honor of Klaus-Dieter Schewe

Alexander Bienemann

Kiel, Germany

`info@bienemann.org`

Abstract. The author has been asked to formulate a paper in honor of Prof. Klaus-Dieter Schewe (KDS). Due to the high importance of KDS to the field of Computer Science and to the friendly character KDS has shown throughout the years, this request could not be denied. The author considers KDS' works on Abstract State Machines, Conceptual Modeling and on the Co-Design of web information systems particularly interesting, although the fullness and depth of KDS' work altogether goes far beyond these areas.

Keywords: misunderstanding, specification, migration project, fun

1 Motivation

Klaus-Dieter Schewe (KDS) has obtained his *venia legendi* at the chair of Prof. Bernhard Thalheim, who was also the doctoral supervisor of the author. We thus share a common professor, so to speak. Also, the author had the opportunity to meet KDS at different venues somewhere between Austria and Australia, including both. This paper is written in honor of KDS.

2 What this paper contains

The intention of this paper is to share some insights regarding how Computer Science and Computer Engineering are lived in the world outside academia, i.e. in a world full of implicit assumptions, imprecise formulations, improper software architectures, and sometimes even impossible specifications.

3 Disclaimer

The reader reads this paper upon his/her sole responsibility. The author is not liable for any consequences nor damage caused by reading this paper or by applying any of the thoughts contained in it.

4 The notion of misunderstandings

At the university, we learn to think precisely. We learn to derive thoughts, to verify claims, to make assumptions explicit. We learn to model data. We learn abstract thinking. Formal thinking. Proper thinking.

In the outside world, one deals with other human beings. Communicating in different ways. Meeting assumptions that sometimes are not verified, or verified too late. Using terminology in different ways, depending on context, on company culture, on personal knowledge.

In the outside world, misunderstandings are common. Even the definition of a process, i.e. of *what to do*, can turn out to have been unclear to the persons involved - even if everyone involved had assured in writing that he/she understood it to the fullest.

Interestingly, misunderstandings are *successful*. We might also say: they are *robust*, or *persistent*. They do not seem to disappear with the next generation of software systems, development environments, or specification frameworks. Instead, we seem to get to a new type of misunderstandings time and again. Misunderstandings increase the amount of work in the IT, i.e. for inspecting and repairing application systems with wrongly designed architectures, based on misunderstood requirements or specifications.

The actual challenge seems to be not to design structures for a clearly defined requirements - but to design structures being able to deal with misunderstandings. Structures being robust enough for the real world. What a tough challenge!

5 Checking if things are good enough

5.1 When is a solution good enough?

Sometimes clear criteria are given, e.g. written down as "Definition of Done". But often the goodness criteria are only vaguely formulated, e.g. "maintainability" of a software system. When it comes to inspecting a system or a given project, the author has experienced many times that the question he is asked to answer needs to be made more precise, that the goodness criteria were too vague - or even misunderstood.

5.2 Testing vs. verification

A common misunderstanding is that tests would help to *prove* that a system is correct. But such a proof would require a formal verification process - which is sometimes done, but usually costly and non-trivial. Tests can only show the existence of an error, if they run into it. This misunderstanding is very common. It is also a great basis for further misunderstandings, e.g. that software unit tests on Java class level would be sufficient to test database procedures used by Java code.

5.3 Integration tests

A non-trivial misunderstanding is related to testing switchovers in a migration project: An integration test should consider all relevant interfaces, involved parties, and neighboring systems. Sometimes long discussions take place between multiple companies involved in the operation of an application system, when it comes to agreeing on what aspects an integration test *should contain*. These discussions can take months, and sometimes involve lots of political issues before an agreement has been achieved. Misunderstandings here take place on multiple layers, e.g. different technical terminology, differing processes, different decision making.

5.4 SSDs vs. indexing

Faster I/O is a blessing for database technology. For instance, relational DBMS can answer queries so much faster now. However, the consequence is that proper database design does not seem that important any more, which is a misunderstanding: the system seems to be fast anyway, at least on test data. It thus requires more discipline now in order to ensure that appropriate indexes are used. The misunderstanding can produce a waste of resources and the need of repairing the system design in production phase.

6 Understanding how to do things

6.1 Analysis vs. solutions brought up-front

A quick-win approach would be to go to the customer, present "the solution" to him, and sign the contract. A solid approach, however, is to go to the customer and *analyze* first what the best solution would be. This costs effort, time and money. Moreover, it does not help to sell a specific solution to the customer, as the analysis result may be different from what was expected. On the other hand, a proper analysis helps to save vast amounts of money thanks to avoiding non-optimal solutions. But somehow it is a common misunderstanding that "analysis is expensive". Well, the analysis and repair of wrongly designed systems later on is usually more expensive.

6.2 Analysis vs. guaranteeing a result

Sometimes the question is asked: Can you please guarantee that your analysis will spawn a performance increase of X% in our production system? No, this cannot be guaranteed, especially as the expert has not seen the system before the analysis, and the result of an analysis as such is open until the analysis is completed. But it is interesting how much this misunderstanding is still present nowadays.

7 Doing things

7.1 Automatic vs. manual migration

People tend to assume that since we have programming languages, we can compute everything. And of course, that we can convert everything to a desired new form. We thus can convert between different graphics, audio and video formats, and nobody is surprised by it. However, this thinking is also commonly used for converting program codes, database structures, configuration files and similar artifacts within a *migration project*. For instance, migrating an application based on a database system A to a database system B, with a differing SQL dialect, at least slightly different performance characteristics, and a different procedural language. It is a common misunderstanding that since we can convert *simple* artifacts, we can also convert artifacts of *full computational complexity*, i.e. Turing-complete ones.

The term *Halteproblem* seems not to be commonly known - or it is frequently forgotten. This results in assumptions like *Oh, I'll write a script that converts all 500 database procedures within a few minutes* - which prove to be wrong as the projects progress. This is sad, but in the end, it also... produces work for IT professionals to clean up the mess later on.

7.2 Data vs. application migration

A common misunderstanding about database migration is that the work is mostly about *data migration*. All the effort related to migrating procedures, queries, database access layer of the application - all this is sometimes not seen initially. It is the job of a migration expert to help to clarify this misunderstanding up front, before a project is planned.

7.3 ORM vs. workload

Using an Object-Relational Mapper (ORM) is a convenient thing, it may even be considered necessary for complexity and project management reasons. However, it is just a tool like any other tool: it can be used in a right or wrong way.

In one of the cases the author was asked to investigate, inappropriate usage of such an ORM resulted of approximately 3000-5000 single SQL queries sent to the database. Each of the queries was highly optimal already, but the overall workload was unnecessarily high. The response times of the GUI of the applications were reaching 16 seconds after a single click. This was more than the approx. 1 second that would have been acceptable.

The misunderstanding was that using the ORM was seen as a *relief* for the software development team, *freeing* them from having to deal with database and query design issues.

7.4 Databases understood as storage

In a 1-day training in database architecture for a very solid and highly qualified software engineering company, one of the participants told me quite bluntly: You know, actually we want to use the relational database as a *storage* where we throw data into, without having to deal with its structures.

Unfortunately, this is a misunderstanding: Using a hammer drill, for instance, also requires to think of *what* is to be accomplished by using it. Just taking the hammer drill in one's hand and somehow placing it at the wall, hoping that it will work out somehow, i.e. the right hole somehow will appear at the right place and in the right quality, is not a proper approach. The same holds for a relational database: If used properly, it will join billions of rows with each other within (tens of) milliseconds. If not used properly, it might be as slow as the file system it is located on.

7.5 Automatic query generation

In one of expertise cases, the author was confronted with a CRM system enhanced by *customization code*. The code was generated (semi-)automatically, along with an additional dynamic view hierarchy put between the database structures and the application.

The generated additional code and the additional views resulted in quite big queries consisting of up to 30-50 outer joins. These outer joins posed a big difficulty to the database system, which was expensive but also well-known for having issues with its query optimizer. This resulted in GUI response times of up to 20 seconds.

It is a general misunderstanding that anything automatically generated will work *just because the computer has generated it*. Automatically generated code is not always optimal. In the end it is always the human expertise which helps to check if the generated outcome has the quality which was expected.

8 Outlook

8.1 Importance of foundations

Teaching Computer Science and related fields nowadays is based on conveying foundations to the students - which is just right. It is the formal stuff, all the LR(k) grammars, functional dependencies, complexity theory, combinatorics, all the different calculi, which help to train a student's mind in order to gain new knowledge, new work methods - and most of all, to recognize patterns among the everyday complexity of the IT.

The best thing we can do is e.g. to give Computer Science students all the equipment they need, formal thinking, as well as plausible and digestible examples, and the ability to search beyond that on their own. In the end, it turns out to be tremendously effective - and useful - if one has been equipped with this foundational knowledge. The same applies to related/combined areas of teaching and research, although here meeting the right choice regarding which foundations are most helpful to fit the purpose of the respective field, and are thus to be taught to students, might be a bit more challenging.

8.2 What comes after the university study

The actual methods, tools and best practices necessary for solving the cases mentioned in this essay, are usually not taught at the university. For instance, analysis of systems, managing migration projects, or performance optimization in distributed large-scale database clusters, are not taught. But a student who received good foundations of Computer Science, will later be able to learn these methods, and to develop his/her own best practices - and thus to move the IT forward.

9 Summary

Best greetings to Klaus-Dieter Schewe, a great member of the Computer Science community!

QoS-Aware Web Service Composition Using Graph Databases

Hui Ma[1], Zhaojiang Chang[1], Alexandre Sawczuk da Silva[1]
, and Sven Hartmann[2]

[1] Victoria University of Wellington, New Zealand
[2] Clausthal University of Technology, Germany

Abstract. Web service composition can fulfil users' requests by composing multiple web services when no existing single service can meet functional requirements of the users. The objective of automatic service composition is to find a composite service that meets users' functional requirements and provides the best quality of service (QoS), in a challenging environment where the number of available services is rapidly increasing. Existing approaches to service composition suffer from excessive in-memory computation on ensuring dependencies between services to be satisfied in order to find a service composition that meets a user's request. We propose an automated QoS-aware web service composition approach using graph databases to store the service dependency information of a service repository. Experimental results on a benchmark dataset show that our proposed approach can efficiently find service composition solutions.

1 Introduction

With the emergence of service-oriented architecture (SOA) [18], web services have gained a lot of popularity in the last few years as a promising technology to implement business processes covering different organisations and computing platforms in a dynamic and loosely-coupled manner [11, 12]. A web service is a self-contained unit with limited functionality that takes input data and produces output data [7, 13]. In order to realize value-added and complex business processes, individual web services having limited functions are often composed together as a workflow which is referred to as *a web service composition*.

Currently there are three main groups of approaches to web service composition. The first group of approaches employs traditional methods such as Integer Linear Programming (ILP) [25] to solve the problem of Web Service Composition (WSC). These approaches lack scalability and are no longer efficient since the number of available web services is increasing extraordinarily quickly. The second group includes various Evolutionary Computing (EC) approaches, such as Genetic Algorithms (GA) [24], Genetic Programming (GP) [14,16,19,21,22] and Particle Swarm Optimisation (PSO) [3,20]. [14,21] are devoted to QoS-aware service composition. [14] apply genetic programming to find optimal QoS-aware service compositions. [21] proposes an evolutionary graph technique to find a service composition with optimal overall QoS values. These approaches are slow, especially when checking the dependencies between the component services, i.e., whether the input of a service can be fulfilled by the outputs of its proceeding services, a task which requires a substantial amount of resources.

The third group is a graph-based approach, such as [8,15]. The approach in [8] uses a graph search algorithm to construct web service compositions. However, this approach does not consider QoS. A graph-based evolutionary computation approach to automated web service composition is proposed in [21]. This approach consists of two steps. The first step is to initialise the population by employing a graph building algorithm based on the planning graph approach described in the literature [7]. The second step is to perform mutation and crossover operations on selected candidates, to generate a new set of candidates and evaluate the fitness of those candidates. This approach needs to search for data dependencies when generating service composition solutions, which incurs expensive computation costs. For each service request, this approach computes and temporarily stores service dependencies in memory, which are then used during the process of searching for service compositions. Approaches proposed in [5, 6] on service composition also make use of dependency graphs. In [6] local search techniques are used to find a path on a dependence graph which preserves lowest response time or largest throughput. However, this approach cannot ensure that the resulted service solutions have the optimal overall QoS value.

Existing web service composition approaches [21, 23] do not permanently store web service dependencies, which means that when running the composition algorithms for any new service requests web service dependencies information needs to be recomputed. This step has a big impact on the efficiency of web service composition approaches. Actually, web service dependencies between services are independent from service requests therefore should be computed only once and stored for computing service compositions for all requests. Graph databases, e.g., *Neo4j* [17], can store data and relationships [17] between data as graphs. They also can efficiently retrieve the relationships between data. According to [10], which evaluates four current types of graph databases, *Neo4j* graph database gets the best performance with loading datasets, finding the shortest path and breadth-first exploration of nodes neighbourhood.

In this paper we propose to use a graph database, i.e., *Neo4j*, to store the information related to services and dependencies of services of a service repository. In particular, we will use *Neo4j* to store all services as nodes and dependencies between the services as relationships (edges). All nodes and relationships have their own individual properties. With each new composition task we can utilize existing dependency information contained within graph database dependency graphs. Our approach also takes quality of service (QoS) into consideration by including QoS attributes with each edge between services.

This paper aims to create a graph database-based approach that generates web service solutions efficiently by reducing the composition costs involved in checking dependencies of services. In particular we will achieve the following objectives:

1. To use a graph database to represent and store services and their dependencies in a service repository.
2. To propose a novel graph database-based approach that can find near-optimal QoS-aware web services composition solutions.
3. To conduct a full evaluation of our approach by comparing its performance against an existing approach using a benchmark dataset.

The paper is organised as follows: Sect. 2 describes the model that represents service composition as instances of graph databases. Sect. 3 presents our proposed approach in detail. Sect. 4 reports on the experimental evaluation of our approach. Finally, Sect. 5 gives conclusions and suggestions for future work.

2 Problem Definition and the QoS Model for Service Composition

We consider a *web service* (*service*, for short) as a tuple $S = (I_S, O_S, QoS_S)$ where I_S is the set of service inputs that are consumed by S, O_S the set of service outputs that are produced by S, and QoS_S the set of non-functional attributes of S, e.g., response time, cost, reliability, and availability of service S. The inputs in I_S and outputs in O_S are parameters that are related to concepts in a taxonomy $\mathcal{T}$.

A *service repository* $\mathcal{SR}$ is a finite collection of services with a common taxonomy $\mathcal{T}$. $\mathcal{SR}$ only contains a collection of services; the potential dependencies between them are not explicitly listed. Instead, these dependencies are to be determined by verifying whether the output types of a given service fulfil the input types of another. This verification is done by querying $\mathcal{T}$. A *service request* (also called *composition task*) over $\mathcal{SR}$ is a tuple $T = (I_T, O_T)$ where I_T is the set of task inputs, and O_T the set of task outputs. The inputs in I_T and the outputs in O_T are parameters that are related to concepts in the taxonomy $\mathcal{T}$ [11,13].

A service composition is commonly represented as a *directed acyclic graph* (DAG). Its nodes correspond to the services in the composition. Two services S and S' are connected by an edge e if some output of S serves as input for S'. An edge between two services represents the *dependency* between them, S' depends on S. Apparently, such outputs and inputs must semantically match to ensure the correct execution of the service composition.

Given a composition task $T = (I_T, O_T)$, we represent a service composition solution for T with services $S_1, \ldots, S_n$ by a weighted DAG, $WG = (V, E)$ with node set $V = \{Start, S_1, S_2, \ldots, S_n, End\}$ and edge set $E = \{e_1, e_2, \ldots e_m\}$. $Start$ and End are two special services defined as $Start = (\emptyset, I_T, \emptyset)$ and $End = (O_T, \emptyset, \emptyset)$ that account for the input and output requirements given by the request.

Each edge e from a service S to a service S' means that service S produces an output $a \in O_S$ that is matched (*exact* or *plugin*) to an input $b \in I_{S'}$ to be consumed by service S' in the composition.

Quality of Service (QoS) is one of the important factors to consider when composing services. For example, a service may be functionally capable of performing a certain task, but might not be sufficiently reliable or efficient to achieve users' satisfaction. With the increasing number of services available, service providers compete with each other to provide services with best possible qualities, such as low response time, low execution cost, high availability, and/or high reliability.

The QoS of the service composition can be obtained by aggregating the QoS values of the participating services. For a service composition with services $S_1, S_2, ...S_n$, QoS of the composite service can be computed with the QoS of its component services. To model the quality of service compositions in this paper we consider four QoS attributes that are often considered by existing works: availability, reliability, execution cost and response time [2,9,19,23,26,27]. Note that other QoS attributes, e.g. security, can be also considered according to users' requirements.

The *execution cost* is the amount of money that a service requester has to pay for using the web service. The global execution cost of a composite set of services is the sum of the local execution costs of all the component services in the composition, i.e., $C = \sum_{k=1}^{n} c_k$.

Availability is defined as the ratio of (1) the time during which a service is ready for use, and (2) the time during which the service exists. The global availability of a composite set of services is the product of the local availabilities of the component services in the service composition, i.e., $A = \prod_{k=1}^{n} a_k$.

Reliability is the ability to guarantee message delivery to the service within the maximum permitted time frame. The global reliability of a composite set of services is the product of the local reliabilities of the component services in the service composition, i.e., $R = \prod_{k=1}^{n} r_k$.

The *response time* is the time of most time-consuming path in the composition, i.e.,

$$T = MAX\{\sum_{k=1}^{\ell_j} t_k |\ j \in \{1,\ldots,h\} \text{ and } \ell_j \text{ is the number of nodes on a path } P_j\}.$$

where h is the number of paths on a graph.

The aim of QoS-aware service composition is to find a service composition with the best QoS. We use the following quality model to measure the fitness (or goodness) of a service composition.

$$Fitness = w_1 \hat{A} + w_2 \hat{R} + w_3 \hat{T} + w_4 \hat{C} \qquad (1)$$

with $\sum_{k=1}^{4} w_k = 1$. The weights can be adjusted according to users' preferences. $\hat{A}, \hat{R}, \hat{T}$ and $\hat{C}$ are normalised values calculated within the range from 0 to 1 using Eq. (2). To simplify the presentation we also use the notation $(Q_1; Q_2; Q_3; Q_4) = (A; R; T; C)$. Q_1 and Q_2 have minimum value 0 and maximum value 1. The minimum and maximum value of Q_3 and Q_4 are calculated across all task-related candidates in the service repository $\mathcal{SR}$.

$$\hat{Q}_k = \begin{cases} \frac{Q_k - Q_{k,min}}{Q_{k,max} - Q_{k,min}} & \text{if } k = 1, 2 \text{ and } Q_{k,max} - Q_{k,min} \neq 0, \\ \frac{Q_{k,max} - Q_k}{Q_{k,max} - Q_{k,min}} & \text{if } k = 3, 4 \text{ and } Q_{k,max} - Q_{k,min} \neq 0, \\ 1 & \text{otherwise.} \end{cases} \quad (2)$$

3 Web Service Composition Using Graph Databases

We propose to use graph databases to store the information of the services in a service repository, the dependencies between them, and feasible service compositions to form composite services. Our approach is outlined in Fig. 1. It consists of four steps:

1. Creating a graph database for a given service repository $\mathcal{SR}$, where the *nodes* represent the services in the repository, and the *relationships* represent the dependencies between services.
2. Generating a reduced graph database for a given composition task $T = (I_T, O_T)$ over $\mathcal{SR}$.
3. Generating an initial population of service composition solutions using the graph database generated in the previous step.
4. Finding a service composition solution in the population with the best fitness (QoS).

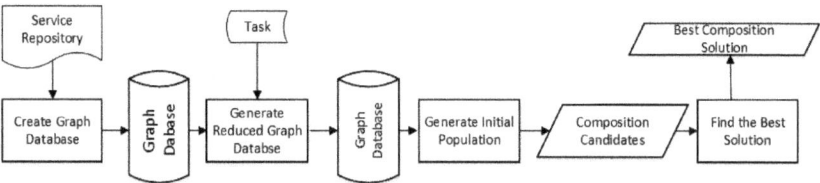

Fig. 1: Conceptual outline of our approach

3.1 Creating a Graph Database for a Service Repository

The first step is to model a service repository which contains a set of services with data dependencies between them. We represent a service as an entity which has a *name* and properties ID, I_S, O_S, QoS_S, *inputServices*, and *outputServices*. *InputServices* is a set of services can provide input to the service, and *outputServices* is a set of services that take the output of the service as their input. The reason why we add these two properties is to simplify retrieval of the connected services and also to reduce the cost of creating relationships between services.

Like service nodes, relationships also have properties *from, to* and a direction. Each relationship is a directed edge that connects the output of a service node to

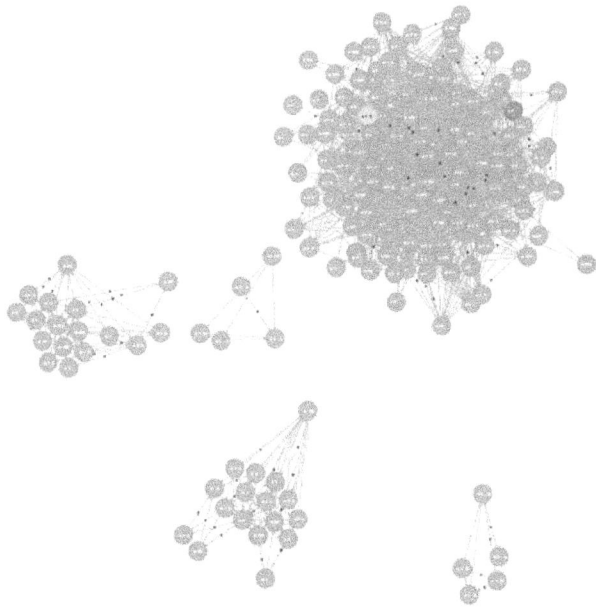

Fig. 2: (Unreduced) Graph database for task WSC08-1

the input of another node according to already established dependencies stored in the taxonomy tree.

We use a 'taxonomy tree' to represent the relationships between the inputs and outputs of services. Any two concepts A and B can be related to one another in one of four possible ways. The first scenario is that A is a generalization of B. The second scenario is that A is a specialization of B. The third scenario is that A and B are not related to each other. And the fourth scenario is that A equals B.

We can now create a graph database for a given service repository in three steps. Firstly, we find the properties of all the services and create a service object for each service in the service repository. We apply Algorithm 1 to find all the *input services* and *output services* for each taxonomy node. Secondly, creating a node for each service using the service objects we created in the first step. Lastly, we create relationships between services based on the *inputServices* and *outputService* properties. Note that taxonomy nodes are not directly included in the composition graph; their sole purpose is to be queried when checking for service dependencies during the graph construction process.

To store service dependencies in a graph database, we need to traverse every service node in the graph database, find all the *input services* and *output services* and then use the *createRelationshipTo* method to create dependencies between the service nodes. Algorithm 2 handles the creation of relationships between services in the graph database. It goes through each service node (N), and loops through the corresponding *inputServices* of each N, and then adds relationships between the services according to the sets of *inputServices*.

Algorithm 1: Population of the taxonomy tree by associating services with the nodes in the tree.

Input : $taxonomyNodes, serviceNodes$
Output: $taxonomyNodes$ with associated $serviceNodes$

1: $i \leftarrow 0$;
2: **while** $i < |taxonomyNodes|$ **do**
3:     $tNode \leftarrow taxonomyNodes[i]$;
4:     $tNode.parents \leftarrow findParentsNodes(tNode)$;
5:     $tNode.children \leftarrow findChildrenNodes(tNode)$;
6:     $i \leftarrow i + 1$;
7: $i \leftarrow 0$;
8: **while** $i < |serviceNodes|$ **do**
9:     $j \leftarrow 0$;
10:    $outputs \leftarrow serviceNodes[i].outputs$;
11:    **while** $j < |outputs|$ **do**
12:       $tNode \leftarrow findTaxonomyNode(outputs[j])$;
13:       $k \leftarrow 0$;
14:       **while** $j < |tNode.parents|$ **do**
15:          $tNode.parents \leftarrow tNode.parents \cup \{serviceNodes[i]\}$;
16:          $k \leftarrow k + 1$;
17:       $j \leftarrow j + 1$;
18:    $j \leftarrow 0$;
19:    $inputs \leftarrow serviceNodes[i].inputs$;
20:    **while** $j < |inputs|$ **do**
21:       $tNode \leftarrow findTaxonomyNode(inputs[j])$;
22:       $k \leftarrow 0$;
23:       **while** $j < |tNode.children|$ **do**
24:          $tNode.children \leftarrow tNode.children \cup \{serviceNodes[i]\}$;
25:          $k \leftarrow k + 1$;
26:       $j \leftarrow j + 1$;
27:    $i \leftarrow i + 1$;

Example 1 *As an example we consider the task WSC08-1. Its service repository consists of 158 services. Fig. 2 shows the graph database that we created in our approach. The nodes represent the services in the repository and the edges are the relationships (dependencies) between services.*

3.2 Generating a Graph Database for a Service Request

For each given task $T = (I_T, O_T)$, only a subset of services in a repository is related to that task. Therefore we created a graph database for each given task. Algorithm 3 describes the process of reducing the graph database for a service repository based on the composition task under consideration.

Algorithm 3 first creates a *Start* node and an *End* node. Secondly, we insert these nodes into the database by creating new relationships between the *Start* node and any node that can consume some of I_T. Similarly, we create relation-

Algorithm 2: Creation of relationships between services.

Input : $GraphDatabaseWithNoRelationships$, i.e. with no edges between vertices
Output: $GraphDatabaseWithRelationships$, i.e. having built the edges between vertices

1: $i \leftarrow 0$;
2: **while** $i < |serviceNodes|$ **do**
3: $\quad sNode \leftarrow serviceNodes[i]$;
4: $\quad j \leftarrow 0$;
5: $\quad$ **while** $j < |sNode.inputServices|$ **do**
6: $\quad\quad inputsNode \leftarrow sNode.inputServices[j]$;
7: $\quad\quad relation \leftarrow inputSNode.createRelationshipTo(sNode)$;
8: $\quad\quad relation.setProperty(\text{``From''} : inputsNode)$;
9: $\quad\quad relation.setProperty(\text{``To''} : sNode)$;
10: $\quad\quad relation.setProperty(\text{``Outputs''} : inputsNode.outputs)$;
11: $\quad\quad relation.setProperty(\text{``Direction''} : incoming)$;
12: $\quad\quad j \leftarrow j + 1$;
13: $\quad i \leftarrow i + 1$;

Algorithm 3: Reduction of the Graph Database.

Input : $GraphDatabase$
Output: $ReducedGraphDatabase$

1: $i \leftarrow 0$;
2: $relatedNodes \leftarrow \{\}$;
3: **while** $i < |serviceNodes|$ **do**
4: $\quad sNode \leftarrow serviceNodes[i]$;
5: $\quad$ **if** $hasRelationship(sNode, startNode) \land hasRelationship(sNode, endNode)$ **then**
6: $\quad\quad relatedNodes \leftarrow relatedNodes \cup \{sNode\}$;
7: $\quad$ **if** $!fulfillInputs(sNode)$ **then**
8: $\quad\quad removeRelatedNodes(sNode)$;
9: $\quad i \leftarrow i + 1$;
10: **return** $relatedNodes$;

---

ships between any node that can produce some of O_T. Lastly, we find all nodes that lie on some paths from the *Start* node to the *End* node by repeatedly using the built-in path finding algorithm available in the graph database platform, while we remove all other nodes. Algorithm 3 shows the process of reducing the original graph database of the service repository by removing all 'superfluous' nodes. Creating a reduced graph database can be relatively computationally expensive, however this only needs to be performed once per task, which means that the cost is amortised over time.

Example 2 *Recall task WSC08-01. The given composition task T has task inputs $I_T = \{$inst1926141668, inst395151449, inst1557679659$\}$ and task outputs*

O_T = {inst1913443608, inst664891780}. *When applying Algorithm 3 to the graph database in Fig. 2 we obtain a reduced graph database, see Fig. 3. This reduced graph database contains only those nodes that represent services which are related to both, the task inputs and the task outputs. In this example, the algorithm shrinks the graph database from 158 nodes to 61 nodes. This illustrates the effectiveness of our approach.*

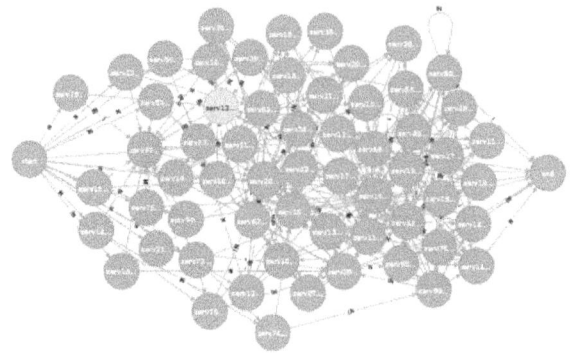

Fig. 3: Reduced graph database for task WSC08-1

3.3 Web Service Composition Candidates

Algorithm 3 allows us to retrieve those service nodes that are related to a particular task. This section explains how we used related service nodes to generate service compositions in order to create an initial set of candidates.

Algorithm 4 is designed to create feasible compositions. It starts from the *End* node and searches internal nodes which are directly connected to *End*. Internal nodes are added into the composition list if they fulfil the input of *End* (this is checked using the *fulfillInputs* function). Then the algorithm recursively goes through each node in the composition list and repeats the above steps until the size of the composition list is stable. When creating a solution, edges are used first and then the corresponding nodes are retrieved by using *fulfilledNodes*. Each set of *fullfilledNodes* makes up a candidate in the composition. This algorithm is also able to calculate the total *response time* for the creation of each composition, thus determining the total *response time* from task input I_T to task output O_T. This total *response time* is then used to determine the quality of the solution, when we generate a QoS-aware service composition.

Algorithm 4 calculates the *response time* during generation of the service composition by setting the duration property which reflects the time taken from *End* node to *Start* node. When adding node N to the composition list, the algorithm first checks the duration property of N. If N's duration property value is less than the sum of N's previous node duration property value and N's execution time (N's QoS time value), then the algorithm sets N's duration property value to the sum of the N's previous node duration property value and N's execution time.

Algorithm 4: Initial Population.
Input : $endNode, relatedNodes$
Output: $candidates$
1: $i \leftarrow 0$;
2: $rels \leftarrow \{\}$;
3: $candidates \leftarrow \{\}$;
4: **while** $i < |getIncomingRelationships(endNode)|$ **do**
5: $\quad rels \leftarrow rels \cup \{tNode.getIncomingRelationships(endNode)[i]\}$;
6: $\quad i \leftarrow i + 1$;
7: $fulfilledNodes \leftarrow fulfilledNodes \cup \{getNodesFulfillCurrentNode(rels)\}$;
8: **if** $|fulfilledNodes| > 0$ **then**
9: $\quad candidates \leftarrow candidates \cup \{fulfilledNodes\}$;
10: $\quad j \leftarrow 0$;

Two service composition solutions are shown in Fig. 4 to illustrate the application of Algorithm 4 to the reduced graph database in Fig. 3. Both compositions show the relationships between task inputs I_T and task outputs O_T. For each node in the composition it is possible to click the service node S to find the inputs of S and also click the incoming edge of S to find the input values related to the service at the other end of the edge.

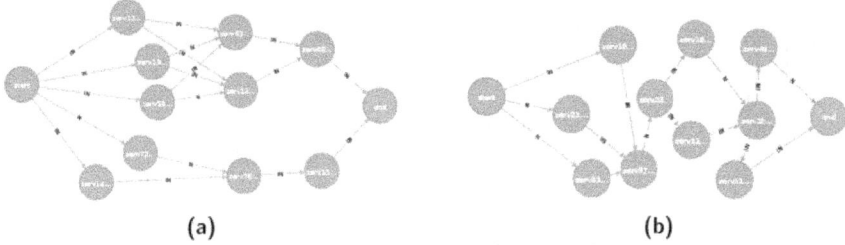

Fig. 4: Two different service composition solutions for task WSC08-1

3.4 QoS-Aware Service Compositions

The next step is to choose a service composition from a set of solutions for a composition task such that its QoS is optimal. Usually there are many service compositions that can fulfil a given task. As discussed in Sect. 1, to meet a user's non-functional requirements we aim to find a service composition with the best overall QoS.

Algorithm 5 is designed to search for the solution with best QoS. Using the set of initial population generated from Algorithm 4, the algorithm goes through every service node invoked in the compositions. It then determines minimums and maximums of *execution cost, response time, availability* and *reliability*. Having found all minimums and maximums the next steps are to carry out normalization to restrict all values to the interval [0,1], and then to apply a fitness function, using normalized QoS properties and user defined weights to each service composition. Lastly, the algorithm returns the service composition which has the best fitness value.

Algorithm 5: *QoS*-aware service composition

Input : *candidates, weights*
Output: *BestComposition*

1: $i \leftarrow 0$;
2: $BestComposition \leftarrow \{\}$;
3: $BestFintess \leftarrow 0$;
4: $Map < composition, normalizedList > compWithNormalized \leftarrow \{\}$;
5: **while** $i < |candidates|$ **do**
6:      $j \leftarrow 0$;
7:      **while** $j < |candidates(i)|$ **do**
8:          $find(maxC, minC, maxT, minT, maxA, minA, maxR, minR)$;
9:          $j \leftarrow j + 1$;
10:      $i \leftarrow i + 1$;
11: $i \leftarrow 0$;
12: **while** $i < |candidates|$ **do**
13:      $j \leftarrow 0$;
14:      **while** $j < |candidates(i)|$ **do**
15:          $NormalizedList.add(normalize(Cost, Time, Availability, Availability))$;
             $j \leftarrow j + 1$;
16:      $compWithNormalized.getKey(composition) \leftarrow NormalizedList$;
17:      $i \leftarrow i + 1$;
18: $i \leftarrow 0$;
19: **while** $i < |compWithNormalized|$ **do**
20:      $fitnessValue = fitnessFunction(compWithNormalized(i), weights)$;
21:      **if** $fitnessValue > BestFitness$ **then**
22:          $BestFitness \leftarrow fitnessValue$;
23:          $BestComposition \leftarrow composition$;

The main purpose of this algorithm to find the best composition solutions, i.e. highest fitness value, among a set of candidate compositions. The fitness values are normalized with the best and worst values of each of QoS attributes considered.

4 Evaluation

In order to assess the performance of our proposed approach we conduct performance evaluation of our graph database based approach, by comparing it with an existing approach, namely, the GraphEvol approach presented in [21]. In particular, we evaluated the effectiveness and efficiency of our approach by comparing the quality of composition solutions and the time taken to generate a service composition.

4.1 Datasets and Parameters

We evaluated the performance of our approach by testing it on an extended version of the benchmark dataset of the web service challenge WSC08 [4] by attaching QoS values to the description of atomic services in the repository. The QoS values have been randomly generated in the range of the values found in QWS [1] collected for web services from real domains. Both our approach and the GraphEvol [21] are stochastic. To conduct significance analysis, we ran each task independently 30 times, for each run recording the best service composition, fitness value and execution time. We used a two-sample t-test to check if the fitness values and execution times for our approach were better than the fitness values and execution times for the GraphEvol approach at the 0.05 significance level. For each run, our approach generated 50 candidates and the best solution from those candidates was chosen. For GraphEvol approach, the best candidate was generated from a population size of 200. For all tests we set the weights for the objective function to the same value as was used with the GraphEvol method, namely, $w_i = 0.25$ for $i = 1, \ldots, 4$. The other parameters for GraphEvol [21] were: a population of 200 candidates, a mutation probability of 0.05, and a crossover probability of 0.5. Individuals were chosen for breeding using tournament selection with a tournament size of 2 [21].

The experiments are conducted on a Macbook Pro Mid 2012 with a 2.9GHz dual-core Intel Core i7 processor, 8GB of 1600MHz DDR3 memory, a 1536 Mb Intel HD Graphics 4000 graphics card, and a 128GB solid-state drive.

Table 1: Comparison of the number of services in the original repository and the reduced graph database

Task	Original service repository	Reduced graph database
WSC08-1	158	61
WSC08-2	558	66
WSC08-3	604	105
WSC08-4	1041	46
WSC08-5	1091	102
WSC08-6	2198	205
WSC08-7	4113	195
WSC08-8	8119	131

4.2 Experimental Results and Analysis

To evaluate the effectiveness of our approach we applied it to the dataset of WSC08. Table 1 shows the number of services for the original service repository and the number of services in the reduced graph database. For example, for task WSC08-2, there were 558 services in the original repository, but only 66 of them were related to the composition task. For task WSC08-8, there were 8138 services in the original repository, but only 131 of them were related to the composition task. Thus, the number of the services was significantly reduced when compared to the number of services in the original repository, which greatly improved the execution time when generating the service composition.

Table 2 shows a full evaluation comparing the performance of our approach and that of the GraphEvol approach [21]. Both approaches were run on the same machine (see Sect. 4.1) to conduct a significance analysis. Table 2 shows two columns for each approach. The *time* column records the average time, over 30 independent runs, which was taken to generate the best service composition. And the *fitness* column records the average QoS fitness value for the best service composition, calculated from 30 independent runs.

Table 2: Average results of the tests for QoS-aware service composition

Task	GraphDB Approach		GraphEvol Approach[21]	
	Time (ms)	Fitness	Time (ms)	Fitness
WSC08-1	2197.90 ± 329 ↓	0.521 ± 0.169 ↓	4845.57 ± 315.42	0.645 ± 0.139
WSC08-2	5347.13 ± 880 ↑	0.48 ± 0.152 ↓	3699.77 ± 364.57	0.906 ± 0.00
WSC08-3	10961.53 ± 790 ↓	0.387 ± 0.1 ↑	17221.53 ± 764.85	0.176 ± 0.045
WSC08-4	3885.70 ± 399 ↓	0.431 ± 0.066 ↑	6076.7 ± 281.58	0.305 ± 0.066
WSC08-5	4510.6 ± 468 ↓	0.403 ± 0.128↑	10444.2 ± 572.59	0.164 ± 0.046
WSC08-6	258503.33 ± 42324 ↑	0.407 ± 0.089 ↑	22183.53 ± 1639	0.228 ± 0.06
WSC08-7	17839.77 ± 763 ↓	0.457 ± 0.097 ↑	20304.37 ± 1257	0.316 ± 0.039
WSC08-8	53003.7 ± 4465 ↑	0.468 ± 0.091 ↑	18567.03 ± 2055	0.315 ± 0.028

In Table 2, it is clear that the P-values from tasks WSC08-3 to WSC08-8 are smaller than the significance level 0.05 for the fitness values set and tasks WSC08-1, WSC08-3, WSC08-4, WSC08-5 and WSC08-7 are less than the significance level 0.05 for the execution times set, as indicated by the upward and downward pointing arrows. This means there is considerable evidence that the sets of fitness values and execution times for GraphEvol approach are lower than for our approach. In other words, the best solutions in our approach are significantly superior to the best solutions produced by the GraphEvol approach for 6 out of 8 tasks. For execution time, our approach is faster than GraphEvol for 5 out of 8 tasks. Only for tasks WSC08-2, WSC08-6 and WSC08-8 the execution time of our approach was poorer than the GraphEvol approach. In summary, our approach successfully generated a service composition for each task of the eight tasks of WSC08. For five tasks our approach was significantly better than the GraphEvol approach in terms of execution time.

4.3 Correctness of the Web Service Composition

We thoroughly verified the correctness of the service composition solutions computed by our approach. For that we checked that all the service node inputs are fulfilled by the output of their preceding nodes in the graph database. Our evaluation has shown that all the solutions were correct.

Example 3 *Recall task WSC08-1. Consider the service composition solution shown on the right hand side of Fig. 4. It includes services S = 'serv976005395' and S' = 'serv283321609'. In this solution, S is the only preceding node of S'. The outputs of S are O_S = {inst722854357 inst1326239605, inst1881697469, inst1437249127, inst1519789560}. The inputs of S' are $I_{S'}$ = {inst722854357,* inst347634243, inst1881697469, inst746203847}. *It is easy to observe that concepts* 'inst722854357' *and* 'inst1881697469' *belong to both* O_S *and* $I_{S'}$. *According to the underlying taxonomy* $\mathcal{T}$ *of WSC08-1, concepts* 'inst1519789560' *and* 'inst1326239605' *in* O_S *are specializations of concept* 'inst347634243' *in* $I_{S'}$, *and* 'inst1437249127' *in* O_S *is a specialization of* 'inst746203847' *in* $I_{S'}$. *Hence, every input of* S' *is matched by some input of* S. *Similarly we can check for every service that all its inputs are matched by some preceding node in the solution.*

4.4 Further Analysis

In this section we analyze the performance of our proposed approach by comparing its performance with other existing approaches [6, 14, 21] that are also evaluated using the same dataset. The experimental results in [6] do not show overall QoS values. Instead, in their results table, response time and throughput values are shown separately, but it is not clear if these are the best results found or the average values. In [6], the number of services used in the composition solutions generated by their approach is presented to show the quality of their results. For easy analysis, we list the number of services used in service composition solutions generated by the above mentioned approaches.

Table 3: Average Number of the Services Used in Composition Solutions

Task	BestKnown	Scalable [6]	Hybrid[23]	GraphEvol[21]	GraphDB
WSC08-01	10	12	10.00±0.00	10.63 ±2.17	10.00 ±0.00
WSC08-02	5	5	5.00 ±0.00	5.87 ±2.12	5.00 ±0.00
WSC08-03	40	42	40.60 ±0.62	41.20 ±0.79	40.00±0.00
WSC08-04	10	14	10.00 ±0.00	10.20 ±0.48	10.00±0.00
WSC08-05	20	26	20.00 ±0.00	22.13 ±2.59	20.00 ±0.00
WSC08-06	40	47	45.80 ±0.92	40.20 ±0.48	40.00 ±0.00
WSC08-07	20	20	20.00 ±0.00	23.13 ±7.33	20.00 ±0.00
WSC08-08	30	39	32.10 ±0.30	32.47 ±3.86	30.00 ±0.00

As we see in Table 3, comparing with the work in [21] and [23], our proposed approach is very stable and can always generate solutions with smallest number of services known so far. Also, for most tasks it outperform the other three approaches in terms of the number of services used in the composition solutions. Except for two tasks, solutions generated by the approach in [6] always have more component services.

5 Conclusion and Future Work

In this paper we have proposed a novel approach to QoS-aware service composition using graph databases. To do that we have proposed to model and store of service repository as graph database. In this way, information concerning services and relationships between services can be stored and retrieved for any given tasks. We have also proposed several QoS-aware service composition algorithms based on graph databases. We have evaluated our approach with a benchmark dataset. Our evaluation results show that the solutions generated by our approach are all functionally correct for all the tasks of WSC08. Further, our evaluation demonstrates that our approach can compute near-optimised solutions with good performance.

Future work can be along several lines to improve our approach. We can investigate using indices to improve the efficiency and effectiveness of querying the database. Also, we can extend our work to semantic web service composition using graph databases.

References

1. Al-Masri, E., Mahmoud, Q.H.: Qos-based discovery and ranking of web services. In: 2007 16th International Conference on Computer Communications and Networks. pp. 529–534 (2007)
2. Alrifai, M., Risse, T.: Combining global optimization with local selection for efficient QoS-aware service composition. In: WWW. pp. 881–890. ACM (2009)
3. Amiri, M.A., Serajzadeh, H.: Effective web service composition using particle swarm optimization algorithm. In: IST. pp. 1190–1194. IEEE (2012)
4. Bansal, A., Blake, M.B., Kona, S., Bleul, S., Weise, T., Jaeger, M.C.: WSC-08: Continuing the web services challenge. In: Enterprise Computing, E-Commerce and E-Services. pp. 351–354. IEEE (2008)
5. Chattopadhyay, S., Banerjee, A.: Qscas: Qos aware web service composition algorithms with stochastic parameters. In: 2016 IEEE International Conference on Web Services (ICWS). pp. 388–395 (2016)
6. Chattopadhyay, S., Banerjee, A., Banerjee, N.: A scalable and approximate mechanism for web service composition. In: 2015 IEEE International Conference on Web Services. pp. 9–16 (2015)
7. Gottschalk, K., Graham, S., Kreger, H., Snell, J.: Introduction to web services architecture. IBM Systems Journal 41(2), 170–177 (2002)
8. Hashemian, S.V., Mavaddat, F.: A graph-based approach to web services composition. In: Applications and the Internet. pp. 183–189. IEEE (2005)
9. Jaeger, M.C., Mühl, G.: QoS-based selection of services: The implementation of a genetic algorithm. In: ITG-GI. pp. 1–12. VDE (2007)
10. Jouili, S., Vansteenberghe, V.: An empirical comparison of graph databases. In: Social Computing. pp. 708–715 (2013)
11. Ma, H., Schewe, K.D., Wang, Q.: An abstract model for service provision, search and composition. In: 2009 IEEE Asia-Pacific Services Computing Conference (APSCC). pp. 95–102 (2009)
12. Ma, H., Schewe, K.D., Thalheim, B., Wang, Q.: A theory of data-intensive software services. Service Oriented Computing and Applications 3(4), 263–283 (Dec 2009)
13. Ma, H., Schewe, K., Thalheim, B., Wang, Q.: A formal model for the interoperability of service clouds. Service Oriented Computing and Applications 6(3), 189–205 (2012)
14. Ma, H., Wang, A., Zhang, M.: A hybrid approach using genetic programming and greedy search for qos-aware web service composition. Trans. Large-Scale Data- and Knowledge-Centered Systems 18, 180–205 (2015)
15. Mahmoud, C.B., Bettahar, F., Abderrahim, H., Saidi, H.: Towards a graph-based approach for web services composition. Int. J. Comp. Sci. Issues (IJCSI) 10 (2013)

16. Mucientes, M., Lama, M., Couto, M.I.: A genetic programming-based algorithm for composing web services. In: ISDA. pp. 379–384. IEEE (2009)
17. Neo4j: Neo4j Developer Manual v3.0 (2016), `neo4j.com/docs/developer-manual`
18. Perrey, R., Lycett, M.: Service-oriented architecture. In: Applications and the Internet. pp. 116–119. IEEE (2003)
19. da Silva, A.S., Ma, H., Zhang, M.: A GP approach to QoS-aware web service composition and selection. In: SEAL. pp. 180–191. Springer (2014)
20. da Silva, A.S., Ma, H., Zhang, M.: A graph-based particle swarm optimisation approach to QoS-aware web service composition and selection. In: CEC. pp. 3127–3134 (2014)
21. da Silva, A.S., Ma, H., Zhang, M.: Graphevol: a graph evolution technique for web service composition. In: DEXA. pp. 134–142. Springer (2015)
22. da Silva, A.S., Ma, H., Zhang, M., Hartmann, S.: Handling branched web service composition with a qos-aware graph-based method. In: E-Commerce and Web Technologies - 17th International Conference (EC-Web). pp. 154–169 (2016)
23. Wang, A., Ma, H., Zhang, M.: Genetic programming with greedy search for web service composition. In: DEXA. pp. 9–17. Springer (2013)
24. Yilmaz, A.E., Karagoz, P.: Improved genetic algorithm based approach for QoS aware web service composition. In: ICWS. pp. 463–470. IEEE (2014)
25. Yoo, J.J.W., Kumara, S., Lee, D., Oh, S.C.: A web service composition framework using integer programming with non-functional objectives and constraints. In: Enterprise Computing, E-Commerce and E-Services. pp. 347–350. IEEE (2008)
26. Zeng, L., Benatallah, B., Dumas, M., Kalagnanam, J., Sheng, Q.Z.: Quality driven web services composition. In: WWW. pp. 411–421. ACM (2003)
27. Zeng, L., Benatallah, B., Ngu, A.H., Dumas, M., Kalagnanam, J., Chang, H.: QoS-aware middleware for web services composition. IEEE Trans. Software Eng. 30(5), 311–327 (2004)

Arcaini, Paolo 242
Banach, Richard 177
Beierle, Christoph 288
Bienemann, Alexander 332
Buga, Andreea 264
Chang, Zhaojiang 338
Dahanayake, Ajantha 3
Ferrarotti, Flavio 92
Gargantini, Angelo 242
Geist, Verena 220
Hartmann, Sven 75, 338
Hegner, Stephen J. 122
Illibauer, Christa 220
Jacquot, Jean-Pierre 202
Kern-Isberner, Gabriele 288
Kirchberg, Markus 75
Koehler, Henning 75
Kossak, Felix 220
Kropp, Yannic Ole 146
Kutsch, Steven 288
Leck, Uwe 75
Link, Sebastian 75
Ma, Hui 338
Mashkoor, Atif 202, 220, 264
Natschläger, Christine 220
Nemeş, Sorana Tania 264
Raab-Düsterhöft, Antje 29
Riccobene, Elvinia 242
Sali, Attila 313
Sawczuk da Silva, Alexandre ... 338
Su, Wen 177
Tec, Loredana 92
Thalheim, Bernhard 3, 44, 146
Turull-Torres, José María 92
Ziebermayr, Thomas 220

www.ingramcontent.com/pod-product-compliance
Lightning Source LLC
Chambersburg PA
CBHW071328190426
43193CB00041B/948